APPLICATIONS OF SUBSEA SYSTEMS

APPLICATIONS OF SUBSEA SYSTEMS

Goodfellow Associates Ltd.

PennWell Books

PENNWELL PUBLISHING COMPANY
TULSA, OKLAHOMA

PennWell Publishing Company
1421 South Sheridan/P.O. Box 1260
Tulsa, Oklahoma 74101

Library of Congress Cataloging-in-Publication Data

Goodfellow Associates.
Applications of subsea systems / Goodfellow Associates Ltd.
p. cm.
Includes index.
ISBN 0–87814–334–3
1. Oil well drilling, Submarine.
TN871.3.G66 1990
622′.3382—dc20 90–44295
CIP

Printed in the United States of America

1 2 3 4 5 94 93 92 91 90

To All Offshore and Subsea Workers in the Oil and Gas Industry, Past, Present and in the Future.

CONTENTS

ACKNOWLEDGMENTS

Goodfellow Associates wish to thank the following for their help and permission in assembling the text and diagrams in this book.

ACB
Amerada Hess Ltd
Amoco (UK) Exploration Co
BP Exploration
Britoil Plc
BUE Hydra-Lok
Cameron Iron Works
Chevron Petroleum (UK) Ltd
Conoco (UK) Ltd
Dunlop Armaline Ltd
Enterprise Oil Plc
Esso Exploration and Production (UK) Ltd
Hamilton Bros Oil and Gas Ltd
MIB International Ltd
Mobil North Sea Ltd
OSEL
Petrobras
Santa Fe
Shell Exploration and Production
Sun Oil
Smit International (UK) Ltd
Stena Offshore
Texas Eastern North Sea Inc
UDI Group Ltd
Vetco Gray UK Ltd
WKM
Willis Flow Control
Webco Marine Services Ltd

Also, we must thank the following personnel within Goodfellow Associates for their time and energy in the production of this book:

D. Appleford, J. Ashe, C. Benfield, L. Brennan, R. Brett, B. Brown, K. Bond, J. Chasserot, L. Clarke, R. Connor, P. Cross, D. Cruickshank, N. Cuzzi, P. Dick, who edited and coordinated the work from various authors, R. Edson, M. Euillet, E. Forde, R. Goodfellow, B. Lascelles, R. Lawrence, K. Lee, C. Mathieson, C. Millum, P. Mitra, K. Morland, L. Morrison, J. Neighbour, R. North, A. Norris, G. Overbury, C. Panagiotopoulos, M. Rimmer, S. Sarshar, D. Scott, M. Scott, D. Smith, M. Strachan, A. Tong, D. Urquhart, G. Vaughan, T. Welch.

EDITOR'S NOTE

The chapters of this book originated with different authors from within the Goodfellow Associates organization, therefore, some alignment of text and, in particular, uniformity in units of measurement was required in editing. These changes should appeal to young and experienced engineers alike on an international basis; however, some explanation is still warranted.

Water depths are quoted in both meters and feet using 3.2808 feet to the meter where conversions were required. Similarly, 0.621 miles to the kilometer is used for distance conversions.

Costs are for the most part quoted in the currency of the country in question. All references to dollars are in United States dollars ($ or U.S.) and where necessary, to standardize units within a continuous text, a conversion of U.S. $1.6 to one pound sterling (£) has been used. The abbreviation NOK indicates Norwegian Kroner.

In line with the international nature of the oil and gas industry, no one system of units has been adhered to. For instance, specialized units have been left as presented by the author, as they are probably easily understood by those interested in them. However, many abbreviations have been avoided in favor of the long-hand version, millions and billions in particular. Within a complex technical text, it is very easy, of course, to miss one or two points. However, it is felt that such anomalies will be easily understood and translated by the reader.

P.D.

INTRODUCTION

The offshore oil and gas industry has a continuing need for field developments to be cost-effective. Against a general background of political and economic uncertainty, subsea technology now offers an acceptable solution to many, if not most, of the problems involved with tie-backs to existing platforms or stand-alone developments.

In the future such technology will offer the only viable approach to stand-alone developments in deeper water, whether smaller fields or in remoter locations. The eventual elimination of the platform as a major cost-center in field development due to new subsea technologies will have far reaching economic consequences.

This book introduces various subsea systems and their components in a number of different scenarios. It will be of assistance to both operators, who have a requirement to produce, and to equipment suppliers, who must continue to supply their requirements.

The general view of any offshore oil or gas field involves a production platform. However, nowadays the subsea development of a prospect will inevitably be one of the scenarios reviewed by a potential operator. Whether or not it is chosen, of course, depends on many factors. This includes operator philosophy and financing, as well as reservoir characteristics, water depth and distance from shore or mother platform.

In engineering terms being subsea can present some complex problems, which *Applications of Subsea Systems* tries to clarify. It leads the reader gently into the subject starting with the reasons for going subsea. It also introduces many examples, reviews the technology involved with templates sitting under floating production platforms and moves toward manifolds sited some considerable distance from central processing facilities. Further, each chapter is complete and can stand alone, thereby satisfying the requirements of more informed readers on specific subjects and helping develop a personalized approach.

Chapter 1

The Expanding Role for Subsea Systems

In November 1986, Dr. Bill Doyle, then Managing Director of Texaco Exploration and Production, pointed out that, "The estimated cost of production from current (North Sea) construction is \$18/bbl. This is the estimated cost to operate, amortize, and provide a 10% real rate of return prior to tax and royalty. Clearly, lower costs are needed to ensure the full development of the North Sea resources. John Jennings' (of Shell) previously set targets for 15% reduction in development costs is inadequate."

Speaking in Aberdeen during the worst point of the slump, Doyle admitted that any predictions were reckless but suggested that "For the next few years I doubt that new developments will proceed unless they are economically sound at \$5/bbl."

"This in turn means that only satellite developments using existing infrastructures are likely to proceed in the medium term. Even here there is a need for new cost reducing technology. This means an increasingly important role for subsea technology over the next few years, and the priority must be on reducing costs and increasing reliability."

Following the oil price crash of 1986—to \$10/bbl and the ensuing major cut-backs and project deferments in oil field development around the world—the price stabilized at around \$15/bbl for 1987 and 1988. New projects once more came forward for development, with a much increased emphasis on low cost subsea satellite developments making use of "spare" capacity in the existing infrastructure of fixed platforms, topside processing and export pipelines.

By the end of 1988 however, following the end of the Iran-Iraq war and continued quota exceedance by many OPEC members, the oil price weakened once more. Oil was trading, in real terms, more cheaply than at any time since 1973. Some forecasts predicted a protracted period with the oil price as low as \$5/bbl.

Figure 1–1 shows oil prices during the period 1978–1988 as expressed in real 1987 U.S. \$/bbl. Official prices for Saudi Arabian Light are shown between 1978 and mid-1985 followed by Brent crude spot price from end-1985 onwards. Historical prices have been converted to 1987 real terms by applying the U.S. Consumer Price Index.

The outlook of continuing volatile oil prices and exchange rates in a now uncertain world economy will, if anything, intensify the drive to reduce exploration and production where such costs are greatest (see Table 1–1).

Even if governments change the tax regimes drastically, few stand-alone offshore projects in the non-OPEC region would be economically viable were prices to slump back into the \$10–\$15/bbl range. Yet the offshore oil and gas regions of the world are important contributors to overall production, with 24% and 20% respectively of the 1986 totals of 55.8 million/B/D and 1,735 billion m^3. An indication of their potential based on sedimentary geology is that offshore regions now account for about half the world's exploration activity and investment.

Due to the increasing economic restriction of offshore oil and gas developments which particularly affect small fields of limited life (i.e., 5 to 10 years), early production systems offer an attractive option to achieving lower capital costs and an earlier return on investments.

With these smaller fields (i.e., less than 100 million barrels recoverable reserves), the limited reservoir data also influences the options in favor of an early development scheme at low cost to minimize the risk. Extended well testing is now being considered, based on floating systems, in order to gain more precise longer term reservoir data before making the commitment to a more permanent longer term production unit.

Traditional offshore developments of oil and gas, after location and definition of a field, have utilized fixed steel or concrete structures. From preliminary engineering through construction, installation, and hook-up, first oil production for this type of structure typically takes from 4 to 5 years. Early production systems aim to reduce this period to 2.5 to 3 years and to date most systems have used floating production vessels. Notable examples are Argyll and Buchan using semi-submersibles in the North Sea and tanker conversions such as Castellon and Tazerka in the less hostile waters of the Mediterranean. Limited

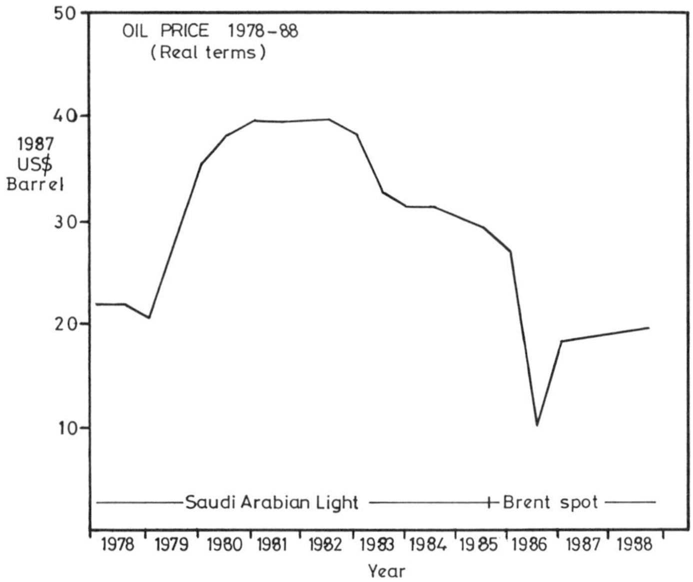

Fig. 1–1 Oil Prices 1978–1988 (in real 1987 U.S. $/bbl)

use has been made of converted jack-ups also, such as on the Ekofisk field in the North Sea and Espoir off the Ivory Coast.

In all such developments, the operator had to choose between converting an existing vessel to a production role or to build a particular design for his specific requirements. There is also an interest in developing purpose-built vessels for use on a series of short-life fields, either under the control of one operator or alternatively financed by a contractor/owner who offers a leasing arrangement to a series of different operators. Such a concept can only be viable if the production system is generalized to cover a range of field characteristics. The Petrojal I ship represents such a concept—owned by Golar Nor and used for extended well testing on the Oseberg Field. A similar concept also applies to BP's SWOPS vessel (currently under construction) which is owned by an operator.

Conversions of semi-submersibles have not been entirely successful to date since the payload constraints restrict the design, and re-classification requirements for production status involve extensive modifications to the existing vessel and its utility systems.

Table 1–1 Oil Prices, 1978–1988 (expressed in real US $/bbl)

	Production Costs $/bbl	Investments $/bbl/d
Conventional oil		
Mean value	5	4,000– 8,000
—Middle East	1	500– 3,000
—Non-OPEC countries	8	3,000–12,000
High cost oil		
Offshore conditions	10–20	10,000–25,000
Deep offshore	10–30	15,000–35,000
Enhanced oil recovery:		
Water flooding	2–10	1,000–5,000
Thermal recovery methods	10–25	8,000–25,000
Miscible gas/CO^2 flooding	10–30	10,000–25,000
Chemical flooding–polymer	10–20	10,000–30,000
–surfactant	20–50	15,000–30,000
Extra heavy crude (1)	20–30	30,000–40,000

(1) Upgrading included
Source IFP/Department of Economics/1986

TREND TOWARD SMALL (MARGINAL) FIELDS

The development of small marginal offshore oil and gas fields presents a challenge to the ingenuity of subsea engineers on a worldwide basis (see Table 1–2).

A wide range of technical approaches has been implemented in the North Sea. Sufficient experience now exists to enable the field development methods to be evaluated for a proposed oil or gas field and the optimum plan selected. Engineering design and construction expenditure can be forecast and maintained within budget. This is essential for smaller (marginal) fields.

For the future of North Sea activities and the U.K. concessional waters, increased emphasis will be placed on the development of small oil and gas fields. For example, forecasts for the U.K. concessional waters suggest that 80% of future developments will be from fields with 100 million barrels recoverable reserves or less. On a worldwide basis, several hundred such small (marginal) fields could exist. For these small fields a different set of parameters apply, and the applications of early production and subsea systems are as follows:

- Semi-submersible exporting through offshore loading systems
- Semi-submersible exporting through submarine pipeline
- Tankers with storage, and shuttle tankers
- Articulated columns
- Jack-up rigs

Factors relating to the commercial viability of developing small fields relate to political/economic factors and technology such as:

- Lower tax
- Lower oil prices
- Lower capital expenditure (CAPEX)
- Lower operating expenditure (OPEX)
- Early production

The application of subsea technology to achieve lower capital expenditure and early production is the theme of this book.

Table 1–2 Distribution of Undeveloped Discoveries and Marginal Oil and Gas Field Alternatives

	Water depth (ft)						
Region	1–100	101–200	201–400	401–660	661–1150	Over 1150	Total
North America	21	34	47	16	11	14	143
South America	33	27	34	14	12	18	138
Europe	38	66	81	46	15	13	259
Mid East (1)	35	16	15	2	—	—	68
Africa	88	102	87	32	12	6	327
Far East (2)	112	80	97	27	14	16	346
World Total	327	325	361	137	64	67	1,281

(1) Includes Mediterranean (2) Includes Australia, N Zealand

Development Alternatives	Field and Location (examples
Steel caisson	Rolf, Denmark
Artificial island	Endicott, U.S. Arctic
Converted jack-up	Espoir, Ivory Coast
Concrete platform with crude storage	Schwedeneck, West Germany
Steel platform with crude storage	Maureen, U.K.
Fixed steel platform	Kotter, Netherlands
Fixed wellhead/floating storage	Bohai, China
Subsea completion/floating storage	Tazerka, Tunisia
Converted semi-submersible	Argyll, U.K.
Converted semi/subsea template	Green Canyon 29, U.S.
Purpose-built semi-submersible	Balmoral, U.K.
Tension-leg platform	Hutton, U.K.: Jolliet, U.S.
Template system	Scapa, U.K.

CAPITAL EXPENDITURE (CAPEX)

The capital expenditure of floating production methods developing North Sea oil fields has ranged from U.S. 120 $ million to 1,760 billion for the existing fields. The expenditure is dependant on the size and location of the field, the availability of existing export systems, and the need to provide oil storage. With the trend in the future to smaller (marginal) fields, the capital expenditure is likely to be in the range of U.S. $ 80 million to 480 million.

An illustrative breakdown of the capital expenditure (CAPEX) for a floating production vessel (FPV) is as follows:

Drilling	25%
FPV	25%
Subsea Systems	20%
Offshore Loading	5%
Management/Insurance/ Certification	10%
Contingency	15%

For the Texaco Highlander field, four development options were considered. Each of the options disposed of the oil and gas through pipeline systems terminating at the Tartan platform. These systems, with their relative project investments costs, including wells, were as follows:

Development Option	Relative Cost
Fixed self-contained platform	2.9
Fixed satellite platform and purpose-built semi-submersible floating	2.8
Facility with template	2.5
Remote subsea production facility	1.0

The substantially lower investment of the subsea facility and the fact that production could be achieved much earlier from the four existing wells both coupled to make this marginal field project economically viable. Texaco's first-hand experience with satellite subsea wells and their associated control systems was deemed sufficient to ensure the technical success of a subsea facility.

SUBSEA SYSTEMS

The technology of subsea systems provides the basis on which future small fields can be economically developed. Subsea technology covers single-well completions, multi-well completions through templates, gas and water injection systems, manifolding, and commingling.

With the development needs of such fields, subsea engineers designing the equipment and facilities need to take into account the following criteria:

- Short lead times
- Simple designs and reliable components
- Ease of installation and maintenance
- Lower costs
- Adaptable to a range of production requirements
- Shorter field life (5–11 years)
- Potential for recovery of the system

From the early 1970s, subsea systems have been used for satellite well (water injection/production), templates, and manifolds.

With only isolated exceptions, production of offshore oil and gas reserves using subsea techniques has been restricted to depths where, at the time the development decision was taken, other more conventional techniques (e.g., fixed structures) were unable to provide a cost-effective solution.

Subsea production technology is now an accepted part of development, compared with the early situation in the 1970s. However, the development has not been as rapid as in other areas of offshore technology, for example, structural analysis. Therefore, this has pushed the anticipated economic depths for subsea from 75 m to 150 m (246 to 492 ft) or more. However, the need for subsea technology has been and will continue to be felt in economic terms, due to the increasing costs of offshore production. In particular, this is the case for the small oil reservoirs (30–50 million barrels recoverable reserves) and especially for oil accumulations in deeper waters. Further factors influencing the growth of subsea technology relate to early production and secondary and tertiary recovery.

To assess the prospects for the next decade, a review of the last 10 years shows the trends for positive and ''sideways'' progress. By 1970 the number of subsea completions (worldwide) was 118, rising by 1984 to a total of 241 completions. This has increased regularly as follows:

Year	Completions
1985	45
1986	43
1987	63
Thru March 1988	42

The distribution of completions by offshore provinces is as follows:

Australia	5
North Sea	151
Gulf of Mexico	52
West Coast Cal.	42
Brazil	163
West Africa	13
S.E. Asia	31
Mediterannean	29
Middle East	26
Arctic Canada	1
Caspian Sea	1
	514

The completions were made by some 16 companies or groups of operators in water depths ranging from 49 to 410 ft. The maintenance of the equipment used on these wells has required a varying degree of reliance on divers. Nearly all have fail-safe valves, operated remotely from a nearby surface installation on a platform or shore, using hydraulic control systems.

Although a few have been equipped for remote connection of flowlines, all but one have required the services of divers to assist in this activity. In summary, to date, no installations have been made in water depths beyond diver capability (150 m).

Operating Expenditure (OPEX)

The operating expenditure requires the greatest attention in the design and selection of the overall subsea systems, particularly the control and monitoring instrumentation. This requires a thorough assessment of the operating life of the proposed field together with an analysis of component reliability. Subsea maintenance must, therefore, be considered in the design from the intitial concept. Annual operating expenditure on these floating production units is approximately 10% of the capital expenditure of the field.

Equipment reliability has been a prime concern with regard to placing equipment on the seabed. As a consequence, alternative approaches to the diverless systems have involved encapsulating the wellheads with seawater at ambient pressure, which can be reduced to atmospheric pressure when manned intervention is undertaken (Neutrabaric system by Vickers-Intertek).

A total subsea production concept is still evolving, and currently, intermediate concepts have been developed incorporating subsea wellheads and floating production facilities. Subsea technology has, therefore, become complementary (not in the greater depths) to field development by increasing production without significantly increasing the capital costs. The main efforts are, therefore, concentrated on satellite wells, flowlines, and manifold centers as well as platform and export risers linking the subsea installation to the surface.

The effectiveness of a marginal-field production system for small fields, based on floating production units, is now completely proven. The principal problem is to engineer an economically viable system. In the North Sea, semi-submersibles give good weather performance and may be an operator's first choice if storage is not required. If storage is required, then the use of a floating tanker-type unit on a weather-vaning mooring would appear to be the first option to be considered.

Working underwater is a lot more difficult than working above water or on land, therefore the incentives to do so must be found in some aspects of project realization, such as:

- Reduced cost
- Advantageous schedule
- Improved technology

In this chapter, these aspects will be considered for a number of floating systems and subsea developments which have been carried out. The North Sea provides examples of floating systems, starting with Hamilton Brothers' Argyll and the development of the buoyant tension-legged platform of Conoco, and the Hutton TLP which was in production in 1984.

Subsea systems are complementary to existing fixed structures and increase the recoverability; for example, BP Magnus has seven satellite wells.

Subsea production began with the Zakum Subsea Production Scheme (1969–72), and now subsea systems have been used in a number of field developments in the North Sea, for example the Shell UMC, Texaco Highlander, and North-East Frigg.

This chapter also discusses some prospective developments in applications, techniques, and equipment, focusing on key research and development projects and outlining current trends in subsea technological advancement.

Most critical of all subsea components, the link between the seabed equipment and the surface, is the riser system. Although top tensioned rigid risers have performed satisfactorily in the North Sea, the flexible-riser system is so inherently superior in performance and easier to maintain and operate, that to date it is considered to be the optimum choice for a production riser system. Examples of production riser systems are discussed in Chapter 2.

Early practitioners of the art of working underwater include marine salvagers and offshore drillers. The underwater techniques which were developed for these activities are being elaborated and improved upon for those more complex activities of offshore oilfield development. Experience, and common sense, has taught that complicated construction work should be done as completely as possible on land before taking units offshore for installation, and that only minimal assembly be required subsea.

Examples of subsea construction activities include well templates, manifolds, pipeline, multiple flowlines, anchor foundations, use of divers, and remotely operated vehicles. Prospects for the future of subsea activities are assured by the present keen interest in floating production systems and subsea satellite developments to be tied-back to existing installations.

The future North Sea development in the U.K. Sector will be primarily:

- Smaller fields
- Satellite fields
- Condensate

Deepwater developments present greater challenges but, with the discovery of sufficient reserves, there is no doubt that these developments will also stimulate new requirements as well as new capabilities for underwater construction. However, as North Sea reserves are depleted, both U.K. and Norwegian markets will turn towards field development using subsea solutions.

Norwegian Subsea Developments

As many as 195 subsea wells, worth U.S. $6–8 billion, may be installed and brought onstream on the Norwegian continental shelf before the turn of the century. Some 39 of these wells are included in field development plans already sanctioned by Norwegian authorities, and of those, the most promising business potential is related to the Snorre and Heidrum early production system (EPS) products.

Saga Petroleum's Snorre project will be the most comprehensive subsea development on the Norwegian shelf so far. The operator has outlined a subsea scheme which brings the overall capex per well down to only U.S. $45 (U.S. $40 million), including drilling. This was accomplished by moving Christmas trees from exhausted wells onto new wells drilled through the same subsea template, with 20 slots but only 10 Christmas trees. Snorre Phase II could include two additional 20-well templates over the northern flank of the field.

The prediction is that several subsea projects will emerge from the 12th Norwegian Licensing Round, which will focus on the prospective areas close to existing infrastructure and on marginal fields. Subsea developments are likely for small prospects tied to existing processing centers and for fields like Smoerbukk, Njord, and Snoehvit.

As indicated earlier in the distribution of subsea completions by province, a major market contender to the North Sea is coming from Brazil where the need to develop in deeper waters is pushing forward the frontiers of technology.

Focus on Brazil

The evolution of Brazil's offshore industry has led to deeper water and as the president of Petroleum International, Wagner Freire, described, "Out of necessity we had to go deeper as our biggest finds were in deeper waters and as we have short budgets we started cost cutting before it became popular. This took us to floating production systems which are for us—and we believe for other countries as well—the best immediately available solution for low cost, cost effective, and reliable development of deep water fields.'' Brazil is the world leader in floating production with 19 such systems installed. However, for future field developments, Procap is considering subsea solutions for water depths of 600, 1,000, and 1,500 m (1,950, 3,280, and 4,900 ft) with emphasis on technical innovations such as subsea multiphase pumping systems in addition to the floating production system.

The most ambitious deep water subsea development is planned in the mid 1990s for the Marlim oil field. Here water depths range from 400 m to 2,000 m (1,300 to 6,500 ft).

Design innovations to be implemented in these deeper waters include:

- Special layaway diverless and guidelineless wet Christmas trees
- Specially designed wellhead and guidebases
- Export monobouy for anchoring in 600 meters (1,968 ft) of water
- Extensive use of flexible hoses
- Permanent deepwater floating production system based on purpose-built semi-submersible

Esso Australia

Esso is at present set to develop a series of marginal finds in the Bass Strait. The range of production schemes chosen to find economic solutions varies considerably—from conventional jackets through gravity based mono tower platforms at Perch and Dolphin to subsea solutions for Tarwhine and Seahorse reservoirs (due on stream early 1990) (see Fig. 1–2).

The Tarwhine and Seahorse fields have the smallest deposits for proposed development at 6.8 million and 4.4 million barrels. Electrohydraulic control is to be provided by the existing Barracouta platform with pipeline and umbilical tie-in. Both flowlines have to be thermally insulated to avoid wax/hydrate formation. Esso is intending to use a reel-lay method to reduce costs during installation, and pipe bending trials are being carried out to qualify a polyurethane foam insulation system.

The tree proposal is a non-TFL design with metal-to-metal seal. The control system uses conductive couplers, and both gas lift and pigging will be provided. The systems also include two subsurface safety valves in tandem to minimize the need for workover.

Esso Australia holds the current record for a step out well with 4,600 m (15,091 ft) in the Bass Strait.

SUBSEA TECHNOLOGY

The development trend toward subsea production is closely linked with developments in technology. Changes in thinking have been brought about by the need to recover field reserves in the most economic method available depending on factors such as field capacity and water depth. Operators now realize the limited potential of conventional field techniques and are looking to subsea for the future. In this respect, most major oil companies are sponsoring research and development projects related to subsea production.

In the North Sea, a significant parameter is the distance of a satellite field from an existing field development. A target would be to increase the distance from 10 km to 30–40 km over which they could be tied into existing facilities.

For subsea systems to improve in reliability and performance, key areas of technology will need to be advanced such as:

- Underwater power and distribution
- Subsea processing/multiphase flow
- Remote controls and robotics (remote maintenance systems)
- Underwater operations and well servicing

On multi-well systems, commingling reduces the need for multiple risers, and an evaluation of the various task requirements highlights the need to consider separate manifolds for production, water injection, and satellite-well production. Whatever the piping and manifold arrangement chosen, the basic design criteria will generally apply. However, consideration of possible future gas lifting of wells must be included in the initial design philosophy, if expensive retrofitting of equipment at a later date is to be avoided.

The key system of subsea production schemes is the remote control system, which represents only 3–10% of the capital expenditure of a subsea completion. However, the system comprises the majority of the complex components and is potentially the greatest source of failure. Although there are five possible types of control systems available, only electrohydraulic multiplexed systems are considered suitable for systems with 10–20 wells or more. This is based on experience from early testing in 1968 by Exxon through successive developments in Cormorant and culminating in the Shell/Esso UMC.

The majority of the installations to date have performed satisfactorily, and even though component reliability is improving, thorough testing at higher than operating pressures and operation in all modes increases the likelihood of service reliability.

Other key components are the trees and wellheads, and recent designs have led to the simplification of tree assembly by omitting the lower master valve which, although being accepted outside the North Sea, has not yet gained acceptance in the United Kingdom continental shelf (UKCS). The tree control system must be laid out efficiently to allow for simple transfer between the production and work-over mode.

In the early stages of the project, a series of initial assumptions will have to be made through discussion with the Department of Energy/Pipeline Inspectorate (DOE/PI) and others concerning flowlines and export pipelines. As the field development is ad-

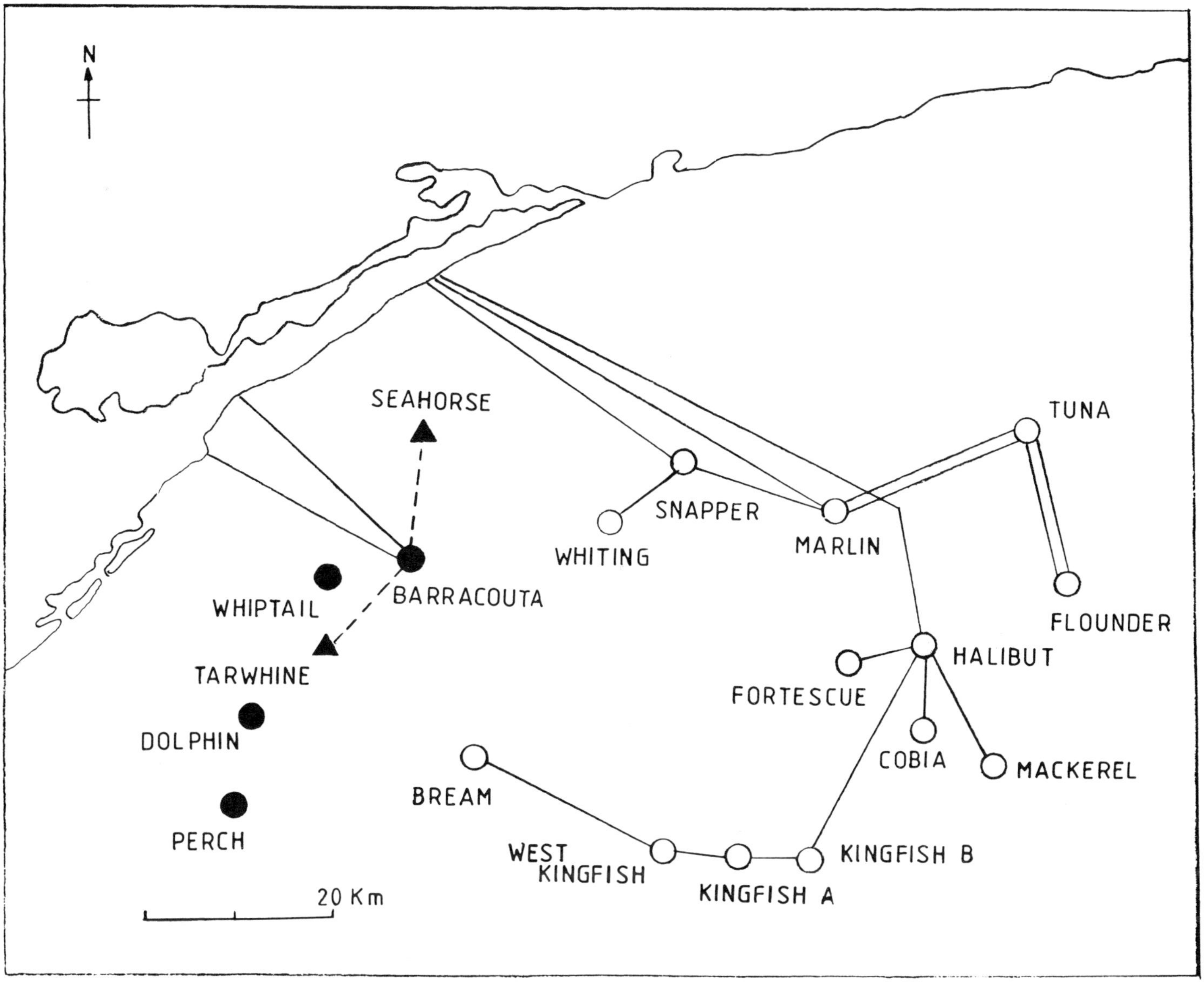

Fig. 1–2 Gippsland Basin, Bass Strait, Australia. White circles show installed platforms, black show those planned. Whiting is unmanned while Seahorse and Tarwhine fields are to be developed as remote controlled wellheads tied back to Barracouta (as of June 1989)

vanced, the cost of the alternative methods must be reappraised continually as it is the most volatile cost item in the budget. The basic difference in flowline costs can vary by 10–20%, if not more.

The expected flowlines throughout the field range from 10–36 in. diameter for the export lines to 4–8 in. for the satellite wells and 2, 3, and 4 in. for the test and service lines. The laying of the larger lines with a conventional lay-barge is the most effective and economical method. However, for the numerous lengths of smaller-diameter flowlines, repeated passes of the lay-barge is expensive and extremely hazardous to the already laid flowlines. The use of dynamically positioned (DP) reel ships laying steel or flexible lines is one cost-effective solution. Alternatively, the bottom-tow method of laying bundles of shore-fabricated pipes may be used when several flowlines and control umbilicals can be combined. Pipe connection is generally performed by divers using rigid or flexible spools.

In the North Sea, fishing is a primary hazard to the flowlines and has led to the practice of burying flowlines. Several different methods are in use, such as jetting, plowing, mechanical-cutter dredging, and rock dumping.

Many crudes have a high wax content which can lead to

hydrate formation or to emulsification, therefore the flowlines need to be insulated. This can be done by burial, but soil conditions will influence the final choice. It is particularly important to note that only flexible pipes can be buried deep enough for effective insulation (more than 2 m deep). However, the bottom-towed method can use preinsulated flowlines and, alternatively, expanded polyvinyl chloride (PVC) foam coating can be applied to provide a high-efficiency insulation.

A fundamental difference between onshore or platform-mounted production systems and floating systems is in well-workover. Well-workover is essential to maintain the maximum output. However, subsea wells present difficult problems. The through flowline (TFL) technique is not considered as yet to be sufficiently economic for marginal fields. Major workovers can be carried out by using an integral riser run through a marine riser and blow-out preventer (BOP). However, multi-purpose tooling can be run on tubing provided that a terminal head is used to recirculate hydrocarbons. Light workover is achieved by entering the tree via the tree mandrel. As drilling operations have moved into deeper waters, wireline operations from floating rigs have become a more common practice.

In the North Sea, there is an increasing need to carry out the well-servicing tasks from less expensive vessels instead of drilling rigs. This is particularly the case in fields with a large number of widely spread satellite wells. Many potential types of vessels are being considered for infield service, including small semi-submersibles and new forms of monohulls.

Development of subsea systems has an important operational interface, namely maintenance. The need to prove the reliability and minimize the costs associated with maintenance has been a major constraint on development to date. Many operators are only experienced in dry access maintenance and have little or no experience with subsea systems. Most of the earlier systems have relied on diver intervention although some use has been made of eyeball cameras on remote operated vehicles (ROVs) for inspection purposes. Where diver depths were exceeded, research and development centered around divers working in the dry. A notable example was the Kvaerner concept developed for Mobil & Statoil, namely the Subsea Atmospheric System (SAS). The main problem with marketability of this system was its reliance on a diver transport system to the one-atmosphere chamber.

Alternative diver-related intervention techniques which have been developed include:

- One-atmosphere suits
- Wet one-atmosphere chambers
- Dry one-atmosphere chambers
- Insert technology/dedicated maintenance vehicles
- Manned submersibles, e.g., Statoil's MASU

Recently, the emphasis has switched to diverless intervention techniques, particularly for the deeper fields where diver-assisted technology is impracticable. This has inspired growth in the application of robotics. Remote vehicles with cameras and lights are an existing part of conventional technology, with the innovation being in the adaption of equipment to facilitate remote maintenance and operation. In particular, tooling packages and module deployment techniques are currently at the forefront.

Operators are testing the latest design developments to gain confidence in the technological advancements prior to full field scenario application. Notable examples are the BP Deep Water Integrated Subsea Production System (DISPS) module deployment system and AGIP's Sisteme Alti Fondali (SAF) study on module handling up to 50 tons. Using small 50-ton maximum modules increases the number of subsea connections but enables regular maintenance to be carried out by small vessels. The larger modules minimize installation costs and the cost of connections but, until handling problems are solved, maintenance costs will be higher. Research and development for handling larger modules subsea (up to 200 tons) is an area for possible future development.

Policy On Intervention

There are two main schools of thought concerning diverless intervention currently being developed in conjunction with subsea projects, however they differ in the use of guidelines. Most subsea research and development projects fall into one or the other of the following categories:

- Guidelineless
- With Guidelines

Operators are aiming to use existing diving support vessels (DSVs) for module deployment. Guidelineless intervention techniques can rely on either remote guideline vehicles or drill strings to run the equipment to the seabed. A common feature is the module stack up arrangement where small modules are interconnected together and stack up usually with the module needing most regular maintenance at the top. This system is seen in the BP DISPS project and similarly in the Goodfellow Associates Subsea Processing System (GA-SP) described later.

For water depths in the range of anchored semi-submersible drilling vessels (SSDVs), guidelines are readily deployed to run equipment packages to the seabed. In deeper water, the technique for guideline assisted deployment would be to use an ROV to install the guidelines, examples are Norsk Hydro's TOGI project and Esso's EDIPS project. However, these systems are expensive because they rely on the use of semi-submersible support vessels.

A novel intervention approach is used on Shell's Diverless Installable and Maintenance Oil Production System (DIMOS) project. Here a small wire-supported vehicle with its own skid-mounted deployment module can be launched from any dynamically positioned support vessel.

Some of these key projects will be described in more detail now.

Mobile/Statoil, Subsea Atmosphere Systems

The SAS is a hybrid wet/dry control system developed by Kvaerner for North Sea application (see Fig. 1–3).

The system comprises a circular template fixed to the seabed by a 1.8 m (5.9 ft) diameter monopile. The template supports a work habitat 18 m high × 8 m wide (59 ft by 26 ft) divided into three compartments:

1. Main nitrogen filled service station incorporating production equipment such as valves, chokes and manifolds;
2. A one atmosphere upper chamber where personnel can work, and;
3. A personnel transfer chamber.

The template itself is divided into 10 bays to provide 9 external wet wellhead trees and one for a flowline connection unit. The system can produce up to 75,000 B/D to a production platform up to 10 km (6.2 miles) away. Split trees are used which can be installed prior to chamber installation. The unit has been designed for diverless installation. Emphasis has been placed on protection from fishing trawler gear and anchor snagging, which results in the overall domed shape.

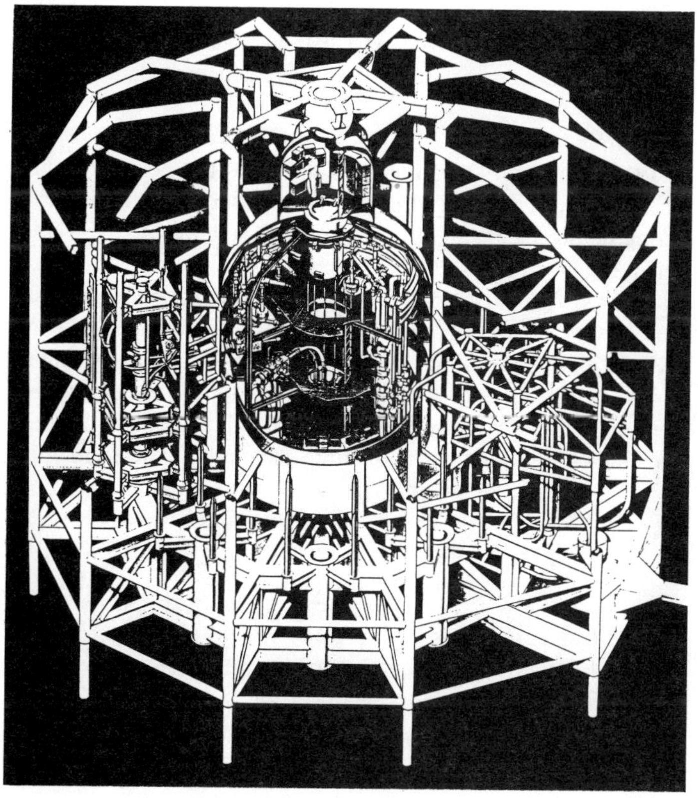

Fig. 1–3 Subsea Atmospheric System

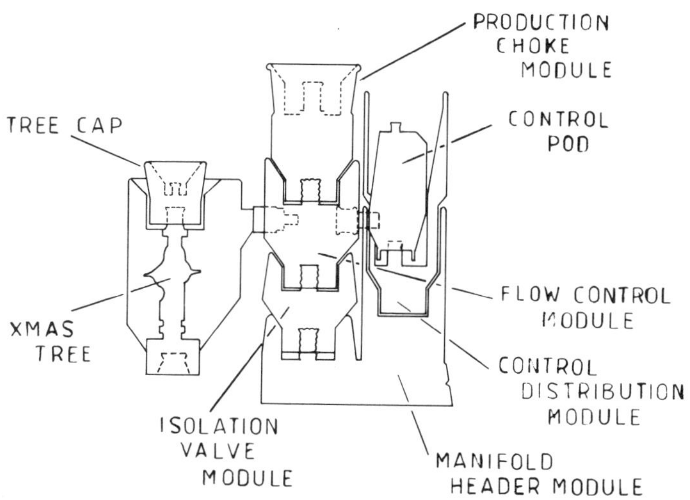

Fig. 1–4 Disps-Module Connection System on the Template (Cross-Section) (Source: BP)

BP Deepwater Intergrated Subsea Production System (DISPS)

The system is based on a template fixed to the seabed. Wells are drilled through both the template and at nearby satellite locations which are tied in by flowlines. A ''slot'' on the template allocated to each well accommodates the production equipment. The equipment is pre-assembled into a number of linking modules which are remotely connected at the seabed (see Fig. 1–4).

The subsea equipment is deployed from an existing dynamically positioned support vessel. Intervention is totally diverless, with ROV tooling developed for minor intervention tasks and module retrieval available for higher level maintenance. The modules are arranged to ensure those requiring the most maintenance are more readily retrievable.

The concept has been taken from R&D to full scale model tests on the representative one-slot test rig. Currently, BP is considering using DISPS's subsea technology for the 23/26 field which is being developed as a floating production system (FPS).

The final decision will be based on the cost competitiveness of the system with respect to existing subsea technology and the availability of essential components being developed for the full scale test program.

Norsk Hydro, Troll Oseberg Gas Injection Project (TOGI)

The TOGI system is a template manifold system with six slots and five monoblock Christmas trees. Gas is piped through a 20-in. flowline 48 km (30 miles) from the template for injection in the Oseberg reservoir. The water depth is 305 m (1,000 ft), and the template incorporates a tubular/cone trawl gear protection frame. Equipment is arranged so that it does not project above the upper framework (see Fig. 1–5).

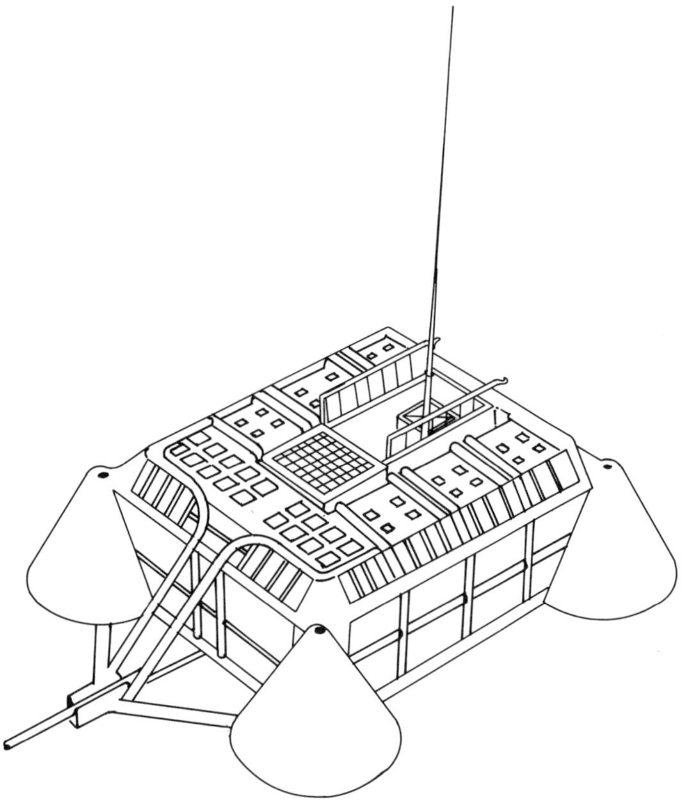

Fig. 1–5 Togi Template

Overall control is provided by Oseberg using an electrohydraulic power supply with provision to transfer control to the workover vessel during well mainteance via control pods on each tree. The all wet controls are handled by ECA pods. The pipeline is concrete coated and includes a safety valve.

Over a period of 12 years, TOGI will boost oil recovery from Oseberg by between 75 and 95 million barrels—squeezing some 7% more from the reservoir. The price to be paid for this—TOGI's cost—is put at U.S. $448 million. However, the switch from water injection to gas injection eliminates the need for some 20 water injection wells, costing around U.S. $299 million.

All in all, the net increase in investment required for TOGI is reduced to about U.S. $194 million, after allowing for other facilities added at Oseberg. Three-quarters of the Troll gas injected at Oseberg can itself be recovered in the later gas production phase.

Esso, Deep Water Integrated Production System (EDIPS)

The system comprises a riser base manifold piled into the seabed linked by a single anchor leg mooring (SALM) to a floating production storage unit (FPSU) vessel. The application is based on water depths of 600 m to 1,000 m, and features which it incorporates are:

- Guidewire deployments
- Dropped object protection
- Removable manifold module
- Porches to support flowlines
- Remote intervention

The oil flow is commingled for piping up the export riser on the SALM. Latest development activities have centered around the subsea maintenance aspects, and some valve tooling is to be developed to prototype test stage.

Processing Complexity

The design of total subsea production and processing facilities removes the dependance on nearby processing facilities. Although developments in multiphase flow are being made, the final barriers to this highly competitive subsea system have yet to be breached. Research and development into subsea separation and multiphase pumping is however coming of age. Both are being put forward as economic solutions for marginal fields.

At the end of 1989, Goodfellow Associates took the detailed design of its GA-SP subsea separation project into manufacture of a prototype system which is to be tested throughout 1990. The GA-SP system is described later in this chapter.

In terms of hardware in operation, subsea separation equipment is already in the North Sea. The Texaco Highlander Slug catcher is a very simple subsea separation unit installed near the Tartan platform. Hamilton and BOET have installed a prototype separation unit at the Argyll field. The full scale BOET system is intended for step out and isolated marginal developments.

Multiphase pumping has yet to become viable offshore, and a large number of pumping manufacturers are designing their ultimate subsea systems. Stothert and Pit and Weir are concentrating on displacement units while IFP/Total is evolving a rotodynamic version.

Another spin off from multiphase flow developments is the multiphase flow meter required to assess reservoir productivity. To save piping costs, flow from individual wells is commingled into a single production line, and the characteristics of production from individual wells can be ascertained by simple by-pass lines into a multiphase flow meter. This avoids the need for a costly test separator subsea.

Multiphase Flow Meters

It has been known for some time that the Japanese industrial giant Mitsubishi Electric has been applying its considerable technological might to the development of a multiphase flow meter. Little information has been available on the project. However at a recent conference in the U.K., some of the gaps were filled in.

Background. Mitsubishi has not been working on the project alone since they gained oil industry input by entering into a joint development and testing agreement with Petro Canada, Inc. in 1984. Set up in 1975 by the Canadian government as the national energy corporation, Petro Canada is the only Canadian-owned, fully integrated, oil and gas company with a national marketing presence.

Since the Mitsubishi-Petro Canada agreement was signed, sufficient progress has been made to begin field tests of a prototype meter in an onshore Canadian oil field. The test began in April 1987.

Meter Details. At the September 1987 IBC Industrial Flow Measurement Conference in London, a joint paper by Mitsubishi and Petro Canada provided further details of the meter and testing program. The paper was presented by Mr. Rafa from Petro Canada.

Mr. Rafa said that the objectives of their work had been to develop a multiphase meter which was both small in size and sufficiently economical to permit them to be installed in the flowline of every well and to measure continuously flow rates of oil, water, and gas without phase separation.

Work has concentrated on the development of a gamma ray oil/water/gas fraction meter. In the field tests, the temperature and pressure of the produced multiphase fluid and the differential pressure across an orifice plate are being measured and used in conjunction with the phase fraction data to deduce flow rates of oil, water, and gas.

Since the field tests are still being carried out, and results therefore incomplete, Mr. Rafa's presentation concentrated on a description of the meter. It covered gamma ray transmission techniques, studies on the technical feasibility of the Mitsubishi-Petro Canada meter, the meter design, and pressure, leakage, wear, corrosion, and content analysis tests.

The meter has multiple measurement paths to distinguish between the oil, gas, and water and is designed to be installed in-line using a standard 50 mm flanged connection.

Other companies are becoming more interested in developing multiphase flow meters, and recently, Euromatic has made some technological advancements. There is still a lot of work to be done before multiphase flow meters are readily available for subsea application.

Since the Mitsubishi-Petro Canada agreement was signed, sufficient progress has been made to begin field tests of a prototype meter in an onshore Canadian oil field. The test began in April 1987.

ALBACORA FIELD DEVELOPMENT

Discovered in November 1984, Albacora was the first giant field discovered in Brazil's Campos Basin which has hosted many world water depth record holes and completions. It lies in waters between 200 m (656 ft) and 2,000 m (6,500 ft) and has reserves estimated well above 1.1 billion bbl. The field is intended to be developed in four phases, each in turn providing information and technology for the next.

Phase 1 started with the field coming on-stream in November 1987 with a world depth well in 419 m (1,374 ft). The installation, from FMC, was diverless with flowlines laid back to a diver-assisted manifold in the shallower part of the field. This was the deepest of six wells installed in phase 1.

From the manifold, production from all six wells was by a dynamic hose system in 236 m (755 ft) of water to monobuoy and yoke to the processing vessel Presidente Prudente de Moreas. Export oil is pumped via a PLEM to a second monobuoy where it is loaded into shuttle tankers. Compressed gas is exported via the PLEM and a 16-in. pipeline some 38 miles to the Garoupa platform and from there ashore.

Phase 1A, which was underway in 1989, is intended to bring an extra six wells on-stream in water depths down to 600 meters (1,970 ft).

Phase 2, due for completion in March 1990, will absorb the fields 12 existing wells and represents an orderly development of reservoirs down to 600 m (1,970 ft), with anything up to 116 wells being brought on-stream.

This involves connections to four semi-submersible platforms, one of which will be a central processing unit with export lines to the Garoupa field. Also involved in this phase is the complete development of the shallow end of the reservoir, involving some 39 extra wells, all directionally drilled from three subsea template manifolds and tied-back to the "PP Moreas."

Phase 3, due to start in 1993, involves the use of some eight semi-submersibles to develop deeper waters beyond 1,219 m (4,000 ft) and to date involves a planned 114 wells and template manifolds. Production is intended to be tied-back to one central platform.

GOODFELLOW ASSOCIATES SUBSEA PROCESSING PROJECT

A significant European Community (EC) sponsored project, with participation by seven oil companies (Agip UK, Chevron, Elf UK, Norsk Hydro, Phillips, Statoil, and Texaco), is under way. These activities cover reliability/risk analysis, prototype trials, and evaluation. The project highlights areas for improvement including interconnections, modularization, multiphase flow, installation, and retrieval.

GA's work on subsea processing arose from a confidential study on subsea systems undertaken for a group of 10 oil companies. This year-long project was completed in July 1986 and involved a detailed and wide-ranging review of subsea components and production systems. It resulted in four comprehensive reports covering:

- Underwater power and distribution
- Subsea processing

- Remote control and robotics
- Underwater operations and well servicing

After completion of this work, GA proposed a project for the participation of oil companies in a stand-alone subsea processing, manifolding, and booster pump exportation system. It is envisioned that the system could have several applications, operating in conjunction with remote satellite wells—either single or arranged as multiwell templates—from an existing platform or below a simplified, low cost, low manned, floating control platform (see Fig. 1–6).

After receiving European Community (EC) funding and oil company participation, GA began pre-engineering work in October 1986. The cost of the project, scheduled over a period of three and a half years, is estimated at £5 million (U.S. $8 million). GA has secured EC support for the first and second phases of the project with additional EC support for later phases. Manufacturers will also make significant contributions to the project with the loan of equipment.

The four objectives of GA-SP are:

1. Development of a 30,000 B/D seabed processing and boosting unit which could operate in water depths of 150–400 m (492–1,312 ft) to produce crude of exportable quality by pipeline.
2. Modularization of individual elements of the system—a support structure for diverless, guidelineless, single datum deployment and connection (using valved multibore connectors) for low cost.
3. Application and proving subsea multiphase flow metering and single phase fiscal metering/sampling devices.
4. Reliability/failure mode analysis of subsea components and systems.

The time scale is:

- Phase 1 (12 months): September 1986–October 1987; pre-engineering and detailed engineering
- Phase 2 (14 months): November 1987–January 1989; detailed development of the system for prototype construction
- Phase 3 (12 months): December 1989; prototype testing

Marginal Field Development

The GA-SP concept is, in essence, a recognition of the growing need to develop cost-effective technology for the exploitation of satellite marginal fields.

For producing over 30 km (18.6 miles) distances, one option—examined in detail in Multiphase Update No. 1—involves using multiphase pumps. Another option is to employ subsea processing; this may still require pumps to boost the oil, gas, and water phases back to existing facilities. In either case, the need still exists to manifold and commingle the production prior to pumping and the requirement to provide power for the fluid transport system.

Although considerable progress is being made in developing multiphase pumps, the application of fully reliable systems in field situations is some time away. On the other hand, processing techniques are well established, and the incorporation of subsea process systems in a subsea scheme would naturally increase the range of exploitation schemes available.

Even when reliable multiphase pumps become available, GA says that their efficiency would be increased by a simple separation process. GA and the participants have identified a number of field applications involving satellite fields with a range of distances away from the main platform or support facilities.

Key Elements

Important features of GA-SP include: subsea booster pumps and separators. Various designs have been and continue to be examined. An important aspect is, of course, the potential for adaptation of components and the system to subsea use. Regarding the separators, both single- and two- or three-stage designs have been considered.

The system also includes a subsea multiphase flow meter and a subsea single phase metering device.

Other important elements of GA-SP include: a pigging system, subsea chokes and manifolds for the production wells, water injection system, gas lifting system, and a multibore connector.

GA emphasizes that their work involves designing the integration of existing components (or those under development) into the GA-SP system—the project, for example, does not cover subsea wellheads and Christmas trees which have an established track record.

GA's work involves surveying what is under development by manufacturers and determining how those can be adapted for GA-SP into a fully modularized system. GA says that ideally, they would like to be in a position to choose from about three manufacturers for each main specialized component.

GA is in close contact with those manufacturers directly involved in developing products in the multiphase area.

Modularization

A study of the various modularization options has formed an important part of GA-SP. Particular aspects covered have included: size, manufacturing ease, minimizing manufacturing tolerance requirements, installation, mating, and mechanization of connections. Other considerations included protection of the system, access by ROVs, and the emphasis on minimal support vessel requirement.

The components which are likely to need maintenance are retrievable and form the primary modules within the system. The secondary modules are mostly made up of pipework serving

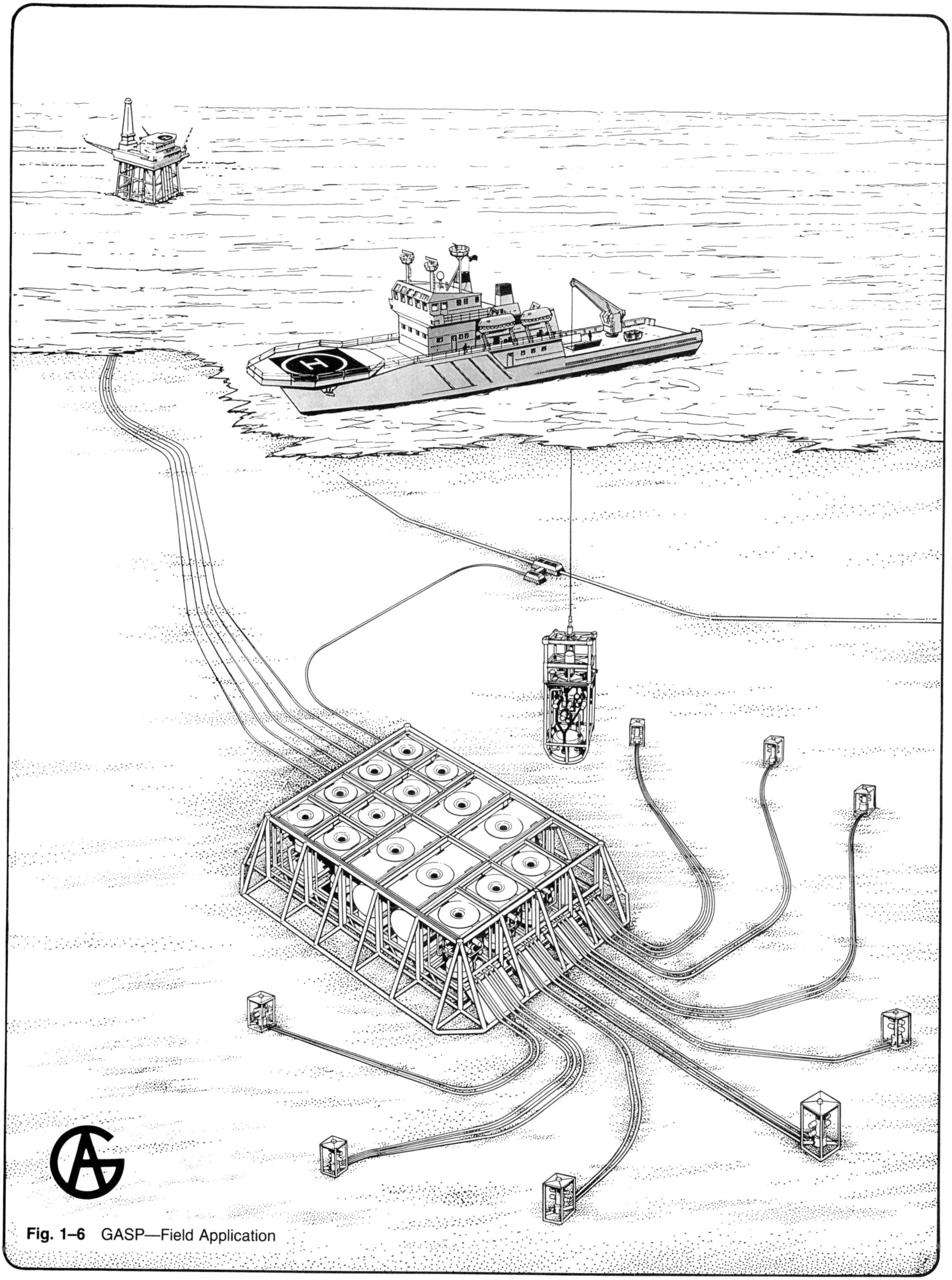

Fig. 1–6 GASP—Field Application

the various components and so require virtually no maintenance. However, the system has been developed to enable the retrieval of the secondary modules if needed.

Reliability Study

An important element of the project is collecting data on the reliability and failure of subsea components and equipment from operators. Information is being sought from a large number of offshore operators, including those not involved in GA-SP. Borregaard Engineering, the Norwegian company with whom GA has set up a joint venture, will be assisting GA in the reliability study.

The information and conclusions resulting from the study will be made available to those companies that supplied a substantial amount of data.

Prototype Testing

The final stage of GA-SP involves testing of a full-scale prototype system based on a single stream processing throughput. This work will involve testing both key components and the integrated system as a whole under a variety of conditions to simulate the subsea production situation (see Fig. 1–7).

For economic reasons, the prototype will not be as fully modularized as in the completed system.

Components for the prototype testing, such as the pumps, separators, and wellhead, will be loaned from a number of manufacturers and operators. After the equipment has been fully tested, the manufacturers and operators will have their equipment returned. This arrangement has helped considerably in reducing the overall cost of the project.

Live Subsea Applications

The discussion on subsea technology has concentrated on the current trends in research and development projects. There are, however, a number of subsea systems which are part of field developments already on-stream. Some of these projects are now outlined in more detail.

EXXON SUBMERGED PRODUCTION SYSTEM

Shell/Esso Underwater Manifold System (UMC)

In 1969 Exxon began a development program which culminated, in 1974, with a pilot test of its Submerged Production System (SPS) in 170 ft of water in the Gulf of Mexico. The success of this U.S. $81 million project included remote flowline connection and robot maintenance. With 18 months of production from three wells, this was a major factor in the development of the Shell/Esso UMC.

Shell and Esso's basic project philosophy was to use proven technology as much as possible in developing a design with emphasis on safety, reliability, flexibility, and ease of maintenance. However, not all existing components met the North Sea project criteria for flexibility and reliability. In-depth design studies and intensive post-construction testing to satisfy these requirements have occupied the two companies since 1974.

Subsea Experience

Shell's input to the UMC project was based on long experience of subsea tree operation offshore Brunei and a pilot test of a 1 atm (101 kPA) encapsulated manifold.

The UMC is similar in appearance to Exxon's SPS but offers additional features—in particular, the ability to serve wells drilled directly through the template and/or remote satellite wells. Essentially, the UMC provides four basic functions:

1. Drilling of wells through the base structure and satellites tie-in
2. Collection of fluids from producting wells for delivery to the South Cormorant platform 45 miles away
3. Distribution of treated seawater from the platform to selected injection wells
4. Maintenance of valves and controls without the use of divers

The basis for the design was a nine-well manifold with each well bay available for template or satellite wells. These wells could be either producers or water-injectors. The structure weighs 2,200 tons and measures 52 m × 42 m × 15 m (170 ft × 138 ft × 49 ft). Production capability is 50,000 B/D of oil and the water-injection capability is 56,000 B/D. The water depth limit is currently 1,000 ft, but it can be readily adapted to suit 1,219 m (4,000 ft) water depths (see Fig. 1–8).

The base structure is a tubular steel frame which supports the other subsystems and assures alignment for connection. (It has 120 funnels for installing guide posts.) It also operates as a drilling and casing guide frame. The structure has nine well bays, four on one side and five on the other, separated by a central track on which the Remote Maintenance System (RMS) runs. Installing choke and "S" spool pieces in any well bay enables satellite tie-ins.

Well Completions

On the Central Cormorant, all of the satellite and template well completions have two 3-in. internal diameter tubing strings, each with a retrievable subsurface safety valve. As a backup, a pumpdown insert safety valve is provided also. Other features of the downhole completion are upper and lower members for circulating fluids. Also, there is a dual detachable packer so

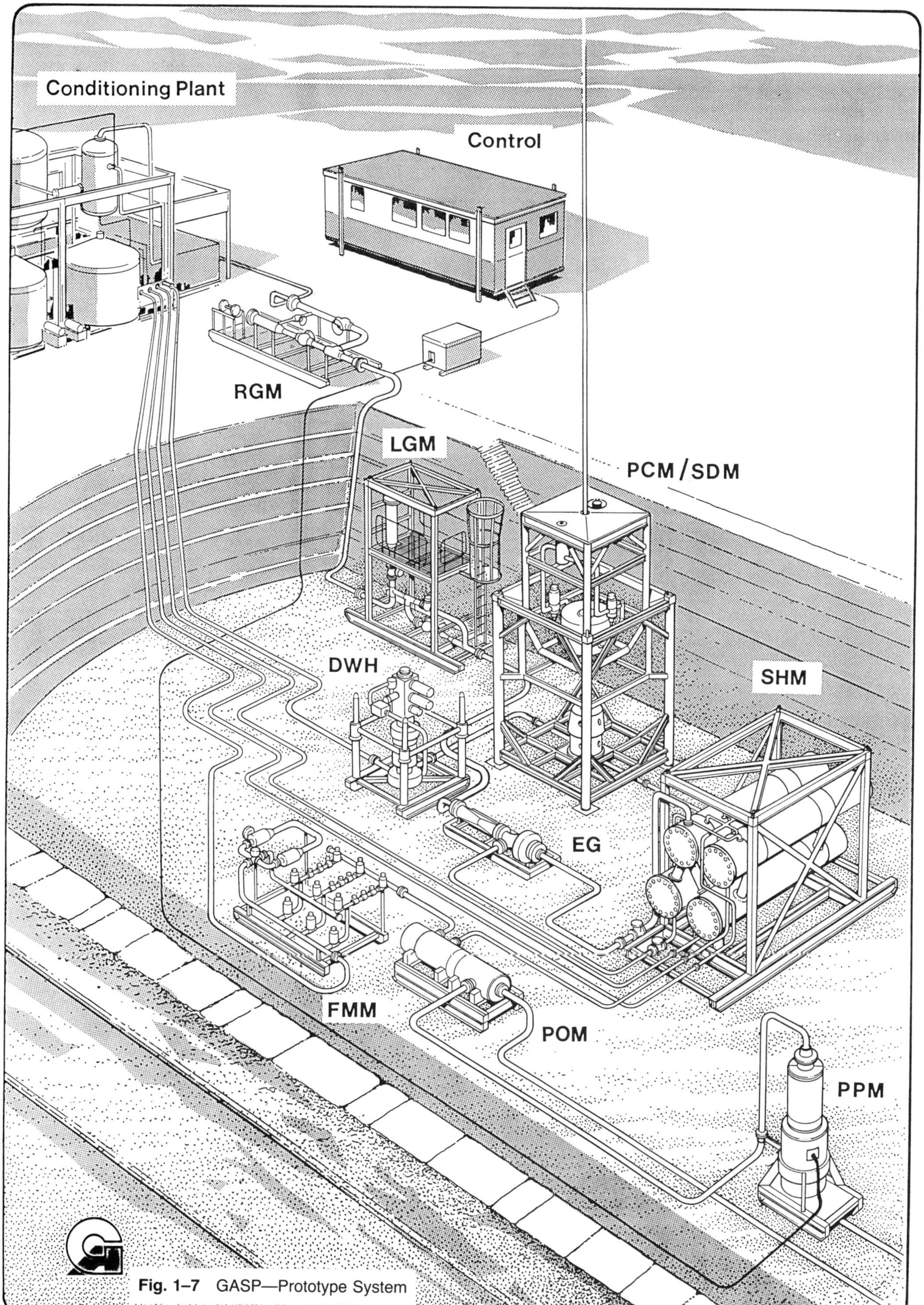

Fig. 1–7 GASP—Prototype System

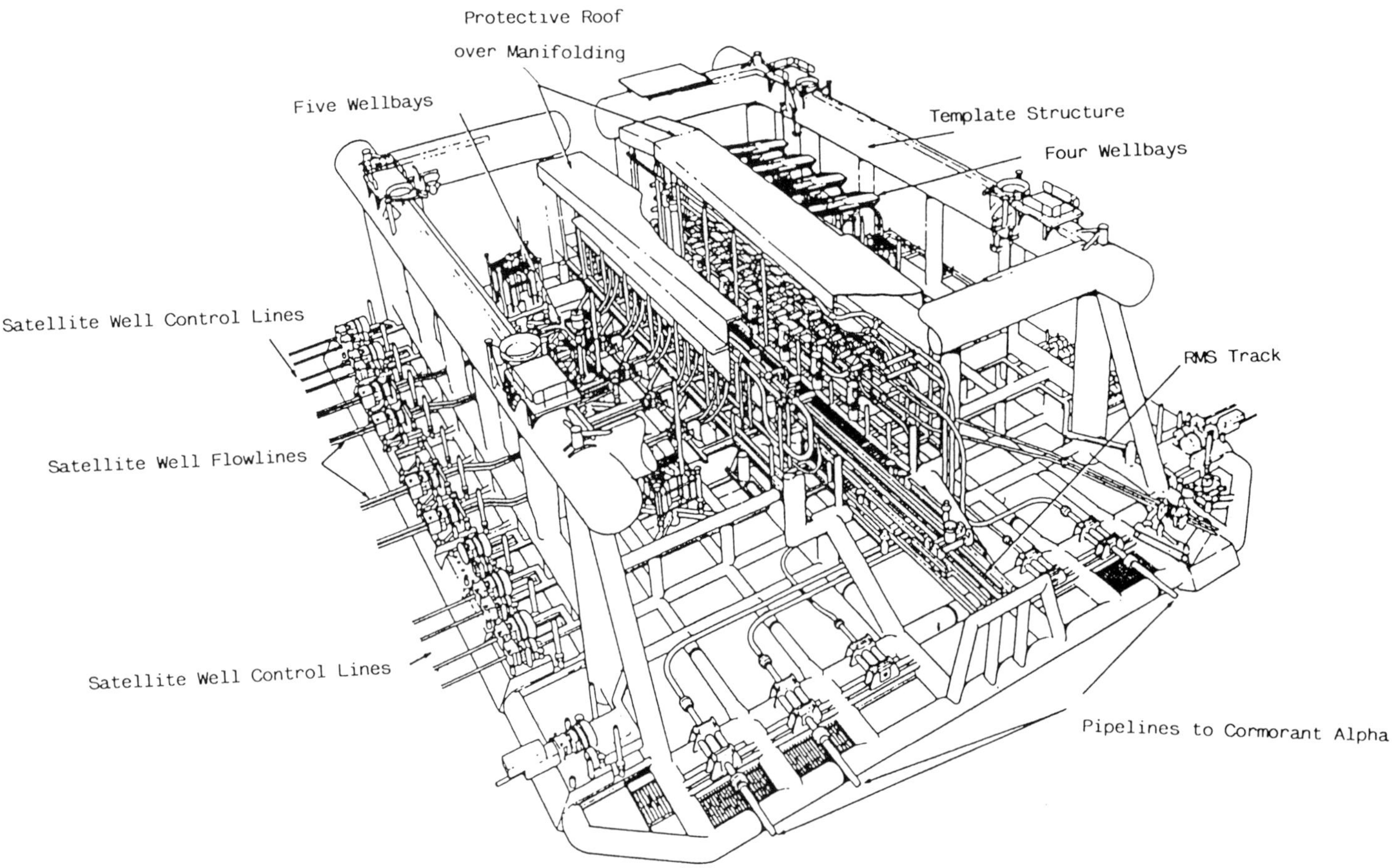

Fig. 1–8 Underwater Manifold Center (Source: Shell/Esso)

that the lower assembly may be left undisturbed during service work. This set-up permits either dual- or single-string flow and allows wells to be used as producers or injectors. All of the completions will be geared for TFL (through flowline) or pump-down tool intervention. In addition to the back-up safety valves, TFL will be capable of changing valves, setting and retrieving plugs, cutting wax, and running downhole surveys.

By using TFL techniques, the need for vertical access for servicing operations, which would be expensive and weather dependent, is eliminated from a semi-submersible rig. Before going offshore, all downhole components have been extensively tested for reliability and performance.

Christmas trees were subjected to extensive interface testing also. The trees feature dual 3-in. strings with dual master valves and a TFL loop. All seals are metal-to-metal and connections are remote diverless. Working pressure is 5,000 psi.

Each well bay on the UMC has a removable cover to protect the Christmas tree, particularly during simultaneous drilling and production operations. The flexibility of the system is such that it is possible to replace a well drilled through the template with a satellite well.

Production Manifold

The production manifold on the UMC is situated on either side of the central maintenance track. On the manifold, there is an 8-in. oil header for collecting oil production, an 8-in. water injection header, and two 3-in. headers for well testing and TFL access. In all there are 110 valves and diverters.

Roofs designed to trap any hydrocarbon leaks cover the manifold and provide protection. In the event of a leak, sensors in the roof sound an alarm in the platform control room. Any hydrocarbons that accumulate as a result of a leak can be recovered back into the 3-in. header system, using valving and blowdown procedures. Packed valve assemblies serving individual well bays and maintainable by the manipulator were used in the SPS project, but in a different configuration.

Two methanol-injection packages are installed on diagonally opposite corners of the structure, with hydraulic accumulator packages for the control system at the other corners. This arrangement ensures 100% redundancy, particularly in the event of damage from falling debris. However, it would be unlikely that debris would hit both methanol and hydraulic accumulator packages.

The UMC is 4.5 miles from the South Cormorant platform and is connected by two 8-in. production lines, one 8-in. water line, and two 3-in. lines. There is a hydraulic/methanol hose bundle together with two electric power and signal cables also. Dual 3-in. flowlines serve the satellite completions. During design of the pipeline system, consideration was given to the effects of cold crude (and line stability during trawl-gear impacts). Wax and hydrate problems could result because of the distance to the platform, particularly for the satellite wells.

Electric Couplings

Inductive couplings have been used because they do not require any mechanical link to transfer the power signal energy. Each half of the coupling houses what is, essentially, half a transformer. This approach was adopted since it was considered more reliable than pin to pin connectors.

OCCIDENTAL SCAPA DEVELOPMENT

The Occidental Consortium's ''Scapa'' field came on-stream as planned some 14 months after Department of Energy approval. The Scapa development consists of a major subsea facility connected to the Claymore platform three miles away by two pipeline bundles and associated control umbilicals. The new topside facilities are connected in turn to the pipelines by a large bundle riser, carrying the produced oil and lift gas and includes a facility for a utility if required. The field is estimated to have recoverable reserves of 40 million barrels of oil with a gravity of 32.5° API to be extracted from a Lower Cretaceous reservoir.

The Scapa development is the first field to build a subsea chemical injection and has a considerable engineering content. The main production riser and template each requires a large proportion of this engineering for the installation design. It is a requirement that all of the subsea work be completed within one season. To accommodate this, the contracts are divided into four groupings:

1. Template — fabricate and install
2. Bundle — fabricate and install
3. Riser — fabricate
4. Riser — install

These packages give the flexibility to maximize the offshore work program involving vessel availability and installation capacity.

This section provides but a brief overview of Scapa's facilities and installation difficulties, together with a detailed description of the riser and template. It should be noted with interest that the total weight of Scapa additions to the jacket is approximately 800 tons, or 2.3% of Claymore's jacket and topside weight. The structure's capability of accepting the extra load requires verification.

Facilities

Pipelines and Umbilicals

The template is connected to the Claymore platform by an extensive pipeline system consisting of two 28-in. multi-line composite bundles, an 8-in. water injection line, and a 3-in. bulk chemical treatment line.

Each pipeline bundle is approximately 2.8 miles long and contains a 10-in. production line, a 6-in. test/service line, and three 3-in. gas lift lines. The bundles are connected to the template manifold and the platform production riser by flexible spools.

The two pipeline bundles, as of 1985 the longest of their type to be towed, were installed using a mid-depth tow method. This involved using the natural bouyancy of the bundle to raise it from the seabed, chains to control its level of flotation and two tugs to tow it. During the tow the tugs kept a set tension on the tow-lines, with the depth of flotation of the bundles varied depending upon weather conditions. The two-lines were attached to the bundles by specially designed tow-heads, which also served to splay the integral lines to accommodate the flange tie-in points. To confirm manual calculations and determine hydrodynamic factors, tests were carried out on a scale model of the tow-head beforehand. Also, a full time-domain computer analysis was undertaken to establish acceptable sea states under which the tow could proceed. A similar analysis was used in failure analysis to establish the results of a major component failure, such as could occur in the leading or trailing tow wire.

The two bundles were laid in a corridor approximately 30 m (98 ft) wide, as were the two control and the chemical injection umbilicals, together with the squeeze scale line. Each of these was sufficiently flexible to be wound onto a carousel mounted on a lay-ship. For the installation, a message line, previously installed in the platform J-tube riser, was transferred over to the lay-ship. Once this was used to pull the various umbilicals in through the J-tube bellmouth, up the riser itself (including the ''S'' bend, described in the next section) to the hang-off point at platform +68 ft elevation. Pull-in resistance was lowered as much as possible by making the radius of each curve on the swan-neck as large as possible. After locking off the umbilicals at the +68 ft level, the lay-ship then steamed along the corridor, laying down the umbilicals as it went. Accurate positioning was achieved using a seabed crawling vehicle, which buried the line as it went.

The water-injection requirements were met by means of a flexible pipeline laid between the South Claymore water-injection template and the Scapa template, two miles distant.

J-tube Riser

The J-tube riser is a 36-in. diameter caisson with four 12-in. diameter internal pipes which are used as supports and guides for the umbilicals during the pulling operation.

It was required that the J-tube riser be protected from vessel impact, so a subsea dog-leg was introduced. The riser was therefore installed in three parts:

1. Section 1 — From seabed to −100 ft, incorporating four splayed bends, each incorporating a bellmouth
2. Section 2 — Four large-radius swan-neck spool pieces from −100 ft to −50 ft to ease pulling forces
3. Section 3 — Top section from −50 ft to +50 ft

The J-tube riser sections were installed using a combination of the MSV Tharos, a semi-submersible vessel with a crane capacity of approximately 300 tons, and platform rigging. The tolerances in the clamp-pipe interface were particularly tight and, together with the clamp neoprene liners, caused a certain amount of obstruction to the installation of the lower section. The upper section was difficult to install, due to limited headroom and the proximity of the support trunnion for the main production riser.

Template

The Scapa production template and manifold (see Fig. 1–9) provides a facility capable of accommodating a maximum subsea design production rate of 33,000 B/D of fluid. The template structure includes the following facilities:

- Drilling support
- Christmas tree support
- Process piping (production, test, gas lift, and water injection)
- Controls and instrumentation
- Chemical injection
- Tie-back of off-template wells
- Protection from external damage and from hydrodynamic loading

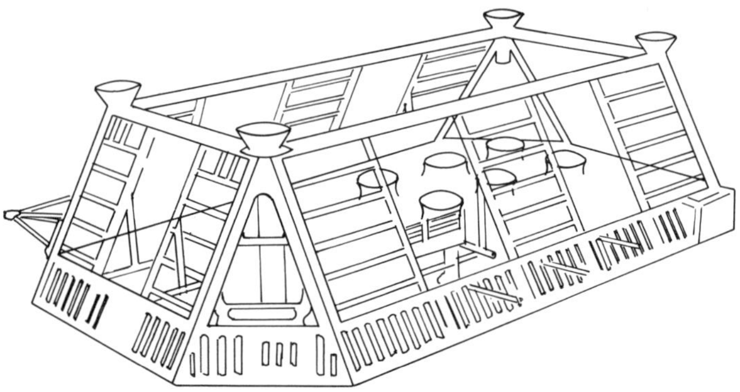

Fig. 1–9 Scapa Production Template

The resulting template structure weighed 920 tons in air and had an approximate in-place operating weight of 400 tons. The template was designed to accommodate eight on-template wells and provided an enclosing protection fence and tie-back for four off-template wells. Three of the off-template wells were existing at the time of installation.

In addition to the above production criteria, the template was designed to be installed as a single item by a four-point lift from a semi-submersible crane vessel. It was required to be accurately positioned over the existing wellheads and fixed to the sea bed within the following tolerances:

- Positioning within +/− 0.5 m (1.6 ft) of design location
- Orientation within +/− 1° of existing well orientation
- Level within +/− 0.3° of the horizontal.

Accurate positioning of the template structure, relative to the three existing wells, was achieved using two docking piles. These docking piles were placed using a temporary docking pile frame, tied back to temporary protection frames on the two extreme existing wellheads.

The leveling system used consisted of three mudmats operated by individual hydraulic cylinders and controlled by a diver-operated console, centrally located on the template. After final leveling the template was secured to the sea bed by four corner piles, each pile fixed.

TENSION LEG PLATFORMS

Tension leg platforms (TLP) pave the way for oil production in water depths beyond the feasibility of conventional fixed structures. It neither floats entirely free nor stands entirely rigid. The Hutton fields' TLP in the North Sea was the world's first commercial application of this new concept in offshore technology. The second is the tension leg wellhead platform (TLWP) installed by Conoco on the Jolliet field, Green Canyon block 184, in the Gulf of Mexico in 1,760 (536 m) ft of water.

Hutton TLP

The Hutton production platform has to be capable of withstanding winds of up to 110 miles per hour and waves of up to 105 ft high. It weighs almost 50,000 tons, and its steel deck structure outflanks a football field. After five years of research, Conoco concluded that the concept was feasible and could be designed to be reliable; equally significant, the cost of a TLP should be relatively insensitive to water depth.

The system allows most of the assembly to be accomplished onshore or in sheltered, near-shore waters before tow-out to the production site. This offers obvious efficiencies over on-site assembly in rough water far from shore.

The hull had to have the characteristics of a normal semi-submersible drilling rig, and yet be capable of remaining in the extreme conditions of the North Sea for 20 years without once returning to dry dock. The design of the steel tension legs was entirely new. A plug-and-socket mechanism had to be designed to allow each leg to be locked into position and held under tension. Yet each leg also had to be capable of release for removal and inspection. Elastomeric technology first developed for the space programs had to be called up to form the special rubber and steel flex joints which would allow the unique mooring system to absorb the natural movement of the waters.

The TLP consists of four basic elements: the deck, the hull, the mooring system, and the well template.

Deck Structure

The deck structure consists of a single integrated dock with a mezzanine carrying all the process and utility equipment plus external modules for the drilling rig, power generations, accommodation, helideck, and flare structures.

Hull Structure

The hull consists of a six column semi-submersible structure which provides buoyancy and contains a ballasting system to maintain platform trim.

Mooring System

The mooring system contains 16 vertical tubular steel tethers, 260 mm in diameter, running from the mooring compartments in the hull to piles and grouted foundation templates on the sea floor. The mooring system is designed to enable the legs to be removed individually for inspection and maintainance without affecting operations.

Well Template

A 32-well-slot template was installed in May 1981 to enable production wells to be drilled before the installation of the TLP.

For final connection, five tugs took the Hutton TLP on its final journey to the field. There to meet it were two large semi-submersible crane vessels, the Hermod and the Balder, whose task would be to hold the TLP in position while the tension legs were lowered from the corner columns and linked to the foundation templates deep below on the seabed.

With the platform ballasted low in the water, the first tension legs were latched into position, guided on their way by engineers using acoustic positioning equipment and underwater cameras. Then, with one "leg" connected to the mooring system from each corner of the structure, the Hutton TLP was gradually brought into tension by releasing the ballast. It was 7:57 A.M. on Friday, 13 July 1984 when vertical heave was suppressed. Within two days the remaining 12 tension legs had been latched and tensioned. Attention could now turn to preparation for first oil production.

As the platform equipment was already fully operational, remarkably little remained to be done. A connection was made to the pre-laid pipeline which was to take Hutton oil via the North-West Hutton platform into the Brent pipeline system and on to the Sullom Voe terminal in the Shetlands.

Next the drill rig was carefully positioned over the first well, one of ten already drilled from a floating rig through the seabed drill template. The well was connected to the platform, the steel casing perforated, and the Hutton TLP was ready to produce its first oil.

Jolliet Field TLWP

Conoco has built on their TLP experience with the installation of a TLWP (Tension Leg Wellhead Platform) in 1,760 ft (536 m), on the Green Canyon block 184 Jolliet field in the Gulf of Mexico in July 1989 (see Fig. 1–10). Compared to Hutton, the TLWP is comparatively small with a displacement of 1,600 tons, only supporting the wellheads (one less than Hutton), well test system, production separators, gas compressors, and 47-person accommodation. Production is piped some 10 miles (16 km) shoreward to a Central Production Platform in 616 ft (188 m).

The mooring system for Jolliet consists of 12 tendons made from 24-in. × 60-in. steel pipeline with 20-mm walls. These are neutrally buoyant, fabricated ashore, and installed in one piece, not joint by joint. This installation was not without some difficulty however, when the first two pairs sank due to various accidents during tow-out. Tendons connect to a 411 m (1,350 ft) deep foundation template, itself placed over a pre-drilling template.

Production from the TLWP passes through 10-in. oil and 8-in. gas flexile lines to the seabed and on to a point some 5.5 mi (9 km) off where they tie-in to the steel pipeline linking to the Central Production Platform.

ELF "GRONDIN" FIELD

The Elf-Aquitaine group has conducted an extensive research and development program to develop subsea production for extreme water depths in the order of 600 m (1,968 ft) beyond diver capability.

The program has included floating drilling and completion of highly deviated wells. Also in the program are the setting and connection of subsea structures without diver assistance, remote production of subsea wells, and automatic flowline connections. Remotely operated vehicles (ROVs) were used for

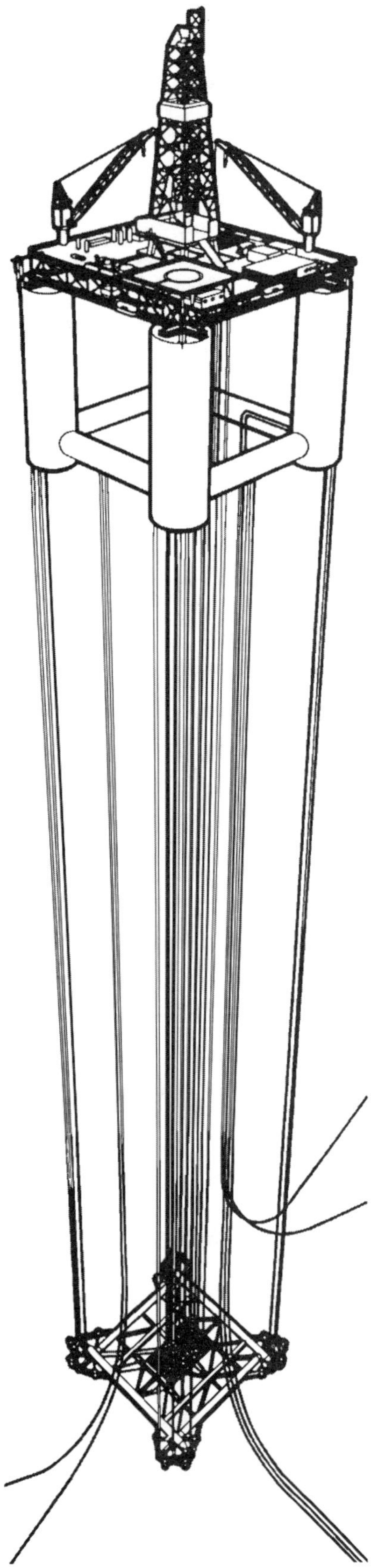

Fig. 1–10 Green Canyon, "Jolliet," TLWP

intervention and maintenance and for reliability and safety analysis.

An experimental station has been located on the northeast extension of the Grondin field, offshore the Republic of Gabon, West Africa. Located 1,500 m (4,921 ft) from the fixed production platform complex, the station is installed in 60 m (196 ft) of water.

The barge Anguille was used as a support vessel with surface equipment. Subsea equipment included template, Christmas trees, production manifold, and the flowlines. All of the equipment is remotely controlled from the Grondin production platform.

Remote control was via a subsea electrical cable with local electro-hydraulic subassembly and surface monitoring facilities. Installation, connection, maintenance, and diverless operations were conducted from a floating service vessel designed and constructed specifically for the station.

From initial engineering in 1974, the Grondin northeast station marked a first step towards establishing the feasibility and reliability of subsea production in deeper waters.

In 1981, Elf-Aquitaine completed its successful remote controlled subsea production system tests. The final phase of the test involved the use of a rail-mounted remote control robot manipulator developed by ACB. Following six weeks of testing of the 13-ton remote-controlled manipulator, TIM (Tele Manipulateur d'Intervention et de Maintenance), two modifications to the system were proposed to increase the capability of handling production from 600 m (1,968 ft) water depths. These proposed modifications first involved substituting an electronic multiplex control system for the electro-hydraulic system to reduce weight. The second involved upgrading of the visual system (mimic diagrams) used to show the work being performed on the subsea production system to the topside operators.

Controlled by operators at a console located on the support barge Anguille, TIM successfully carried out 104 different operations. These operations were during 24 separate dives which involved a total of 160 hours bottom time. During tests, the system's manipulator was brought onto the barge between trials for tool change-out. However, this work could be done underwater, with the robot selecting tools needed for its tool store and performing long periods of maintenance on the sea floor.

Although the depth at the experimental site was only 60 m (196 ft) the design limit of 600 m (1,968 ft) was selected for the experimental performance of the manipulator package. This criterion imposed care in selection of the components to be immersed as well as a conceptual study excluding any direct human intervention.

Mission of TIM Robot

The 11-step mission of the TIM robot consisted of:

1. Connection of hydraulic jumper pipes between a wellhead and a flowline on the manifold

2. Hydraulic supply for operation of connectors
3. Laying of electric cables and hydraulic hoses
4. Assistance for electric jumper cable positioning and connection
5. Installation of operating tool for valve opening or closing (manual safety valves on manifold and wellheads)
6. Releasing and connecting the guide line
7. Brushing, grinding, cutting, tightening, and loosening nuts and bolts
8. Handling of cleaning water hose nozzle
9. Handling of cleaning suction head
10. Handling of TV camera and projector
11. Handling of miscellaneous tools

NORTH-EAST FRIGG/GAS PRODUCTION

Elf's development in the Norwegian North Sea is located just on the border between Petronord-Statoil's 25/1 block and Esso's 30/10 block. The gas field, with recoverable reserves of 8,000 million cubic meters has a planned five-year production life at 6 million cubic meters/day. This field development is the first subsea development of a gas field and the first use of articulated tower concept as a service platform.

Development Plans

The design called for six subsea wells installed on a template, to be controlled from an unmanned articulated tower located about 150 m (492 ft) away. Production is remotely controlled via radio link from the TCP2 platform on Frigg.

The close proximity of the giant Frigg Field's gas accumulation provides adequate compensation for possible shutdowns on North-East Frigg. This compensation was an incentive for the unconventional development scheme. In particular, the scheme would be more suitable to the small size of the reservoir. The approach resulted in a subsea production system consisting of a template to support and protect six wells, a manifold, and a 16-in. gas line which links the manifold to the Frigg treatment platform (TCP2). An articulated field control station (FCS) supports the main function of controlling the subsea equipment and treatment. Metering and boosting modules are installed on the TCP2 platform to process the gas.

Apart from processing 6 million cubic meters/day of gas,

North-East Frigg Field—Key Information Summary

Operator, Partners	Elf Aquitaine Norge A/S Esso Exploration Production Norway, Inc. Norsk Hydro Produksjon A/S Total Marine Norsk A/S Den Norkse Stats Oljeselskap A/S
Water depth	100 m (328 ft)
Recoverable Reserves (Original)	8,000 million cubic meters
Production Start	8 Dec 1983
Production Rate	6,000 cubic meters/day
Field Control Station	
Concrete base dimensions	45 m x 42 m x 6.5 m (147 ft x 137 ft x 21 ft)
Concrete base weight	6,800 tons dry 4,500 tons submerged
Universal Joint weight	150 tons
Column (steel) length	127 m
Column (steel) diam	8 m
Column dry weight	4,500 tons (net buoyancy 500 tons)
Head house weight	630 tons
Accommodation	6 persons (12 in emergency)
Number of decks	4 Methanol/main/quarters/helideck
Heli deck	Sikorsky S 61 N
Gas Line	
–diam	16 in.
–length	18 km
–working pressure	170 bar
Transport facilities	16-in., 18 km (11 miles) subsea pipeline to Frigg field platforms
Service	From Frigg
Development cost	NOK 2,000 million (1984)

the treatment modules on the TCP2 platform also have a daily throughput of 40 million cubic meters of water/methanol. Then the gas is transported to the United Kingdom through the submarine pipeline system via St. Fergus.

As the reservoir characteristics were well known, the wells were not individually tested and no control chokes have been installed. Of the six wells, five are directional and one is vertical. The wellheads are conventional 13 ⅜-in. 5,000 psi working pressure wellhead housings and a 9 ⅝-in. casing hanger.

The Christmas tree is a wet monoblock unit including the wing valve. The template measures 17 m × 30 m × 8 m (56 ft × 98 ft × 26 ft) and weighs 350 tons. It accommodates six wells plus two spare slots (see Fig. 1–11).

The control system is designed for control of subsea equipment during normal production operations, such as routine maintenance and special operation and testing of subsea valves. It is automatically controlled and incorporates automatic safety valves.

Articulated Tower

The articulated tower has a concrete gravity base (45 m × 42 m × 6.5 m/147 ft × 138 ft × 21 ft high) which supports the universal joint. It has a dry weight of 6,800 tons. As the soil conditions are good, no specific anchoring or grouting is necessary. The universal joint can be connected or disconnected from the base of the tower by a remotely controlled hydraulic locking system. An 8-m diameter cylindrical steel column is the largest component in the tower. In the lower sections of the columns are heavy ballast material and buoyancy and stability tanks. Water systems are housed inside the column. The column also supports the risers for the umbilicals, a fendering system at sea level, and the field control system on top. The station consists of four decks: a methanol deck with methanol pumps and storage tanks, a main deck with the well control manifold, a living-quarters deck, and a helideck. Within the remote control station are the hydraulics for the subsea control and monitoring and communication systems between the main Frigg Field and the control station.

NORTH-EAST FRIGG/SKULD SYSTEM

North-East Frigg Control—Field Control Station (FCS)

An articulated column installed 150 m from the subsea station supports the field control station (FCS). The FCS houses the equipment required for:

- Converting electric control signals from the Frigg Field into hydraulic pressures for operating the subsea gas production valves
- Closing control of wells through individual 2-in. kill lines
- Periodic leak testing of the production tubing safety valves in each well
- Continuous injection of hydrate inhibitor into wells during gas flow

Umbilicals

Six specially designed umbilicals, each consisting of one 2-in. tube and twenty ½-in. tubes, link each Christmas tree to the deck of the FCS.

Links to the Frigg Field

The North-East Frigg facilities are linked to the TCP2 Frigg Field platform via a 16-in. gas line. A 1¼-in. hydrate inhibitor supply line and an electric cable provide the FCS with both electric power and remote control signals. These signals are also transmitted through a back-up radio link. Treatment, metering, and boosting modules are installed on TCP2. This allows the North-East Frigg gas, after treatment, to be injected into the Frigg gas flow.

SKULD System

Elf-Aquitaine Norge A/S planned to develop subsea remotely controlled clusters as part of the SKULD program. The program was detailed in two phases:

- Phase 1. Qualification of a subsea production remote control system
- Phase 2. Manufacturing of complete subsea systems for marginal fields in the North Sea

Main Objectives of the Project

These are the main objectives:

1. Produce over a long distance (20 km/12.4 miles) between subsea cluster and control room.
2. Diverless installation and retrievable control system
3. High reliability (five years to a shutdown)

Main Principles

Seven principles are followed:

1. Use of Grondin's experimental results
2. Trials with two Christmas trees: a live one (the production tree in this case is simulated) and a dummy
3. Use of either qualified or field-proven products
4. Electro-hydraulic control system
5. Electronic multiplex transmission

6. Closed hydraulic circuit using subsea pumps, with hydraulic power supply from the platform in back-up
7. Back-up for electric power supply and transmission through a buoy

Control System

The SKULD experimental subsea station is located 1 km (0.62 miles) offshore in 100 m (328 ft) of water. The control room is onshore, with back-up via the buoy.

Subsea Station

The subsea station consists of two wellheads on a template and a manifold controlled through an electro-hydraulic system. The items on the template manifold are:

- Two subsea wellheads
- A common service container (CSC) with a bundle head
- Two production and control jumper units
- A connecting frame (to connect CSC and jumper units)

On each tree, the following functions are remotely controlled (eight commands):

- Two fail-safe valves
- Two DHSVs

Also remotely controlled are pressure measurements, valve position indicator and gas detectors. On each jumper unit (three commands) there are:

- Two fail-safe valves
- One choke

There are also redundant pressure measurements on each line (flowline, annulus line, and methanol injection line), temperature transducers, valve position indicators, gas leak detectors, hydraulic fluid pressure and flow measurement, and solenoid valve position indicators.

On the manifold (nine commands) there are:

- Six fail-safe valves
- Two close-assist valves
- One bowl valve

The common service container (CSC) consists of subsea hydraulic power supply, control, and monitoring. Control and monitoring cover subsea pump start-up, reservoir management, hydraulic pressure measurements, flowmeters, water and gas detection, and solenoid valve redundancy management.

Electric supply control and monitoring covers installation measurement, platform/buoy switching, solenoid valve position indicators, and subsea/on shore hydraulic power switch monitoring. To connect the several packages together, automatic connectors are provided through what are defined as "multiconnectors."

Surface Equipment

The SKULD electronic surface unit consists of:

- Control desk equipped with two-color remote control terminals and a functional keyboard
- Central processing unit with a minicomputer system and associated interfaces

The subsea remote terminal units (TESUMA) are connected to the central connecting unit via two alternative transmission links:

1. Hard-line transmission through the main platform umbilical
2. Radio transmission from the main field to the buoy, then a hard-line transmission from the buoy to the cluster

All of the subsea information (valve positions, hydraulic status, pressures, gas detection, etc.) are locally gathered by the remote terminal unit. Then the information is transmitted upon request via one of the transmission links to the central processing unit in the form of serial data bits (scanning cycle). This information is processed and presented to the operator at the control desk. The operator is automatically alerted of all alarm conditions at the subsea level on critical events. Shutdown actions are automatically elaborated by the system.

From the control desk the operator can operate the valves and pumps using the remote-control terminal (CRT) and the functional keyboard. Corresponding local actions are completed by the remote terminal unit concerned.

Also, all surface information (electrical power status, hydraulic, or methanol production status, etc.) is gathered by the system and displayed on the control desk.

East Frigg

Recently the SKULD technology has been put to use by Elf Aquitaine Norge on the East Frigg Gas field. The modular subsea development was designed for diverless installation (see Fig. 1–12) and incorporates flowline bundles.

FUTURE TRENDS

With future trends towards smaller fields (less than 100 million barrels recoverable reserves) and lower oil prices, the term "marginal" takes on a significant meaning. With current emphasis on floating production, subsea systems have an assured future; however, to increase their importance in the future, subsea technology needs to produce a low cost processing/production system. Before this can be achieved some significant advancements in multiphase technology must be made. The record of achievement to date and developments in subsea technology have been covered briefly in the earlier part of this book.

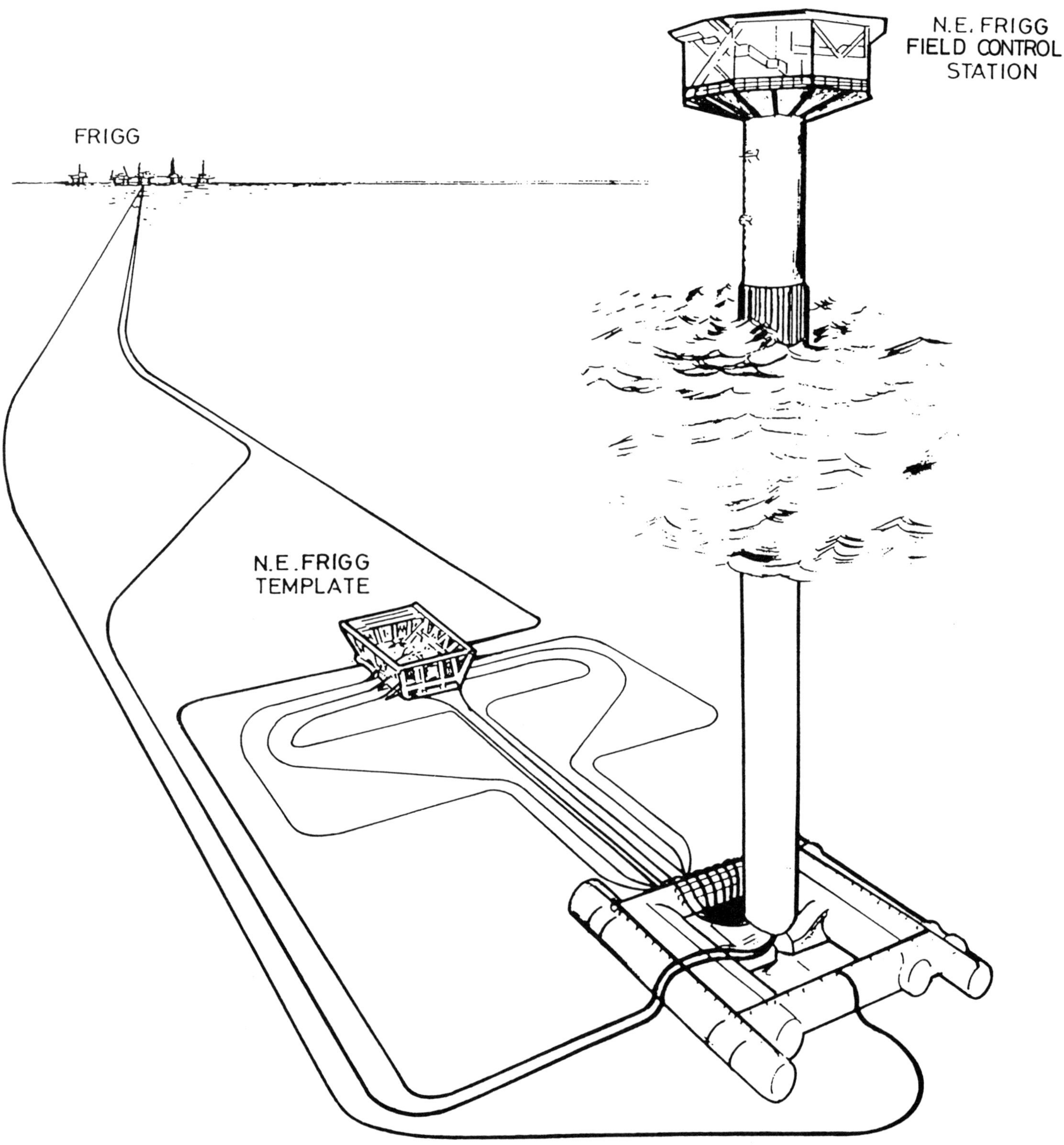

Fig. 1–11 North-East Frigg—Field Layout

The following chapters discuss the individual components of a subsea system: production risers, templates, controls and chokes, wellheads and trees and the way in which they interface through commingling and manifolding, flowlines, and export pipeline systems.

In conclusion, the reliability of subsea systems will be determined by the minimal operation and maintenance expenditure during the life of the field, and it is probably in this area of subsea that most of the efforts of the subsea engineers will be directed in the future.

Chapter 2

Production Riser Systems

ALTERNATIVE RISER DESIGNS

When there is a need to connect the mobile/floating platform to the seabed system, four types of risers could be given serious consideration for analysis and design: rigid integral, rigid non-integral, hybrid (buoyant rigid) and flexible (dynamic).

The rigid integral riser design comprises a structural pipe to support the flow carrying tubulars. The riser is usually fabricated in 50-ft (15-m) lengths with mechanical connectors at both ends of the structural pipe. When the connector is made up, all the tubular flow paths are made up simultaneously also. Current drilling risers utilize the integral design concept.

The non-integral rigid-riser design consists of a central structural pipe supporting vertically spaced guide funnels or other means of guiding tubulars. Flowline tubulars are individually run down the riser through the funnels and latched to the template or manifold. The structural pipe must be installed and tensioned before running the flowline tubulars. This concept has been employed successfully in the North Sea on the Buchan and Argyll Fields, as shown in Figures 2–1 and 2–2.

Structurally, the two risers are, for all practical purposes, identical in performance. The non-integral design, however, is lighter than the integral design in water, and therefore the tensioning requirements are reduced. Also, it is more difficult to test and fabricate the integral riser. Furthermore, the close tolerances, required of the design to compress the metal-to-metal seals between flowline joints, require extensive acceptance testing.

Operational considerations of the two designs are decidedly different and provide the ultimate selection criteria. Both risers can survive a 100-year storm, either connected or in the hang-off mode. However, for operational reasons, they are both disconnected very early before the storm, with a direct impact on loss of production.

The hybrid-buoyant-rigid riser, as used on Placid's Green Canyon Block 29 development in 468 m (1,534 ft) of water in the Gulf of Mexico, is a more recent innovation and suited to deeper waters. Studies have shown the concept, based upon a floating production system, as economically advantageous below 350 m (1,148 ft). Essentially, the design utilizes the rigid riser concept as the most economical means of recovering well fluids to the surface and oil and gas production back down to subsea pipelines, being stationed over a template supporting most field wells. Operationally, it would have a 20-year life and be able to survive a 100-year storm. Furthermore, it is free standing when disconnected from the floating production system, having buoyancy chambers within and syntactic foam externally mounted on its upper section. Other features include individual production and annulus connections to each well (no subsea chokes) and no interference between production and drilling/workover riser operations.

The flexible-dynamic-riser design consists of continuous flexible pipe or bundle of pipes, with a specific structure adapted to the production requirements, connected at the surface (together with a possibility of quick disconnection) and on the seabed. Dynamic risers have significant technical and economical advantages, probably the most important being its capability to survive a 100-year storm without disconnections and thus save a tremendous amount of production time.

Rigid Integral Risers

The production riser system spans the distance between the floating semi-submersible and the seabed template or manifold (see Fig. 2–3). It consists of three major sub-systems: the lower riser assembly, riser joints, and the upper riser assembly. Only the integral design is developed in this chapter, the work being based on non-integral risers used for years on BP Buchan and Hamilton Argyll fields.

The lower riser assembly (see Fig. 2–4) is the interface between the riser and the template or manifold. It provides mechanical, flow, and control interface connections between the manifold and the riser. It contains four flexible joints to allow horizontal motion of the surface vessel without bending and overstressing the structural riser pipe. A mechanical surface-actuated connector is provided to allow remote connection and disconnection of the lower riser assembly and of the entire riser. When the mechan-

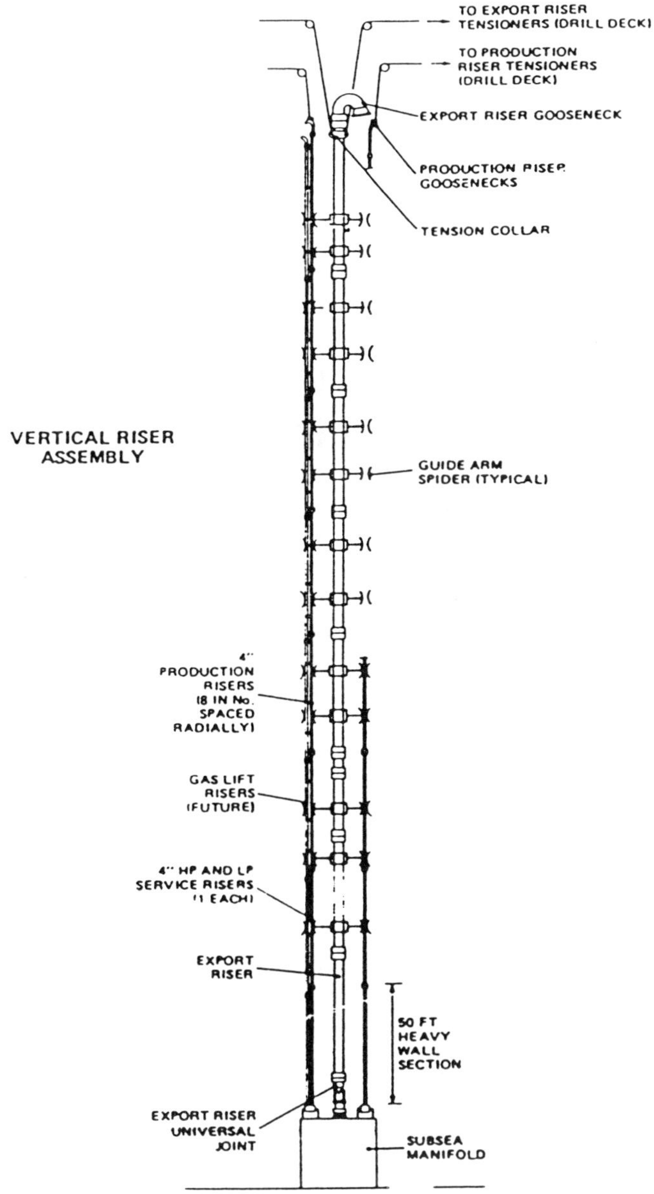

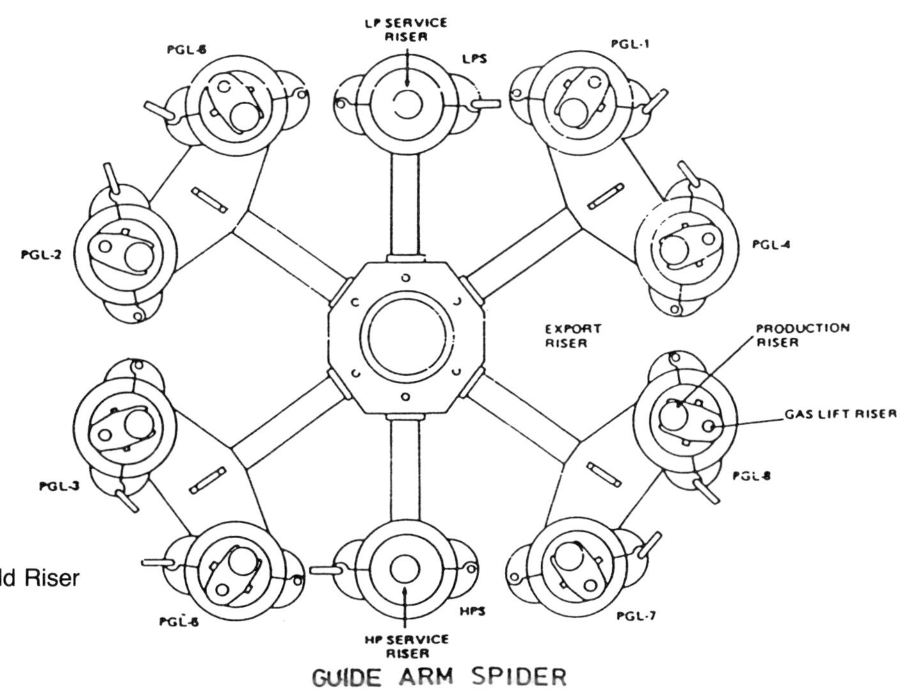

Fig. 2–1 Buchan Field Riser

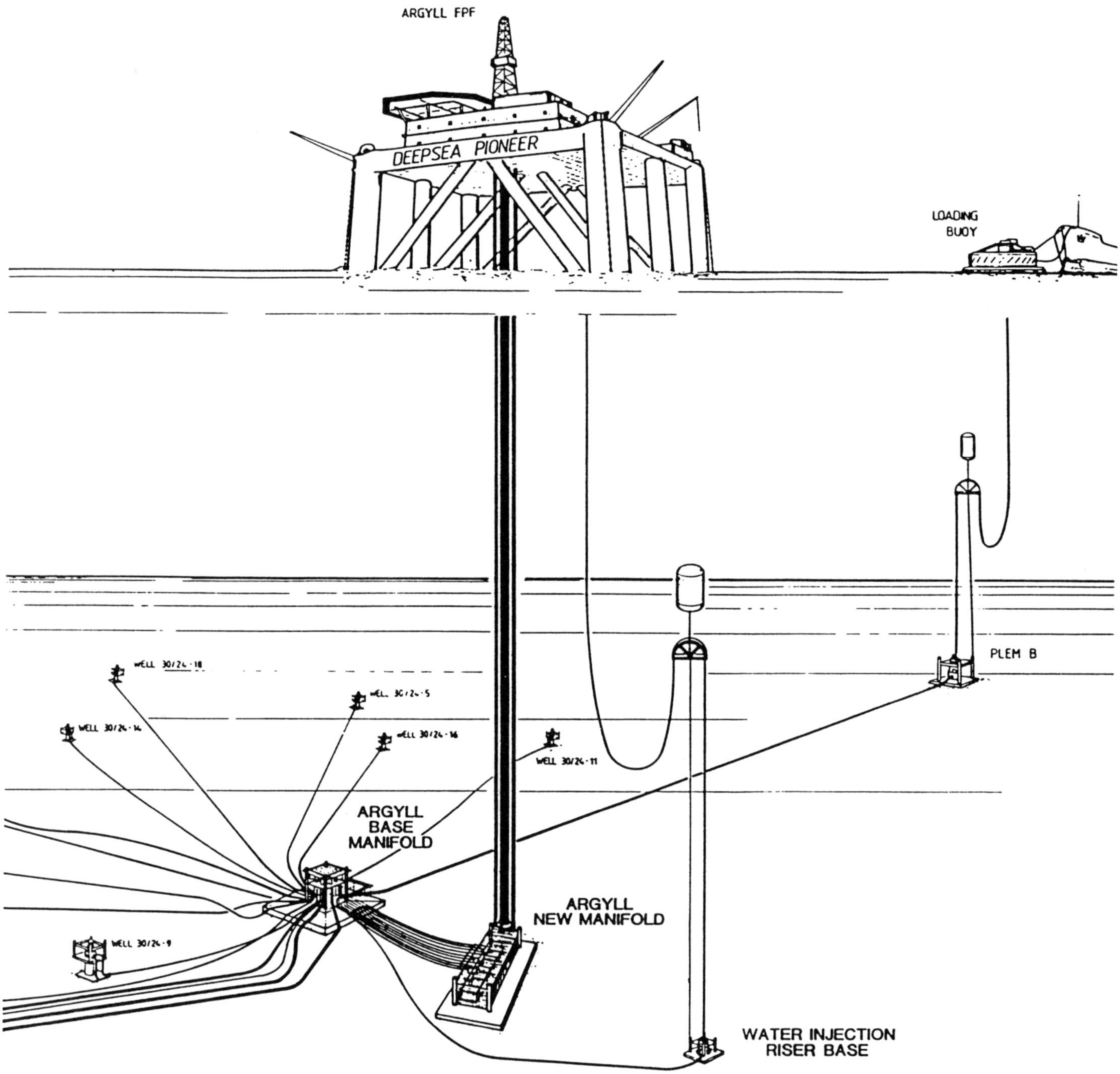

Fig. 2–2 Argyll Field

ical connector is guided into place and locked to the manifold mandrel, all the tubulars supported by the riser are simultaneously connected to their respective manifold points through stabs.

The preferred design uses a central structural 12-in. pipe. This also serves as the conduit for water injection, pumped from the semi-submersible to the template for distribution to the injection wells. The riser is fabricated in 50-ft long sections or joints with mechanical connectors at the end of each joint. These connectors are designed for a rapid connection and make up of the complete riser and are welded to the 12-in. structural pipe. Each pair of connectors then supports the nine flow carrying tubulars: six for production fluid, one for crude export, and

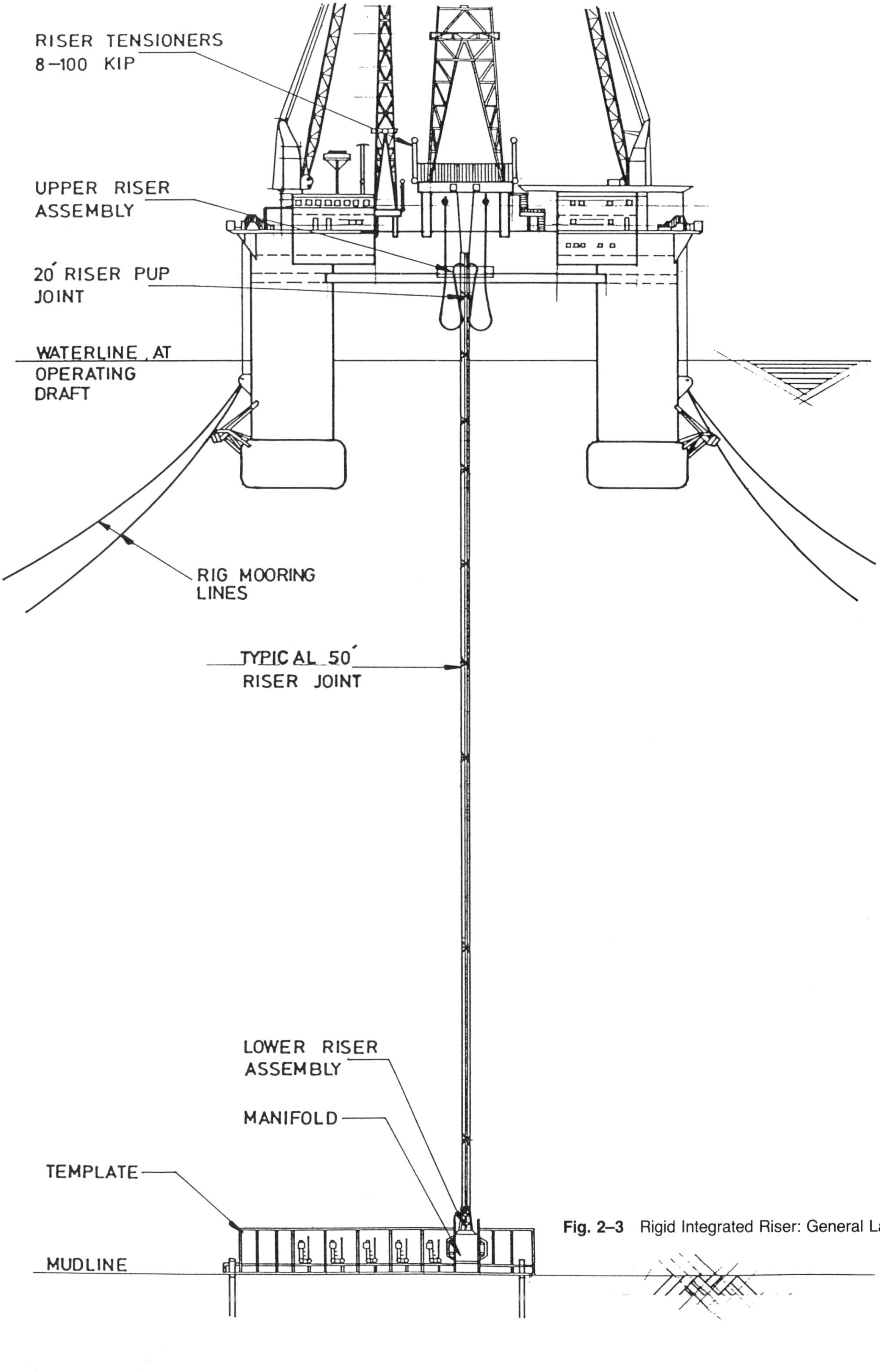

Fig. 2–3 Rigid Integrated Riser: General Layout

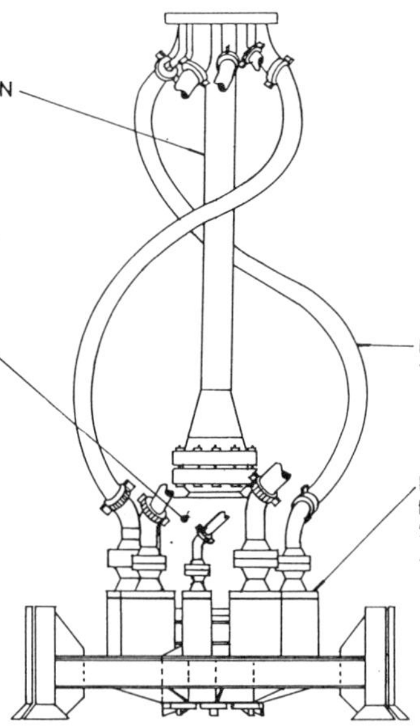

Fig. 2–4 Rigid Integrated Riser: Typical Lower Assembly

one each for well test and annulus access as shown in Figure 2–5.

The upper riser assembly is the interface between the riser and the floating vessel. The riser tubulars terminate into steel goosenecks and then continue to the vessel through high-pressure hoses draped in the moonpool area. Immediately above the goosenecks, the riser tensioner wire ropes terminate in a tension ring attached to the riser pipe. Tension is applied to the riser pipe through this ring.

For the riser to survive a major North Sea storm in the connected mode, it must be supported by eight, large capacity tensioners. These tensioners control the riser and allow for relative motion between the riser and the floating vessel.

The riser joints terminate to flange type couplings at each end. The joint diameter is sufficiently small to allow them to run through a standard 49½-in. rotary table. However, it is recommended that the drill floor be fitted out with a rotary and a hydraulic spider assembly to handle the riser joints during running or retrieving.

The production riser is connected to the subsea template through the template well manifold, which is the main subsea flow and pressure control unit for the field. The manifold provides the secondary means for field shutdown, which allows the riser flowlines to be flushed clean prior to disconnect and thereby prevent pollution. In addition, the manifold structure transmits the relevant riser tension and bending loads into the template.

Lower Riser Assembly

The lower riser assembly is the interface between the riser and the template well manifold. Its main functions are to provide a means of disconnecting the riser and to allow for vessel offset without inducing high stresses in the riser and flowlines.

The riser is connected and disconnected at the manifold by a 16¾-in., 5,000 psi, high angle release hydraulic connector attached to the bottom of the main structural member. The connector is capable of handling the maximum tension and bending moment imparted to it by the riser. Through this connector passes the water injection at 1,000 psi into the manifold for distribution to the various injection wells. The riser connector is attached rigidly to a structural guide frame that stabs over the manifold guide posts. This provides initial alignment for the connector to interface properly with the manifold mandrel.

Spaced in a radial pattern around the connector are nine non-load bearing flowline stabs with radial seals that interface with the required manifold circuit. These stabs are hydraulically operated and can be extended to complete the flowpath to the manifold or retracted to disconnect from the manifold. Being retractable prevents binding between the choke and kill lines and the main stack connector in a high angle disconnect situation. This is a similar situation to that between the production manifold and

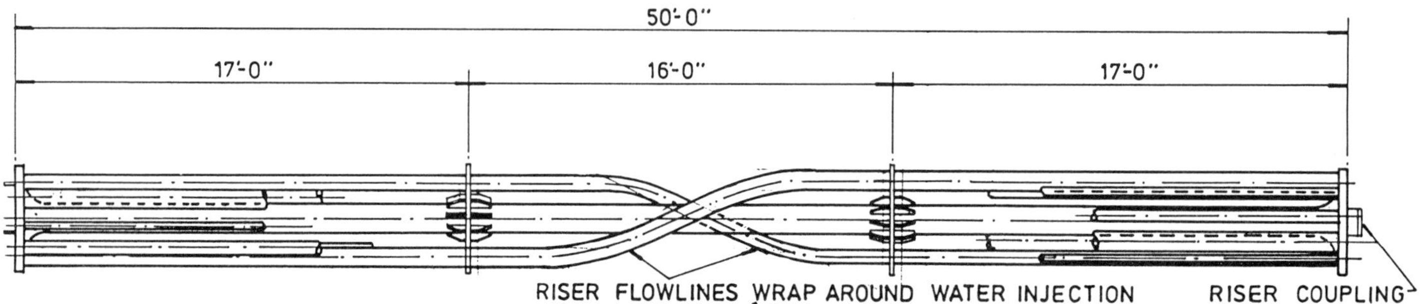

Fig. 2–5 Rigid Integrated Riser: Typical Riser Joint

the lower riser, although in this latter case there are nine stabs instead of two.

Located above the connector is a 16¾-in., 3,000 psi, 15° flex joint also capable of handling the maximum riser tension bending moment. The purpose of the flex joint is to allow riser angular displacement without imparting potentially damaging bending stresses to the riser. The lower riser assembly is completed with a section of 12¾-in. pipe, approximately 10 ft long, and terminating in a riser coupling half. Flexible jumper hoses are connected between the riser coupling half and the retractable stabs to relieve the bending stress on the flowlines and to compensate for the angular motion of the flex joint. These flexible lines wrap around the central water injection core 180° and also compensate for the increased flowline radius from the riser flange to the retractable stab.

Riser Joints

The riser sections are designed by using a similar concept to that of an integral drilling riser. A typical riser may comprise nine equally spaced flowlines on a 17-in. radius around a 12¾-in. central structural member. These nine flowlines coil 180° around the central member in each 50-ft joint to prevent buckling from thermal expansion. Because of this flowline rotation, it is necessary to design right- and left-hand joints which then are connected to each other in an alternating sequence. Two intermediate spaced plates are provided in each section to keep the flowlines properly spaced around the central core and to prevent vibration.

Each joint weighs approximately 22,000 lbs in air, and no added buoyancy was provided for operation in water. For particular riser use, buoyancy can be considered, but it was discovered it did not significantly reduce the tension requirements due to increased dry weight and wave loading.

The riser joints are connected to each other by couplings. Each flange uses eighteen 1¾-in. diameter high-strength bolts to make and preload the seals in each connection. These bolted flange couplings are rated at 1,000 tons axial operating load or about three times the maximum expected riser tension. The locking bolts for the flanges are self-contained in one flange by a snap ring which secures the bolts in the retracted position ready for make-up. This design feature protects the bolt threads from damage and also prevents the rig floor personnel from having to handle and store the bolts between running and retrieving operations.

The riser flowlines are sealed against the flowline in the adjacent joint with a metal AX gasket. A patented, adjustable locking collar draws the flowline against the gasket and allows for adjustment of the gasket position. The central riser pipe is sealed at each flange with a metal AX gasket clamped in place by the flange bolts. All seal grooves and sealing surfaces are inlaid with 316L stainless steel, or equivalent, to prevent corrosion to critical seal surfaces and to ensure long seal reliability. The 10 riser flowpaths are of different size, function, and pressure rating.

Upper Riser Assembly

The upper riser assembly is the last joint in the system, and it also provides the interface between the platform process equipment and tensioners to the riser.

All the flowlines terminate at this joint with hard pipe goosenecks. High pressure drape hoses connect the goosenecks to piping on the deck.

These flexible hoses compensate for rig motion relative to the riser and complete the flowpath from the riser to the vessel. They are all flexible high pressure hoses except for those carrying the injection water. All hoses are rated to at least the same working pressure as the riser to which they are connected. High pressure flexible pipe, such as Coflexip or Dunlop, is recommended in all lines containing oil or gas because even small percentages of sulphur-bearing contaminates will chemically attack and drastically shorten the life of rubber hoses. The main problem to overcome in using the flexible type hose is how to allow for the large bend radius inherent with its design. Because of the limited moonpool space it is not possible simply to allow the hoses to hang in a 180° horseshoe bend. It is necessary to limit the lateral space between the vertical hose sections to 10 ft or 2 ft less than the desired space. This will mean that the hose will have a tendency to bell out at the bottom and seek its natural bend radius. Coflexip was consulted and confirmed that this is acceptable practice and should cause no unexpected hose failures.

The 12¾-in. central core water injection line terminates at the upper riser joint at a cross weld fitting. Two 8-in. 90° elbows with hose connections attach the 8-in. inner diameter high pressure rubber hoses to the upper riser. It is necessary to use flexible hoses for this application because of the bend radius of 8-in. flexible would not fit inside the moonpool.

Above the water injection cross, a tension collar is located to provide eight pad eyes for attaching the tensioner wire ropes. This collar is the structural interface to the platform tensioners and can rotate around the riser pipe to compensate for rig yaw motions. The collar must be provided with either a thrust bearing or a lubricated shoulder to react against when the wire ropes are tensioned.

Riser Analysis

The production integral riser is designed to meet flow and pressure requirements and analyzed to evaluate stresses, flex joint deflection, riser structural configuration, and top tension requirements. A maximum wave, corresponding to environmental conditions for the field, is used for the connected mode analysis.

In addition, a static offset (approximately 10% of water depth) must be imposed at the top of the riser. In order to satisfy the requirements for the computer program analysis, the composite riser could be so modeled as to account for the added weights of flow pipes and production fluids and for increased stiffness provided by the additional flow pipes. It is also important to note:

- The flowlines make a 180° expansion loop around the central member but do not contribute to the tension resisting cross-sectional area.
- The bending stiffness of the flowlines should be added to the bending stiffness of the central member through the calculation of an equivalent modulus of elasticity.

Riser Installation/Retrieval Procedures

Production riser installation operations can be considered very similar to those of a drilling riser. The template well manifold takes the place of the drilling BOP, and the production riser replaces the drilling riser. In the mid-water depth of marginal fields, the riser could be run on guidelines permanently attached to the floating production vessel.

When the floating production vessel is brought and permanently anchored above the template, the first step is to establish guidelines to the template guideposts. The first guidelines can be established by guidelineless drilling techniques and the subsequent three by use of a special guide frame. Conversely, divers could be used to establish guidelines.

Next, the lower riser assembly is brought over the moonpool. The procedure at this point assumes that the riser manifold has been run and connected to the template piping. Then the guidelines are inserted within the assembly's guide funnels. Next, the control system unbilicals are connected to their respective terminators, and the assembly is ready to be lowered to the manifold already installed on the template.

The first riser joint is made up to the lower riser assembly, the spider beams are spread apart, and the assembly is lowered within the moonpool opening. Successive riser joints are added to the string and run through the rotary table. The joint couplings are made up on the rig floor at the top of the substructure while the string is suspended on a spider. Below the substructure, control umbilicals are unreeled from their storage drums and firmly strapped to the riser joints to prevent them from wrapping around the riser under the action of waves and current.

When the riser/lower riser assembly is within 20 ft of the manifold mandrel, the last riser joint is picked up with a long handling tool that can reach from the rig floor to the moonpool spider beams, approximately 40 ft below. The spider beams are brought back in place and a split riser hang-off spider is skidded over the beams. The riser is hung-off on the spider to allow installation of the upper riser assembly stored underneath the substructure. The upper riser assembly could be handled either by an overhead trolley system or by an articulated arm.

The riser tensioner wire ropes are connected to the tensioner ring. The flowline drape hoses may be connected now or after the entire riser has been connected to the manifold and tensioned.

Now the entire riser/upper riser assembly is picked up with a handling tool, and the handling tool is attached to the motion compensator through the hook. The riser tensioners are extended to their full stroke. The make-up assembly is slowly lowered towards the manifold mandrel with increasingly more weight being taken up by the tensioners. When the lower riser assembly connector is on top of the manifold mandrel, the connector is locked in place and the tensioners are immediately set to the required tension.

Now the production riser is installed and tensioned. The drape hoses, if they were not connected previously, can be installed now. Final preparations can be made to verify pressure integrity of the manifold valves and proper functioning of the control system for the manifold functions.

Normal riser retrieval procedures are an exact reversal of the installation procedure with the exception that the production manifold stays in place and all lines containing oil are flushed out. In an emergency riser disconnect, the riser is disconnected at the connector between the lower riser assembly and the manifold. The disconnection is activated with the control system and, before it is carried out, all the subsea wells and manifold valves must be closed since all control functions are effectively lost by disconnecting the riser.

Riser Onboard Handling

The onboard handling of the production riser joints can be accomplished by one of several methods used to handle drilling risers. However, in some cases, where at most 10–12 riser joints will be installed and speed of installation is not too critical, it is recommended that the joints be handled with a simple onboard crane and sling.

During installation, the joint is picked up from the pipe rack by the crane with slings, and one end of the joint is laid down on the rig floor through the derrick V-door. This end is picked up with the riser handling tool suspended from the hook, and the crane handles the other end until it is well within the rig floor. The method is simple and does not require additional handling equipment other than that normally provided on a semi-submersible.

Retrieval of the riser also would be similar to the retrieval of a drilling riser. The joints would be stored vertically in the derrick, if the vessel's motion characteristics can accommodate the high center of gravity. If not, the joints would be laid on deck immediately next to the moonpool.

The riser/vessel interface recommended is a concept that permits the pulling of the riser during rough weather. The purpose is to permit operations personnel to delay pulling the riser as long as possible to minimize downtime. Thus the upper riser assembly handling system requires careful design to allow safe operation in rough weather.

The system is essential for the speedy retrieval of the production under emergency conditions. It should be able to handle the entire upper assembly without having to disconnect the riser tensioners at the tensioning ring or the flowline drape hoses at the gooseneck. An overhead trolley or an articulated arm system can be visualized to accomplish these goals. This approach minimizes the manual work required to pull the riser, thus reducing the effect of vessel motion or weather on operational personnel.

The main moonpool dimensions of 25 ft × 25 ft (7.6 × 7.6 m) are necessary to accommodate the upper riser assembly's drape hoses rather than for passage of any subsea equipment.

Operational Criteria

Riser installation is a motion-sensitive operation and should be carried out with as good environmental conditions as possible. The manifold and lower riser assembly landing and locking in place are the most critical operations. In general, the landing operations should not be attempted with more than 3–4 ft of vessel heave which, dependent on the semi-submersible vessel, will correspond to 10–12 ft waves.

Retrieval of the riser with the manifold will require similar environmental conditions to those of the installation due to the close fit of the manifold through the production moonpool.

In the previous riser analysis section, it was mentioned that the riser can remain connected during a 100-year storm from the point of view of riser structural integrity. In reality, however, the riser cannot remain connected during these conditions because semi-submersibles normally change into a shallow draft in anticipation of major storms. Some semi-submersibles decrease draft by as much as 20 feet, which effectively removes 20 ft from the riser tensioner 50-ft stroke capability. Such a draft reduction permits a total semi-submersible heave of only 30 ft before the tensioners bottom out or the drape hoses are pulled apart. For a major storm, the 30-ft vessel heave will probably be exceeded.

This problem could be solved by installing hoses with an additional 10 ft of drape length and by readjusting the tensioner wire rope length before the storm arrives. However, hoses with such a long drape length are undesirable because during normal operating conditions they hang in the water. Readjusting tensioners while a storm is developing is by no means an easy task. The conclusion is that although the riser can structurally remain connected, it will probably be hung-off or retrieved for major storms.

Hang-off analysis indicates that hanging off does not present any structural problem to the riser. Operationally, to accomplish a hang-off, the spider beams at the moonpool level must be brought together and firmly anchored to the deck. The riser would be hung-off from the moonpool spider. It is essential that this hang-off spider be able to gimbal; otherwise, excessive riser bending stresses may occur at the top, caused by the action of the storm waves and currents.

Operationally, the decision of whether to retrieve or to hang-off the riser before a storm is a difficult one. Approximately 8–10 hours could be necessary to retrieve the integral riser. Additionally, the riser should not be disconnected with total vessel heave exceeding 40 ft because of the danger of collision between the lower riser assembly and the manifold during disconnect. Therefore, when vessel heave approaches 30 ft with signs of worsening weather, the riser should be disconnected and hung-off. It is assumed here that the mooring system can maintain the vessel offset so that the bottom riser angle will not exceed 5–6°. Otherwise, the connector will not disengage.

To be practical, riser retrieval should probably not be carried out with vessel heave past 20 ft. The vertical and horizontal vessel motions during such weather prevent personnel from working safely, even with the recommended system. Therefore, if operating personnel decide that the riser should be retrieved, the retrieval process should begin when total vessel heave is 20 ft, or less if weather forecasts indicate a fast moving major storm.

Hybrid Riser Systems (Semi-sub and Loading Columns)

Mobil-IFP Compliant Production Riser Systems for Deep Water

The Mobil compliant riser uses both drilling technology and the compliance of flexible pipe to handle high production rates and remain connected in the most severe environment (see Fig. 2–5A).

The design precludes the use of divers and is heavily dependent on the use of flexible pipe having a reasonable operating life. After some six years of bench testing followed by a successful sea test of a large diameter high pressure flexible pipe, the viability of the flexible pipe for such an application was successfully demonstrated.

This riser has been designed for application to 2,500 ft water depths. The lower riser section is made of rigid tubulars and is installed using typical drilling methods. This lower section includes a central core and some 11 peripheral lines which are hydraulically connected to a riser base and held in tension by two subsurface buoys, termed the auxiliary and main buoy. Guide funnels are rigidly attached to the core at various intervals for guiding the peripheral lines, typically in double lengths of threaded tubulars with premium threads. Each line is pretensioned atop the subsurface buoy. A tubular transition piece cou-

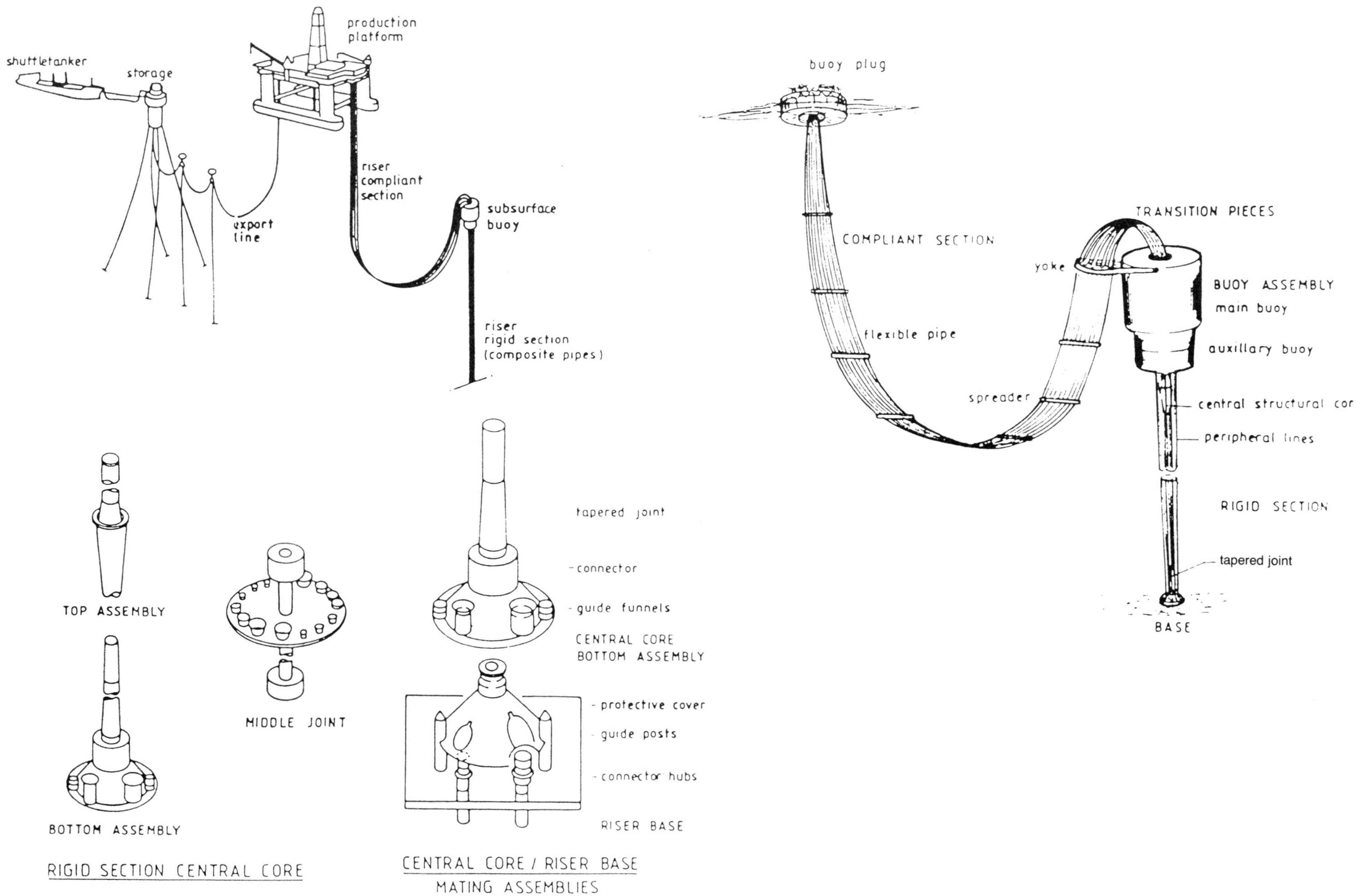

Fig. 2–5(1) Mobil/IFB Compliant Riser System

ples the peripheral lines with the catenary configuration of flexible pipes for the upper section.

The upper section consists of multiple lines of flexible pipe terminating at the floating production facility. This termination design depends on the floating vessel selected.Since the subsurface buoy is some 91 m (300 ft) subsea, the flexible section is some 219 m (720 ft) in length in order to accommodate the floating platform's watch circle in a severe environment 100-year storm condition. The upper section can be either a constrained bundle, which utilizes spreader beams, or unconstrained. The unconstrained approach is preferred because of easier installation. Each flexible is individually installed, connected at both ends hydraulically, and pressure tested.

The flexible pipe termination for a shipshape vessel is into a "buoy plug," mechanically latched in the ship turret. This buoy plug has a number of buoyancy compartments and guide tubes for the flexible pipes. Each flexible is hung-off atop the buoy plug in the dry. This assembly is designed for immediate release and descends to a neutral buoyant depth comparable to the sub-surface buoy. A deadman anchor maintains this assembly within a prescribed watch circle, thereby preventing structural damage to the flexible and contact with lower riser section. Recovery is by a tether line released with the assembly.

Although the number and size of individual lines depends on the specific on-site application, the design includes 10 lines from 3 in. to 12 in. plus a hydraulic bundle and two cables.

Distinguishing features of this riser design include the following:

- The lower section is of conventional tubulars, made up on site and installed using typical floating drilling tools and procedures.
- The hardware is an adaptation of equipment with field demonstrated performance.
- The area of complex installation is atop the subsurface buoy within easy diver access depth (91m/300 ft).
- Although the design is intended to stay connected in the most severe environment, a method is provided for ready release of the upper section from the floating platform.
- Each tubular line and flexible is individually retrievable from the floating production platform, thereby not interrupting other flow paths.

A second example for very deepwater field development using this compliant riser is the IFP design. This differs principally in the replacement of conventional tubulars by composite tubes for peripheral lines. These tubes are made of epoxy resin reinforced by carbon and glass fibers.

Weight becomes a factor of major concern if the application of the system is extended beyond 914 m (3,000 ft). Increase of the buoy assembly size incurs larger wave and current forces. Light weight tubular materials offer an advantageous alternative which provides, in addition, a high resistance to corrosion.

Prototype tubes have been designed, fabricated, and tested by Aerospatiale and IFP under an EC-supported program for the range of diameters and working pressures considered in the riser development program.

Based on these experimental results, it has been found that a system designed for 762 m (2,500 ft) water depth could be enhanced up to 1,828 m (6,000 ft) without any increase of the buoy size, if peripheral lines are made of composite tubes.

Placid-Green Canyon Bouyant Riser System

On the Green Canyon project in the Gulf of Mexico, Placid has used a buoyant riser to which tension is applied to control its position (see Fig. 2–5B). The main components are noted below from the riser templates upwards:

- The riser base is welded to the template which is piled into the seabed.
- The connector locating the riser on the template is of the collet type.
- Above the connector there is a "one-piece" titanium stress joint.
- The riser is assembled from a series of lengths each with air bouyancy chambers at the joints. These provide the support for the riser.
- The sales and annulus lines are supported by the joint structures.
- Air tanks at the top of the riser provide additional lift.
- There is a top riser connector package which allows emergency release of all the elements.
- Flexible lines link the riser to the vessel.

The riser does not act under tension from the vessel but as a bouyant structure held in position by steadying tension.

Lower Riser Package. The lower riser package serves several functions:

- To connect the reservoir and flowline to the template
- To transfer the riser loads into the template
- To provide the necessary flexure between the fixed template and the relatively mobile riser

The stress joint assembly, which is located immediately beneath the lowest riser joint, provides the flexiblity needed for the relatively abrupt bend in the riser near its foundation. The stress joint provides a smooth displacement curve for the production riser, producing manageable static loads and a lengthened fatigue life for both the stress joint and its included production, annulus, and export lines.

The lower collet connector is flanged to the bottom of the stress joint and lands on the connector hub which is welded into the riser base. This connector hydraulically locks the riser

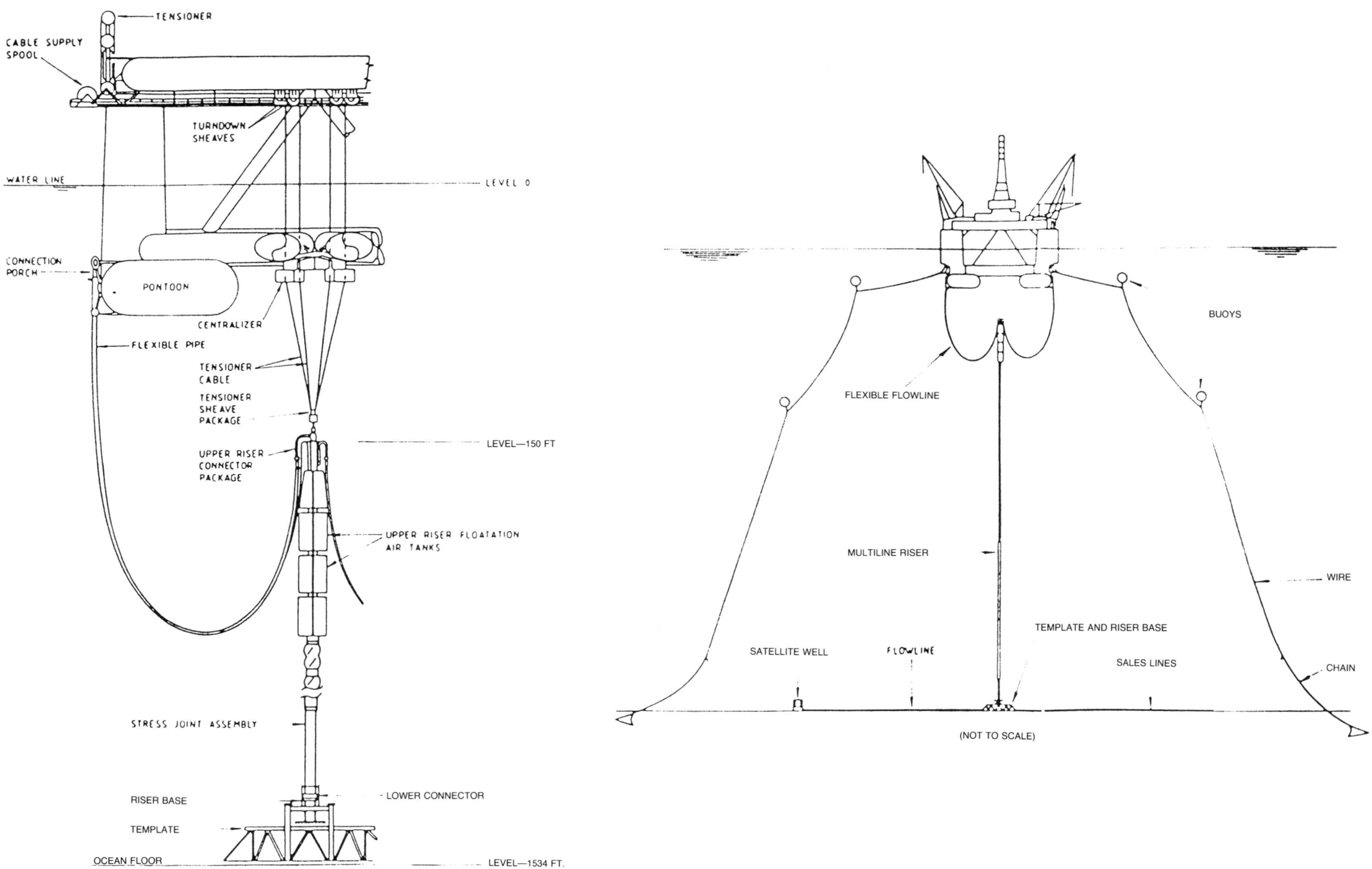

Fig. 2–5(2) Placid Green Canyon Buoyant Hybrid Riser System

to the template, thereby transmitting the riser loads into the riser base.

The riser base is welded to the template and transfers the riser loads to the template. It acts as a fixed connection point for the riser and flowlines.

The flowlines are run after the riser has been connected. The production and annulus tubing anchors are stabbed into the mating receptacle located in the riser base and are locked in through weight setting of an anchoring latch containing elastomeric seals like those used to set tubing in downhole packers. The tubing can be retrieved using either right-hand rotation or overpull. The export tubes are locked into their riser base receptacles using tieback connectors containing elastomeric tip-type seals.

Intermediate Riser Package. The riser consists of twenty-six 50-ft joints of flanged and cathodically protected high strength steel pipe. Each joint has an open bottom internal air tank with a cascade tube welded inside to air the joint above it. These joints are encased in syntactic foam modules that provide passive buoyancy and provide pathways molded within them to guide and support the lines.

An external air valve also is provided on each joint for an ROV to increase or decrease the amount of air in each joint as desired. Control umbilical troughs or guides and vortex shedding strakes are attached to the outside of the foam module during the running of the riser.

Upper Riser Package. The upper riser package serves several functions:

- To connect the flowlines to the rig while simultaneously decoupling the relative motion of the rig and riser
- To maintain the riser in the proper relative position with respect to the rig
- To allow the rig to disconnect quickly from the riser if desired
- To individually disconnect each flowline for maintenance.

Floating Production Vessel. Production facilities on Pernod 72 are installed in a prefabricated module which sits in a space created by removing the original helideck. It adds approximately 450 tons of new deck load.

The vessel is to remain on station for the entire life of the development—some 15 years. To survive maximum operating conditions, its mooring system, the largest ever used in the Gulf of Mexico, comprises (for a typical mooring line):

- 77 kip flat fluke twin shank drag imbedment anchor with 7 kip chain depresser
- Approximately 2,450 ft of 4⅛-in. oil rig quality plus 20% stud-link chain (Anchor section)
- 1,000 ft of 4¼-in. oil rig quality plus 20% stud-link chain (Dip section)
- 1,250 ft of 4½-in. spiral strand wire rope, spin resistant polythene jacketed (Riser catenary section)
- 18-ft diameter riser buoy of carbon steel with ballasting system and outer polythene protective coating, providing 130 kips of bouyancy
- Intermediate catenary section of 1,300 ft of 4½-in. spiral strand wire rope, spin resistant and polythene jacketed
- Intermediate buoy of 13-ft diameter in carbon steel, polythene jacketed with a balasting system, providing 50 kips of buoyancy
- 1,200 ft of 5-in. multistrand wire rope, spin resistant, non-jacketed but galvanized

On board Pernod, the mooring system is handled by equipment in new sponsons added to each corner leg. Each sponson handles two of the eight mooring lines.

Operational Considerations. The hybrid-buoyant-rigid riser installed by Placid is intended to remain in place for the field life (20 years). Its lower assembly (connected to the template) is rigid, while the upper assembly consists of flexible hose. Any disconnection would separate the flowlines, which then would be drawn up under the production systems' pontoons by the on-board tensioners, while the rigid-buoyant lower assembly stayed connected to the template and free standing. Disconnection was planned as "hurricane protection," however, tank tests indicated that it would probably never be required in the field.

Flexible Dynamic Riser

The five technical and economic advantages of incorporating flexible risers in subsea production systems are the following:

1. Significant reduction of downtime costs or loss of production
2. Relative simplicity in system design
3. Built-in flexibility: system can be easily adapted to changing economic objectives as field development and life progresses
4. Easy and rapid installation: use of relatively lightweight and economical D.P. vessels for minimum dependance on weather and other vessels in the field
5. Minimum impact on floater design

The flexible dynamic riser concept has been used successfully for several years in various parts of the world in relatively temperate weather conditions and is now accepted by the industry for applications in the most harsh environments. Flexible dynamic riser systems can be designed for most types of floating production systems, but of the many different possibilities, five basic configurations are currently considered most suitable (see Fig. 2–6).

Free Hanging Riser. This system is the most suitable for temporary production (in rougher seas) or for permanent field develop-

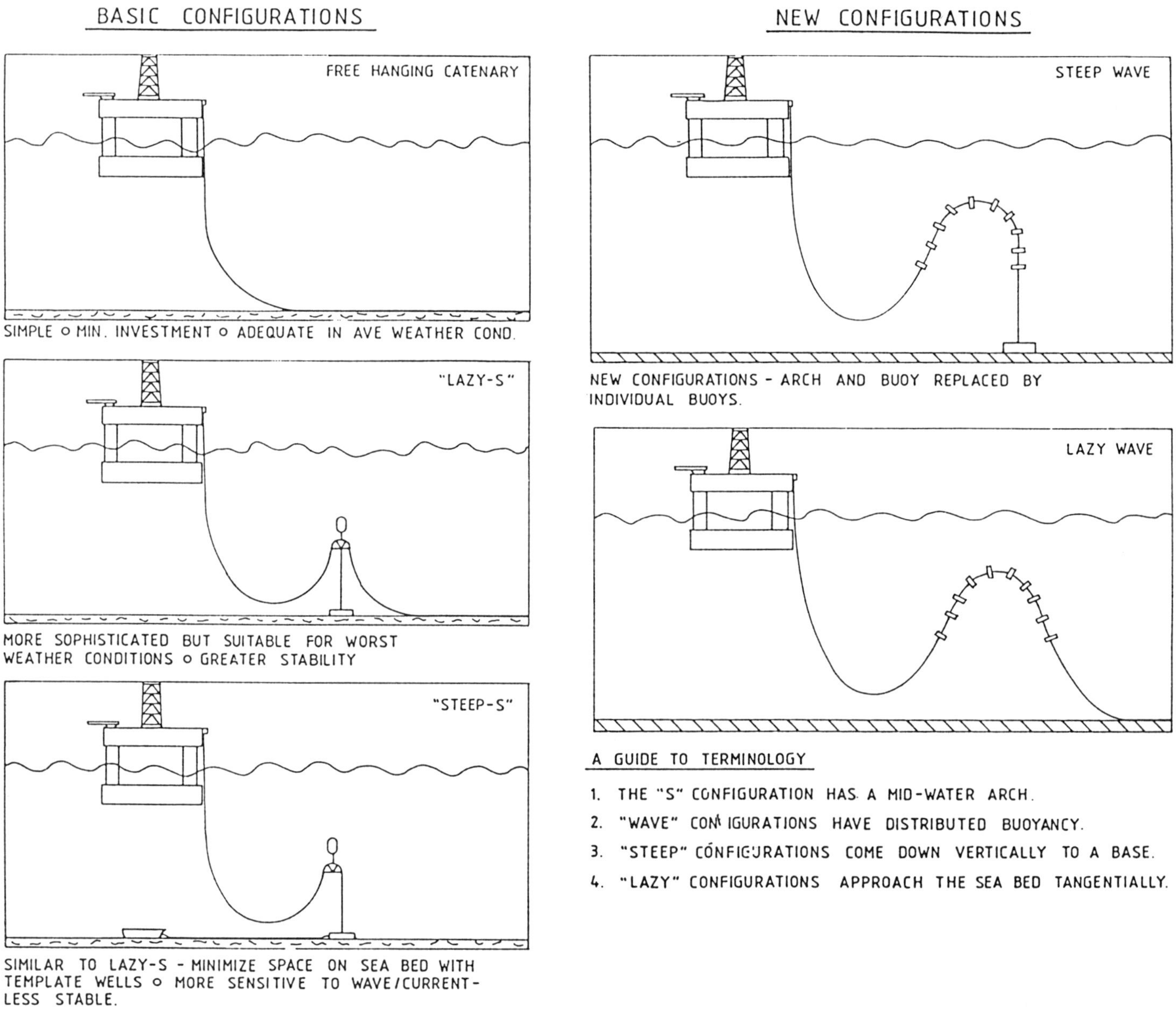

Fig. 2–6 Flexible Riser Configurations

ment in average weather conditions. Being the simplest of the five, it requires a minimum investment in equipment. The flexible riser runs in a catenary configuration from the upper connection point on the floater straight down to the seabed where it can be connected to any type of subsea equipment.

Although the dynamic behavior of the riser in this configuration does not compare with the performances of the other four systems described, results have proven the system to be perfectly adequate in average weather conditions. Since the first free-hanging risers were introduced in 1975, the system has seen service in various parts of the world, notably in Indonesia, the Mediterranean, and Brazil, in record water depths of 250 m (820 ft) and over. This system is now fully accepted as standard equipment for what has come to be a typical scheme of field development (see Fig. 2–7).

Double Catenary Lazy-S. More sophisticated than the free-hanging riser, the double catenary configuration is suitable for all types of field development, even in the worst weather conditions. The system is used to best advantage in production schemes involving satellite wells, or whenever the subsea equipment is located at a distance from the floater.

The system comprises a flexible line running down from the upper connection on the floater via a sub-surface buoy and mid-water arch, in a double catenary. The first catenary, suspended between the floater and the buoy, absorbs most of the motion induced by current and waves, and the second catenary runs directly from the mid-water arch to the sea bottom for connection to the subsea equipment. The sub-surface buoy is tensioned by means of a chain and deadweight.

The double catenary system is usually installed in a diverless operation from the laying vessel. This type of dynamic riser is currently operating with success in the Mediterranean and offshore Brazil and soon in the North Sea.

Simple Catenary Steep-S. The Steep-S is the most advanced system of the five. It is being developed as a safe solution to the problem of over-crowding on the sea floor for production systems where the subsea equipment is located directly beneath the floating facilities. Although being suitable for all weather conditions, this system is best adapted for harsh environments.

The simple catenary system differs from the double catenary in that the flexible riser itself is used to tension the sub-surface buoy, terminating in a riser base or PLEM (replacing the deadweight) at the subsea connection point. Although the second catenary is eliminated, the dynamic behavior of the riser is very much the same. This configuration provides the most direct liaison between the surface facilities and the subsea equipment. With the elimination of tie-back loops, the system occupies a minimum of space on the seabed, thus reducing the risk of interference with other subsea equipment and accidental damage from anchor chains and other hazards. The simple catenary riser system has been operating with success since 1978 on installations in the Far East and Brazil and is now being incorporated into systems in the North Sea by Hamilton Brothers for the Argyll field and by Sun Oil for the Balmoral development (see Fig. 2–8).

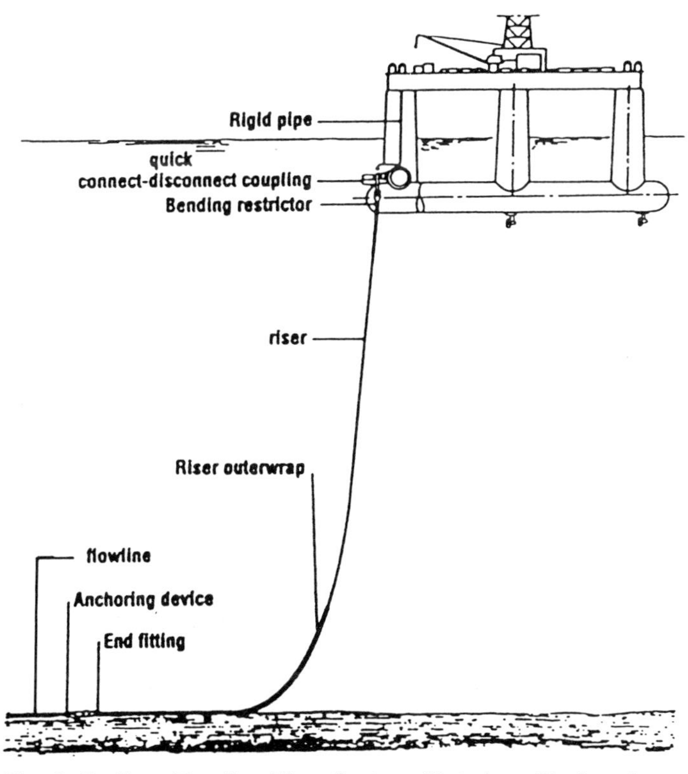

Fig. 2–7 Free Handing Riser System (Petrobras/Enchova)

Steep Wave. The Steep Wave is a straightforward extension of the classic Steep-S configuration. The mid-water arch no longer figures and the associated buoyancy tanks are replaced by smaller buoyancy modules strategically distributed along the flexible riser itself. Bulky equipment is eliminated and, as a result, installation techniques can be simplified. These features make the Steep-Wave system particularly attractive for deep water applications. With regard to dynamic behavior, the Steep Wave is very similar to the Steep-S riser. Buoyancy is distributed so as to accommodate floater excursions while maintaining optimal flexible pipe curvature and acceptable loads on the surface facilities. This system is especially suitable for incorporation into single well, early oil production and/or temporary testing systems.

Lazy Wave. The Lazy Wave riser system is a result of the application of the distributed buoyancy concept to the Lazy-S configuration. The mid-water arch, deadweight and connecting sling are eliminated. By determining the most appropriate distribution of the buoyancy modules for each individual case, it is possible to obtain precise control over the loads exerted on the floating facility by the riser system. This is obviously of considerable importance in deep water applications. Since a Lazy Wave riser system requires less hardware than the Lazy-S, installation procedures are simplified. While this system is appropriate for most types of field development, it is applied to best advantage in production schemes involving satellite wells and/or floating export systems.

Dynamic Riser Typical Configuration

Apart from the flexible pipe itself, the key components are shown in Figures 2–8 and 2–9 which illustrate a typical design selected for a project in the North Sea—Balmoral.

The Upper Riser Connector of QCDC. A key element in the safety of the system is the quick-release hydraulic coupling. Although the riser system itself is designed to be fully operational in the roughest weather without disconnection, provision has to be made for quick disconnection in emergencies arising from broken anchor chains, blow-outs, fire on the platform, etc.

The connector is in two parts, held face-to-face by means of a hook-and-spring system. A ball valve on each part avoids

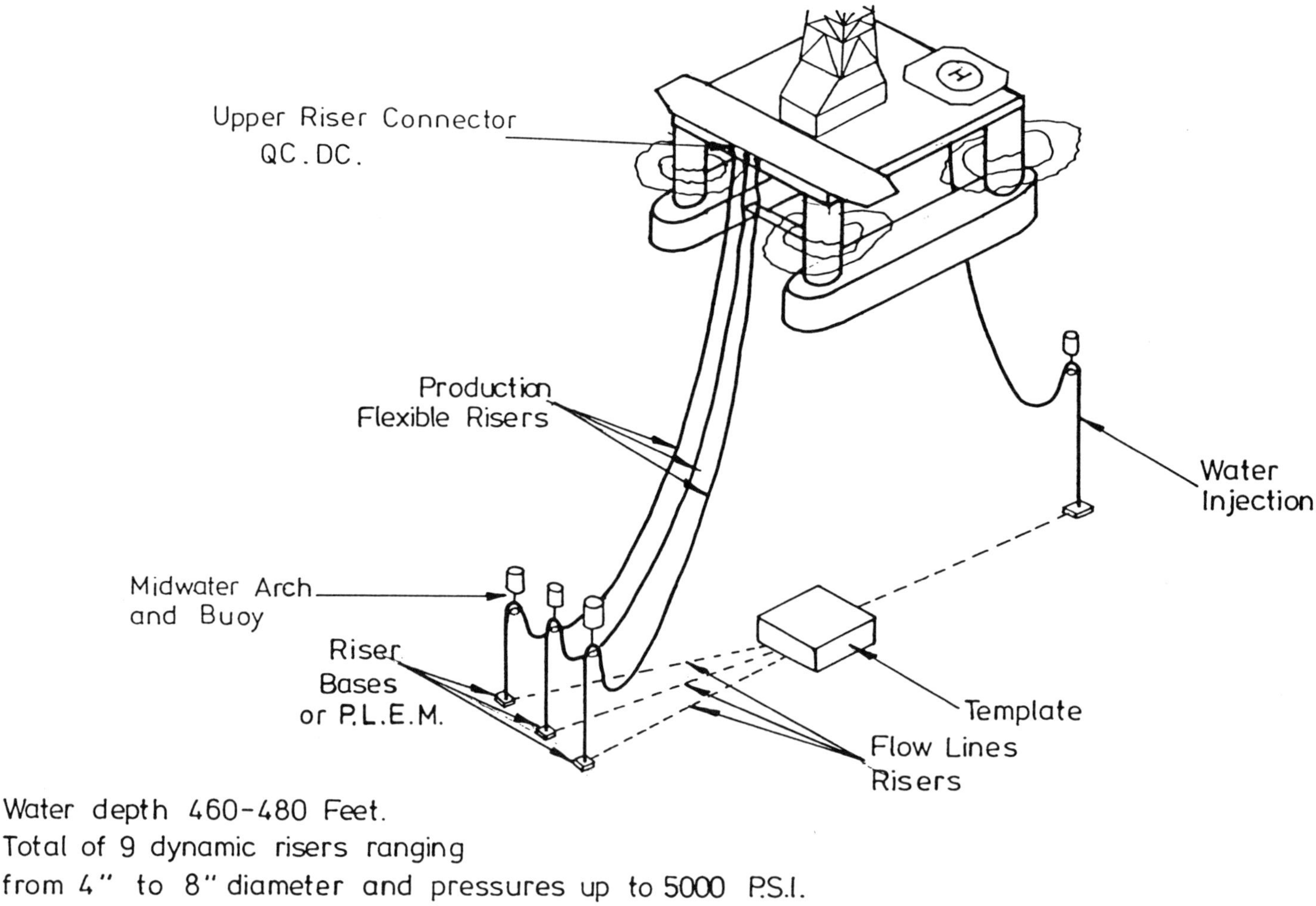

Fig. 2–8 Simple Catenary Design

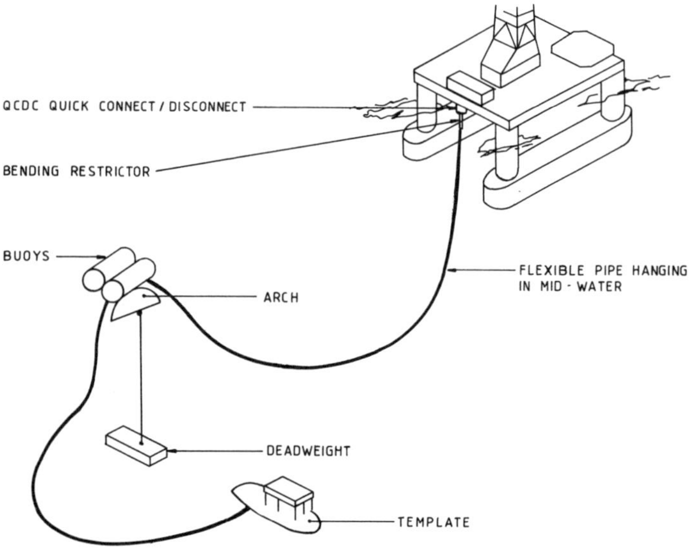

Fig. 2–9 Typical Key Components

any risk of pollution. A special guide enclosure system provides easy alignment and connection of the male and female halves of the coupling by the introduction of the mandrel's guide post through the main body guide funnel (see Fig. 2–10).

The Bend Restrictor. The movements of the floater induce cyclic bending adjacent to the riser end-fitting. To limit the movement of the riser and increase the stiffness at this point, a plastic-molded bending restrictor is fitted behind the end-fitting.

The Sub-surface Mid-water Arch and Buoy. Depending on the specific operational requirements and water depth, the buoy can be made of either steel or syntactic foam. The flexible pipe is clamped over the curved pipe tray and moves freely over the rest of its length.

The Riser Base or Pipeline End Manifold. The subsea termination of a riser in this configuration consists of a simple steel and

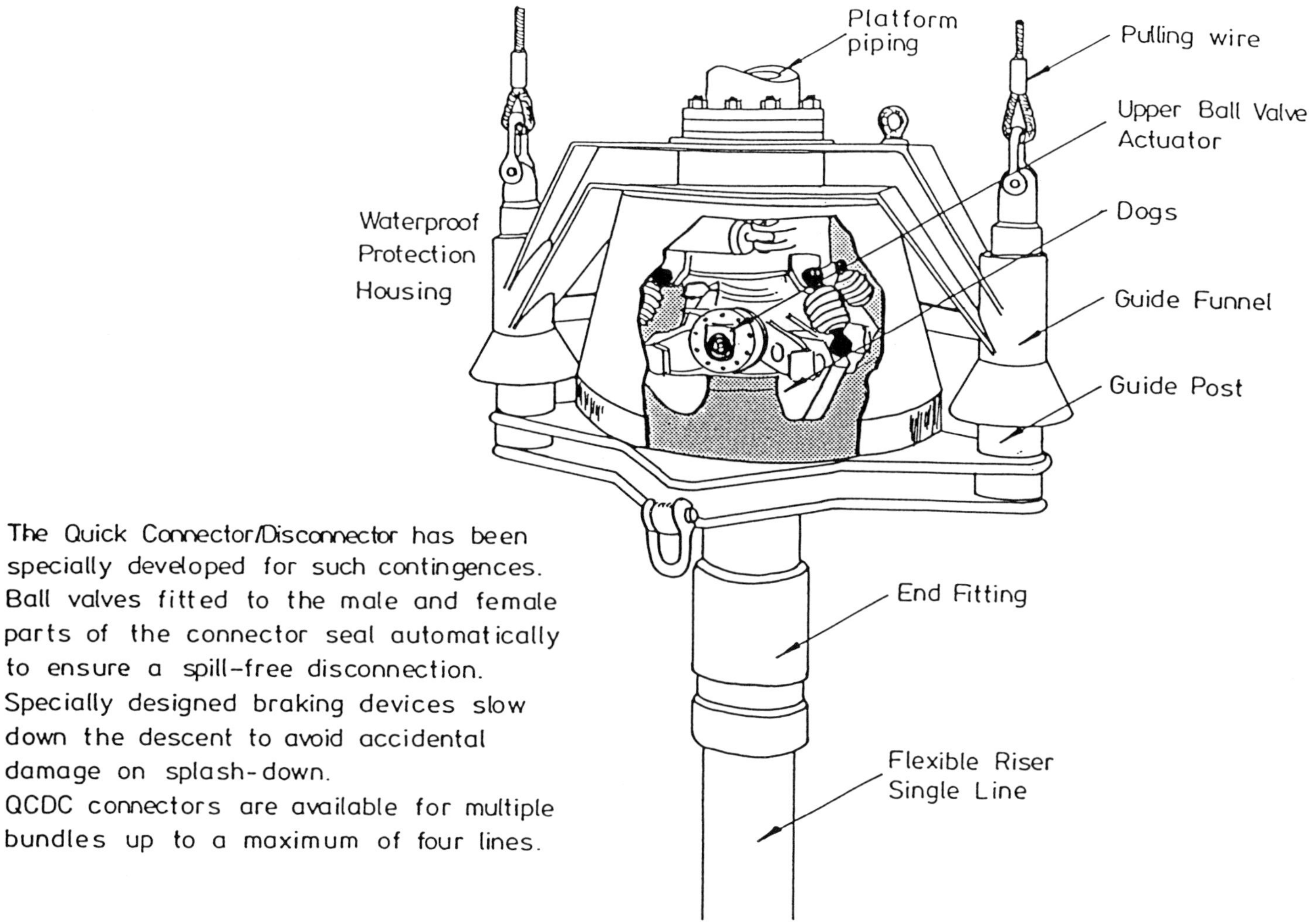

Fig. 2–10 Topside Connector

concrete structure which is stabilized by its own weight. It can be piled if required by soil conditions. Refer to Figure 2–11 for a three-line typical riser base.

Design, Computer Simulation, and Model Testing

The engineering of each dynamic riser system should involve a comprehensive study covering the following five key topics:

1. Definition of the most appropriate pipe structure, based on the specification fo the project in question. Refer to a typical structure in Figure 2–12.
2. Selection of the most suitable riser configuration based on the following criteria:
   - Weather data
   - Anticipated motion at the surface
   - Facilities
   - Water depth
   - Extreme loads
   - Required service life

   Using these data as input, a comprehensive computer package should be used to predict both the dynamic behavior and the limits of the service life of the system.

   Service life can be expected between 8 and 10 years with existing pipe structure. Refer to Figure 2–13 for a typical computer simulation print out.
3. Design of the necessary ancillary equipment, riser platform interface, sub-surface buoy and mid-water arch, riser bases, etc.
4. Definition of the most appropriate installation method and handling equipment. A typical installation connection procedure is shown in Figure 2–14.
5. Simulated model basin tests, if necessary. Regular and irregular waves up to 100-year storm height should be generated from different headings to simulate the behavior of the system

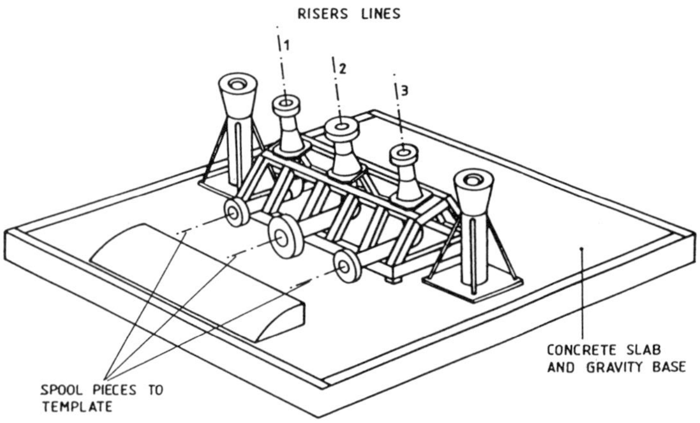

Fig. 2–11 Typical Riser Base for Flexible Risers: Gravity Type/ Triple Line

in all possible situations. Several model tests have been carried out up to date, and one of the most remarkable results to emerge from these tests is the overall stability of the dynamic flexible riser system in the harshest conditions—even in the splash zone, normally a critical area. Simulated emergency disconnections, with quick connect/disconnect couplings (QCDC) connectors, have been thoroughly tested and proved to be perfectly safe for both the flexible risers themselves and for the platform—as well as for the subsea equipment beneath.

Operation and Maintenance Advantages

The incorporation of dynamic flexible riser systems in a production scheme results in major advantages for the operator:

1. Generation of early cash flow and minimum investment in equipment.
2. Easy extension of field capability by simple successive upgrading of the riser system.
3. The ability to remain fully operational even in centenary weather conditions.
4. No or very little downtime is attributable to the riser system.
5. Marginal impact on overall rig design.
6. The central moonpool is left free on the floating production platform enabling re-entry and workover operations without shutdown of other wells when a subsea/template production system is located underneath.
7. The lightweight engineering of the system totally eliminates the need for heavy tensioning system.
8. Routine disconnection during temporary shutdown for platform maintenance and refurbishing is a simple operation. The risers are winched down to the seabed where they are placed to avoid interference with other subsea equipment. Reconnection is a simple operation performed in a reversal of the disconnection procedure with ROV or diver assistance.

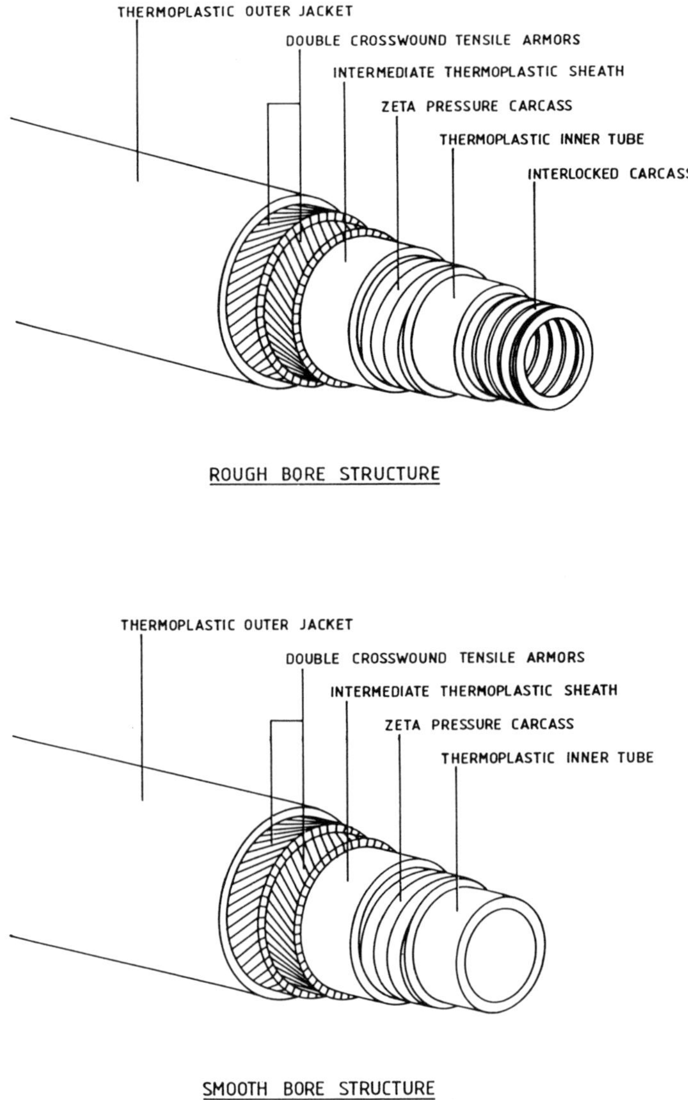

Fig. 2–12 Typical Coflexip Riser Structure, Rough and Smooth

9. Emergency shutdown is completed in a matter of seconds with QCDC connectors (60 to 120 seconds). As with routine disconnection, the risers are easily recovered for rapid reconnection when the emergency is over.

From a maintenance point of view:

10. Flexible dynamic riser systems require no maintenance other than regular ROV inspection of the flexible lines and subsea equipment and routine checks of the connectors and their associated equipment—this up to their in-service life of 8 to 10 years.
11. The flexible risers have built-in corrosion resistance, provided by the external polyamide sheath or other material. In addition, all metallic components of the subsea part of the system are cathodically protected.

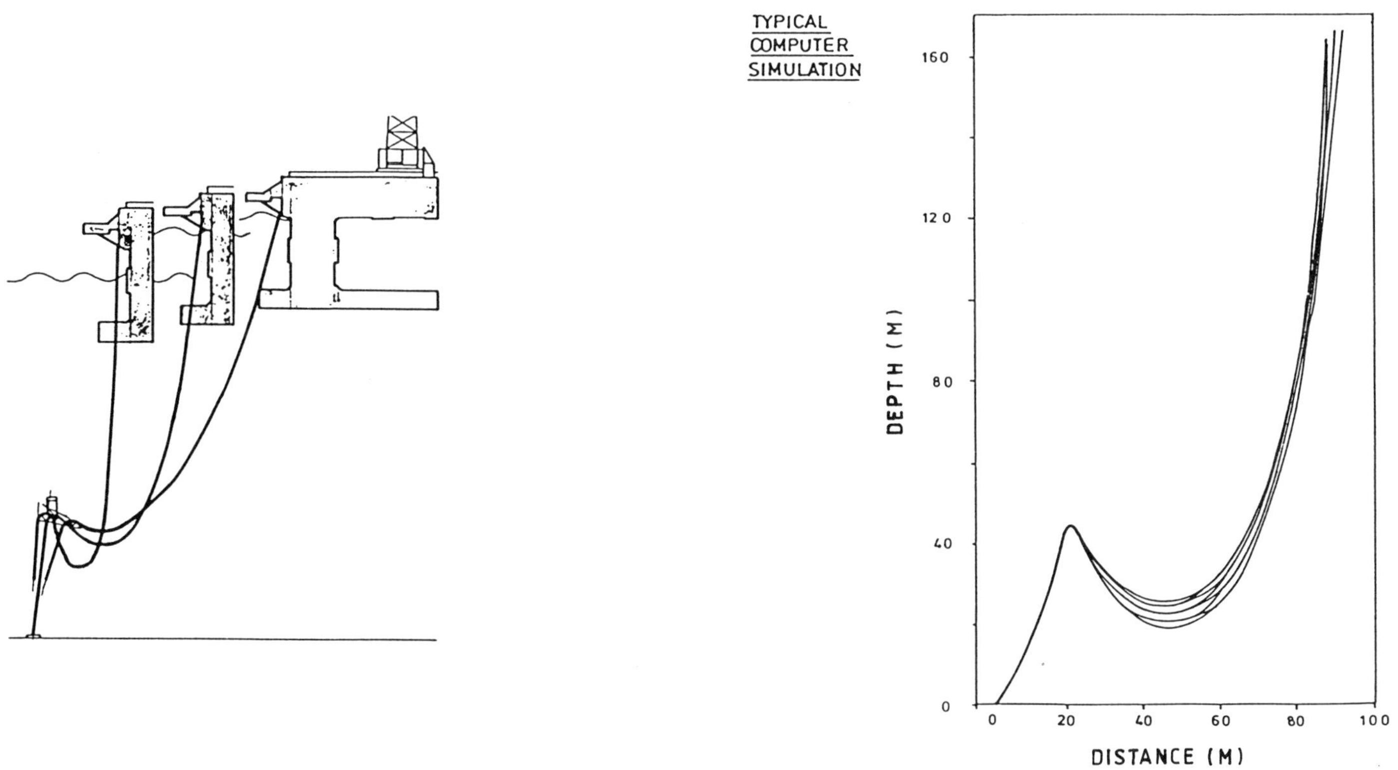

Fig. 2–13 Flexible Dynamic Risers

Comparison Between Rigid and Flexible Riser Systems

Many of the problems highlighted below are overcome by a hybrid riser system, which is not included in the following comparison.

Rigid Riser System.

- A rigid riser system is very likely to be hung-off or retrieved in bad weather. A strong riser base/manifold is required.
- Considerable diving effort (mostly on the subsea manifold) has been necessary on existing systems, Buchan and Argyll. For water depths in excess of 304 m (1,000 ft) diverless systems would have to be designed.
- Riser running and retrieval times for deeper water depths would be lengthy.
- A constant load tensioning system is required to allow for vessel heave taking away a lot of deck load.
- The moonpool equipment can be extensive, with many winches, and have compensated systems for riser and umbilical guide wires, television camera, etc. Moonpool modification of an existing semi-submersible can be very complex.
- Once the system is installed, it is not easy to incorporate additional risers unless these have been specifically allowed by the original design.

Flexible Riser System.

- Flexible risers are more likely to meet platform survival environmental conditions, with platform motions taken up by the catenary of the risers and umbilicals.
- Particular care is necessary to ensure that flexible risers can withstand all the design conditions. For example, the durability under the effects of the temperature, pressure, and chemical constituents of the produced fluids must be ascertained.
- Quick release connectors on the umbilicals and risers need only be used in an extreme emergency.
- The lower end of the risers terminate at a PLEM or a flowline connector/anchor. These are simple structures which experience relatively low loads from the risers.
- A conversion may require very limited modification to the moonpool.
- The moonpool is free for workover equipment. A derrick is not required unless one is necessary for workover. Drill crew accommodation can be released for other use.
- The riser system should have good operational availability. Riser maintenance or changeout may be more expensive than with a rigid riser. Equipment for riser installation/retrieval does not need to be incorporated on the platform.
- Once in production, it can be relatively easy to tie in extra wells and introduce additional risers. However, for pontoon

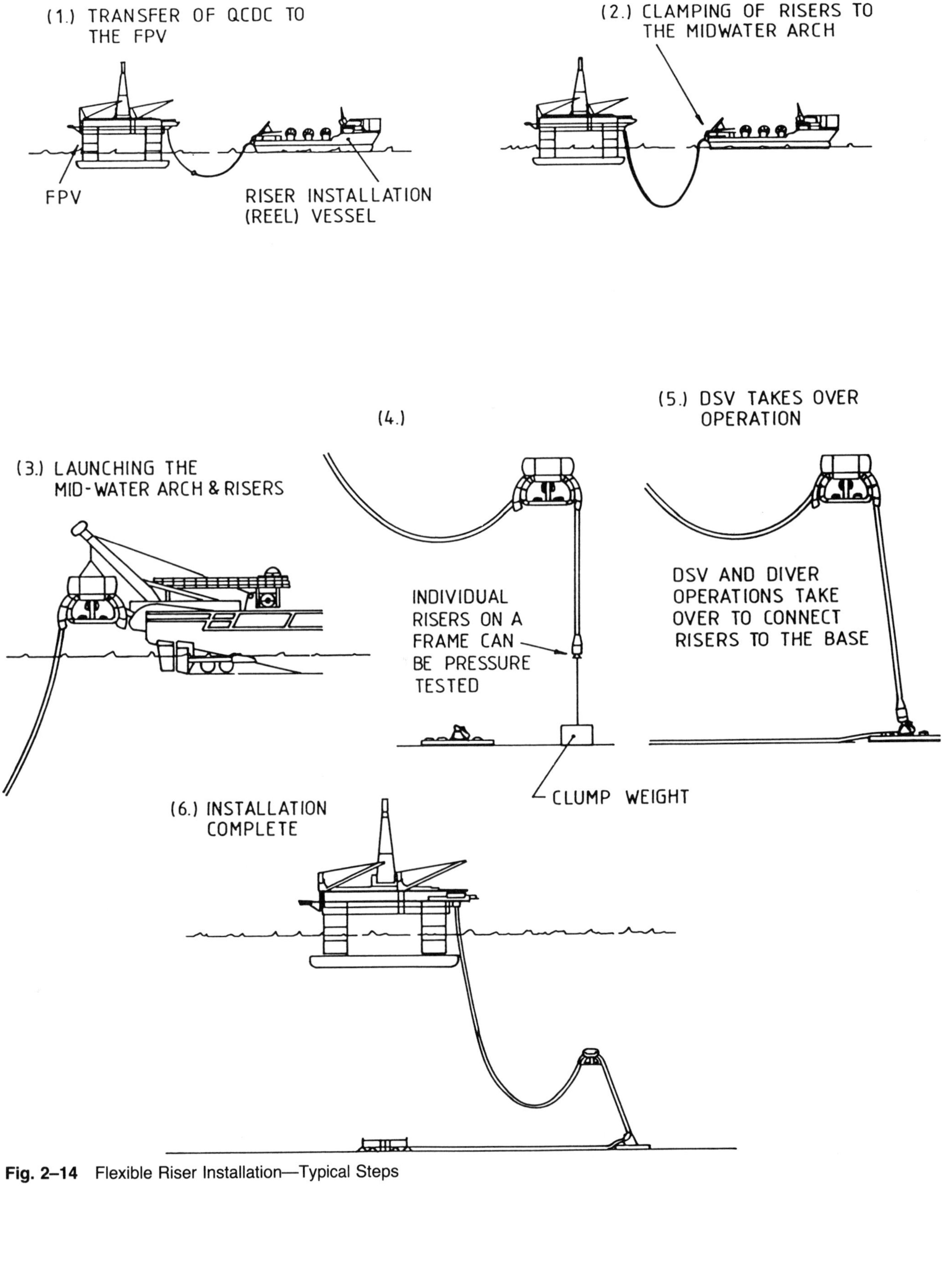

Fig. 2–14 Flexible Riser Installation—Typical Steps

mounted connection, additional connectors would have to be incorporated during initial construction or conversion.

- However, the potential gains over a rigid riser make flexible risers very attractive when used with semi-submersible and tanker based production schemes. A trend toward their use, in even the severe weather conditions of the North Sea, can be discerned already.

R & D PROJECT "FLEXIBLE ARCH," NORTH SEA TEST (1983)

The Flexible Arch was an experimental project of which half was financed by Mobil Exploration Norway, Inc. and the remainder by Elf-Aquitaine Norge, Total Marine Norsk, and the Institut Francais du Petrole (IFP), together with an EC subsidy. The IFP acted as project manager. The purpose of the experiment was to verify the behavior of a flexible pipe in a seabed-to-surface liaison, in deep water and under the severe weather conditions of the North Sea. The test was carried out over one year (1982–1983) on a 135 m (443 ft), lone 10-in. I.D., 4,500 psi working pressure (WP) flexible pipe, filled with water under pressure, and suspended between a surface buoy and a subsea float submerged in a catenary mode. Sensors located at each end and at the lowest point of the "arch" provided a continuous measurement of the parameters through which it was possible to know at any given time the configuration of the arch, the accelerations it underwent, and the internal pressure. The data were transmitted to a nearby platform for recording and processing (see Fig. 2–15).

The threefold aim of the test was to:

1. Carry out an endurance test on a 10-in., 4,500 psi working pressure flexible riser pipe under severe conditions
2. Determine the loadings and movements to which the pipe was subjected by means of appropriate instrumentation
3. Compare and analyze the components of the flexible pipe before and after the endurance test for comparison

Components of the Installation

1. A 135 m (443 ft) length of nominal 10-in. I.D. flexible pipe
2. A moored surface buoy designed to have movement characteristics comparable with those of a large tanker
3. A subsurface buoy submerged at 60 m (196 ft) depth
4. A deadweight to anchor the subsurface buoy
5. Instrumentation at selected points along the length of the flexible pipe and on the surface buoy, as described hereunder
6. A telemetering system to transmit data from the acquisition system on the surface buoy to data-recording equipment on the nearby Frigg platform

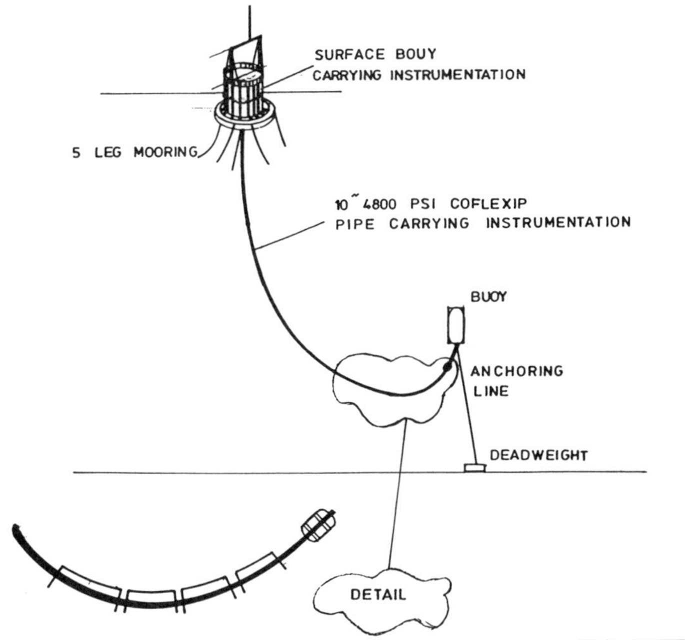

Fig. 2–15 Mobil Flexible Arch

Environmental data from platform instrumentation were collected also.

Test Site

North Sea environmental conditions provided a severe test of the flexible pipe. The proximity of the Frigg platform provided a convenient location for routine inspections of the test installation and for the installation of data-recording equipment. It also gave access to existing environmental recording devices. A flat seabed topography and 106 m (348 ft) water depth were suited to the dimensions of the test configuration.

Installation and Retrieval

Installation was accomplished in three phases in the summer of 1982. These included:

1. Laying out of surface buoy mooring-chain lines
2. Mooring of the surface buoy
3. Installation of the flexible pipe, subsurface buoy and dead weight.

The test installation was retrieved in mid-July 1983.

Instrumentation

The test installation included 41 sensors measuring:

- Movements of the upper end of the flexible pipe
- Movements of the lowest point of the catenary and the subsurface and of the flexible pipe

- Wave and tide height
- Tension of the flexible pipe
- Curve radius of the pipe at the lowest point in the arch, by 12 pressure sensors
- Pressure within the pipe
- Relative (pipe-to-buoy) angular deflection

Environmental data available on the platform included wave height and direction, wind speed and direction, and surface and bottom current velocities.

All data were recorded at a rate of 17 minute sequences every 3 hours. During periods of rough sea states, when present levels were exceeded, sampling frequency was increased to hourly and, at the roughest sea states, to a continuous basis. Over 10 million complete data sequences were recorded.

Data Analysis

The data collected in the test were analysed in three phases. Initial processing was devoted to collating data and converting it into engineering units. Time series of all parameters could, therefore, be obtained from this phase of processing.

Intermediate processing developed correlations between various parameters, with emphasis on significant values of the environmental data and the response data at the upper and lower ends of the flexible pipe and at the lowest portion of the catenary.

Final data processing produced two principal results:

1. Frequency analysis providing detailed response characteristics, such as movements and forces experienced by the flexible pipe for a wide range of environmental conditions representative of the test period
2. An estimate of the number of bending cycles experienced by the flexible pipe by classes of curve-radius variations

Physical Analysis of Flexible Pipe and Conclusions Drawn

After its retrieval, the flexible pipe was cut into sections for analysis to determine fatigue and wear rate. The characteristics of all components of the flexible pipe were measured and compared with those of a reference sample. The general conclusion was that the flexible pipe survived the test and the objectives were attained.

A large diameter Coflexip type flexible pipe had been demonstrated to maintain high internal pressure in a severe environment over the endurance test period. The number of cycles, in conjunction with results of the post-test examination of the pipe, provided a basis for estimating the effective lifetime of flexible pipes in an application similar to the test configuration. Data are also available to correlate computer programs that predict the response of flexible pipes, which are a key component of floating production systems for hydrocarbons.

PETROBRAS/ENCHOVA EXPERIENCE

Preliminary Studies

The purpose was to investigate the feasibility of three different riser system designs. Each design consisted of a bundle of four flexible lines to enable the connection of the subsea wellhead to the semi-submersible platform. The three different riser systems studied were the double-catenary riser, tensioned riser, and free-hanging riser.

The double-catenary riser system consists of four flexible risers suspended from the deck of the platform. These are run in a catenary curve to four wheel-type supporting structures from which they continue in a catenary curve to the sea bottom. The four supporting wheels were suspended from two floating cylinders and anchored to a base plate on the seabed by means of a chain.

The tensioned-riser system has four vertical flexible risers connected to a base plate on the bottom and tensioned by a floating buoy. Four flexible risers were suspended in a catenary curve, one end being connected to the semi-submersible deck and the other to the floating buoy.

The free-hanging riser system consisted simply of four free-hanging flexible risers running to the bottom in a catenary curve.

Computer Calculations and Model Testing

Before proceeding with model tests, simplified computer calculations were carried out to ascertain the basic feasiblity of the project and the geometric data for each of the three systems.

The riser systems were tested under different sea conditions (irregular waves), current and wind, and orientated in three directions with respect to environmental elements. Some test were carried out with risers pulled in through J-tubes and other tests where the risers were connected directly to the semi-submersible deck.

Tests performed on the three flexible riser system designs resulted in the following main conclusions:

1. No J-tube was required; the risers remained vertical at the sea level and no contact was observed.
2. Measured forces in the pipe connected at the point of attachment on the semi-submersible and at the base plate showed that the dynamic effect was very low (40%) for all the riser systems.
3. The tensioned riser system showed a very stable behavior, and the low point of the catenary curve never touched the sea bottom under the platform movement.
4. There was no essential difference in behavior between the double-catenary and the tensioned riser systems; slight movements were observed at the bottom end of the short catenary riser.

5. The free-hanging riser system showed a behavior slightly less favorable than the other two, but the movements of the risers on the seabed appeared, nevertheless, acceptable for the riser structure. Therefore, the system was considered to be feasible for the final evaluation.

Riser Configuration Selection

From the different riser configurations studied during the model tests, Petrobras chose the free-hanging riser system for the production of the Enchova field, as shown earlier in Figure 2–8.

The double catenary and the tensioned riser systems offered the advantage of very good riser behavior even under survival conditions, but the incorporation of an intermediate buoyancy tank complicated the laying operations. These systems were to be retained for the permanent production facilities.

Although its behavior appeared slightly less favorable than the other systems, the free-hanging riser system had the primary advantage of easy installation and retrieval. It was concluded that this system seemed the most suitable for the early production system. A complete computer study of the free-hanging riser behavior was conducted under both survival and operating conditions including wave and current effects, confirming the validity of the choice.

The typical riser structure designed for the Enchova field included five principal layers, the characteristics and dimensions of which were determined according to the requirements of the service involved.

SUN OIL BALMORAL SYSTEM

The originality of the Balmoral field development lies in the fact that it is the first ever purpose-built floater and flexible riser system designed for North Sea conditions.

The maximum environmental design conditions for this area of the North Sea are:

- Significant wave height —15.9 m (52 ft)
- Maximum wave height —29.1 m (95 ft)
- Period (zero up-crossing) —14.3 sec.
- Current speed surface — 1.25 m/sec.
  bottom — 0.45 m/sec.
- Wind speed (3 sec. gust) —52.2 m/sec.
  (1 hour mean)—38.1 m/sec.

Summary of Flexible Riser System Requirements

The riser system consists of four bundles—two production bundles and one sales line on the aft side and one water injection bundle on the fore side of the FPV.

- Bundle 1 & 2: 1-in. x 8-in. + 2-in. x 4-in. production/annulus/test/kill/gas lift risers
- Bundle 3: 1-in. x 8-in. sales line
- Bundle 4: 2-in. x 8-in. water injection lines

The design pressure of the risers ranges from 1,650 psig for the production/annulus and test/kill lines, to 2,300 psig for the water injection, and 2,900 psig for the oil sales lines.

A 4-in. production riser, initially intended for Devonian production test, may be eventually converted to gas export service if required.

The riser system is designed to permit:

- Production operations over the full range of weather conditions, including the above specified centenary conditions
- Release of the riser bundles for emergency in any weather conditions, with the FPV in operating, survival, or damaged mode
- Reconnection of the riser bundles of the FPV

In addition, the riser system can tolerate a horizontal movement of the FPV within an area of 26 m × 28.8 m (85 ft × 94 ft) from the central axis of the template which, taking into account the dynamic response of the FPV to the wave, wind, and current action, means a total maximum offset of more than 25% of the water depth.

The minimum required design life is 10 years for the risers themselves and 20 years for auxiliary riser equipment.

Trade-Off Comparison: Flexible Versus Rigid Risers (See Fig. 2–16)

An extensive trade-off study was made to compare a flexible riser system with a rigid system. Candidate riser systems were developed, and comparisons were made between operating performance, operational flexibility, and costs. The flexible riser system was eventually selected for the following reasons:

- A flexible riser minimizes the influence of weather on production downtime, as it is capable of remaining connected in all operating and survival weather conditions. Rigid riser systems would not be able to safely withstand these same conditions, requiring disconnection from the template and recovery as a storm developed. This was a critical consideration for the Balmoral development because, as a small field, its economics were very sensitive to production downtime. Balmoral uses a pipeline export rather than a more weather-sensitive shuttle tanker off-loading system, hence, the risers become a particularly critical element in maintaining field production.
- Flexible risers do not require handling in severe weather conditions and can be located on the fore and aft ends of the vessel rather than in the vicinity of the derrick. This frees the drilling rig and center moonpool of the FPV for workover, diving inspection, and subsea maintenance operations.

Fig. 2–16 Comparison of Riser Types (Source: Sun Oil)

OPERATIONS

	RIGID		FLEXIBLE
	INTERNAL	NON-INTEGRAL	
NORMAL	CONTROL VESSEL POSITION AND RISER TENSION. LIMITED WORKOVER OPERATIONS.		MONITOR VESSEL POSITION. UNRESTRICTED WORKOVER.
SEVERE WEATHER DISCONNECT	DISCONNECT & RETRIEVE.		NOT REQUIRED.
EMERGENCY DISCONNECT	DISCONNECT & HANG OFF.		DISCONNECT AT VESSEL & DROP RISER TO SEABED.
RECONNECT	OPPOSITE TO PULLING PROCEDURE. TEST FLOWLINES AS RUN.		REQUIRED DIVER ASSISTANCE. TEST AFTER PULL-IN.
CRITICAL COMPONENTS	RISER COUPLINGS SUBSEA FLEXIBLE CONNECTIONS.	SUBSEA FLEXIBLE CONNECTIONS.	END TERMINATION DISCONNECT/ RECONNECT SYSTEM.
ADAPTABILITY	DIFFICULT TO MODIFY. CAPACITY MUST BE BUILT IN.		LIMITED NUMBER OF ADDITIONAL RISERS CAN BE INSTALLED.
RELATIVE COST	1.2	1	1
DOWNTIME (Weather & repair)	LESS THAN 10%	MORE THAN 10%	LESS THAN 6%

Fig. 2–16A Operations

- The flexible riser system can be installed in phases. The Balmoral system is designed to accept future installation of two additional riser bundles for development of other fields with minimal initial capital investment. Similar expansion to the design of a rigid riser system is more difficult without imposing extensive initial commitment.

The capital cost of a flexible riser system is greater than a rigid riser system. But over the field life, this added cost is largely compensated for by its superior weather capability (production uptime).

The more complicated construction of flexible pipe makes it inherently more difficult to inspect than a comparable steel pipe. Moreover, the flexible riser system extends out from the vessel, putting much of it out of reach of direct inspection, maintenance, and repair from the FPV. Visual inspection of certain sections of the flexible riser system can be performed by an ROV operating from the FPV, but maintenance or repair work will normally require auxiliary vessels with diving systems and flexible pipe handling equipment. However, if properly designed and constructed, the flexible riser system requires little maintenance, and its design life will equal or exceed field life.

Selection of Riser Configuration

The free handing configuration was immediately excluded for Balmoral in view of vessel motions. This would have led to excessive topside riser loads and angles, excessive motions at the touchdown point, as well as compression modes and possible buckling of the lower part of the catenary.

The two possible configurations remaining, therefore, were the Lazy-S and Steep-S. Both are similar in behavior, with the Lazy-S having slightly better stability against mid-water arch weathervaning, but having a poorer seabed layout as the risers require more area to loop back to the template. The Steep-S configuration was therefore selected as the most appropriate for the Balmoral field conditions, subject to further substantiation which was to be gained from the model tests.

MAINTENANCE

	RIGID		FLEXIBLE
	INTERNAL	NON-INTEGRAL	
ROUTINE	SERVICE TENSIONERS AND SURFACE EQUIPMENT.		SERVICE SURFACE EQUIPMENT.
REPAIR	SHUT IN AND PULL COMPLETE RISER TO REPAIR FLOWLINES	FLOWLINE CAN BE RETRIEVED INDEPENDENTLY FOR REPAIR. SHUT IN NOT REQUIRED	SHUT IN AND REPLACE. RISERS ARE NOT FIELD REPAIRABLE.
INSPECTION	NDT INSPECTION PERFORMED WHEN RISER IS ON DECK.		VISUALLY INSPECTED BY ROV/DIVER.
EXPECTED LIFE	20 YEARS WITH MAINTENANCE.		11 YEARS ESTIMATED REPLACE RISERS.
SAFETY PERSONNEL	REQUIRES MOONPOOL AND RIG FLOOR OPERATIONS DURING SEVERE WEATHER DISCONNECT.		MINIMAL OPERATIONS FOR DISCONNECT DIVER OPERATIONS FOR RECONNECT.
SYSTEM	RISER FAILURE MAY CAUSE EXTENSIVE DAMAGE TO TEMPLATE.		RISER FAILURE MAY CAUSE LIMITED DAMAGE TO TEM-PLATE

Fig. 2–16B Maintenance

Model Test

The Steep-S configuration was model-tested in 1983 in the SSPA model basin at Goteborg (Sweden). The test was performed at a scale of 1:50 and simulated the field conditions with regard to:

- Water depth 144 m (472 ft).
- Support vessel (model of the GVA 5000).
- Outside diameter, length, and weight of the risers.
- Regular and irregular sea state corresponding to a typical 100-year storm in the North Sea using a Jonswap Spectrum, and a uniform 0.3 m/s current colinear with the waves. Wave headings of 0, 90, 135, 180, and 270° relative to the vessel were studied.

Disconnection tests under 100-year storm conditions were simulated and recorded on video tapes also. Several disconnections were made at different wave headings.

In each test, wave statistics, vessel motions (surge, sway, heave, roll, pitch, yaw), and top and bottom riser tensions were measured and recorded.

The riser motions were recorded on video tapes and high speed movies. No interference problems were observed between the riser in each bundle. The tests indicated a potential for interference between the mid-water arches which was easily avoided by increasing the spacing between the riser bundles.

These tests demonstrated the total feasibility of the system in terms of motion envelopes and riser tensions.

Engineering of the Riser System

The engineering of a riser system is twofold and consists of:

1. The engineering of the configuration by optimizing its different parameters (i.e., riser length, position, and buoyancy of arch and position of riser base) in order to obtain admissable top/bottom tensions and motion envelopes
2. The engineering of the hardward components themselves, particularly the risers

It should be noted that the engineering of a riser system is an interactive process: the configuration is a function of the selected riser structure (weight, O.D.) and in turn the riser structure depends on the loads obtained from the configuration engineering.

Engineering of the Configuration

Refer to the General Design Flow Chart in Figure 2–17.

Essentially, this is based on one computer program, FLEXAN-C, which has been specifically adapted to the analysis of the

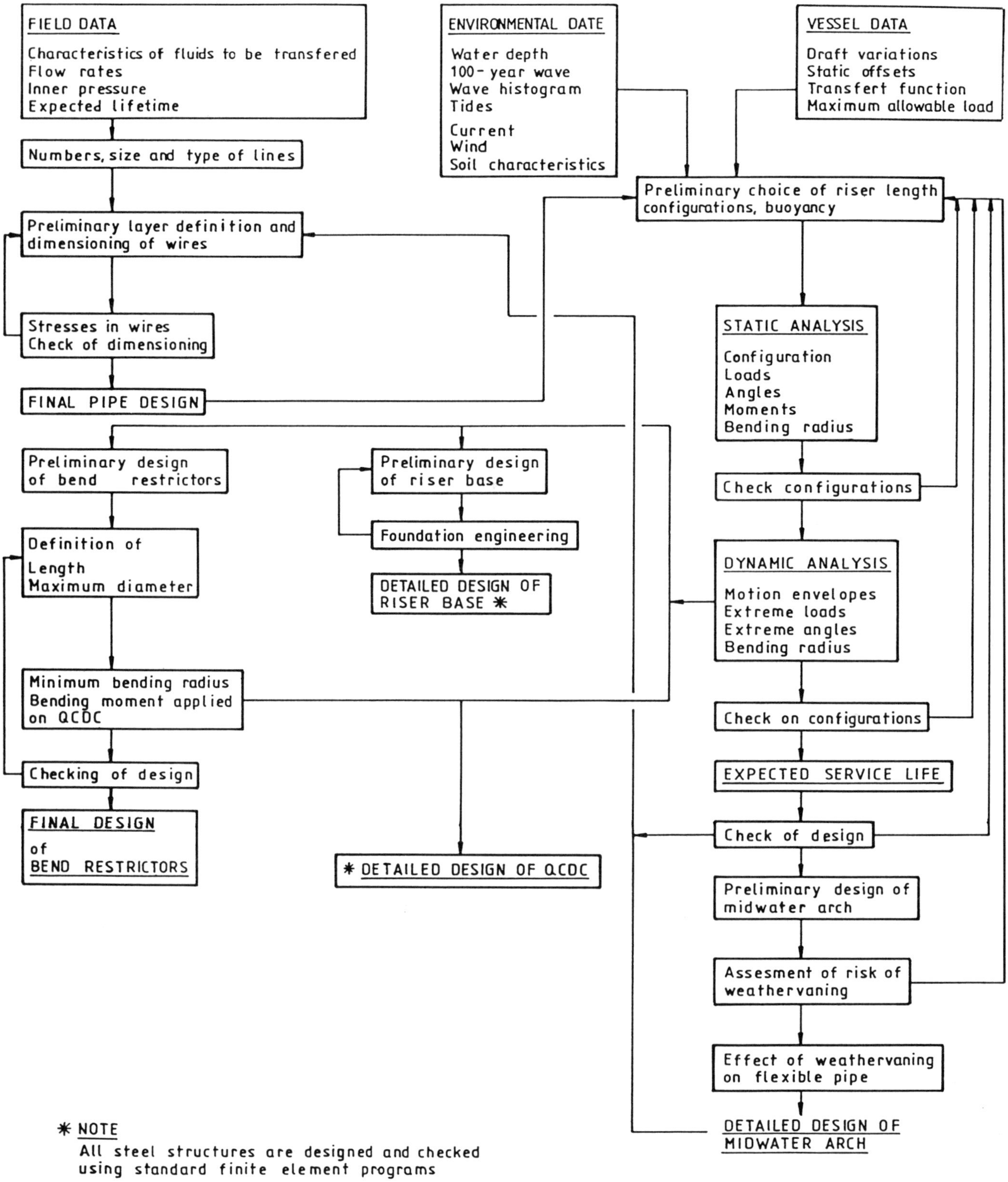

Fig. 2–17 Dynamic Riser Systems, General Design Flowchart (Source: Coflexip)

flexible riser. The configuration analysis goes through model test analysis and static dynamic analysis.

Model Test Analysis. The intent was to validate the computer program by comparing the motion envelope and tensions predicted by the program with the model test results.

The riser models did not exactly duplicate the true dimensions of the full scale risers. This was taken into account in the model test analysis by simulating the model conditions in the program.

The conclusions of this analysis were that:

- A riser bundle may be treated as a single riser of equivalent weight and diameter for the purpose of predicting tensions.
- The total tensions predicted by FLEXAN-C in regular wave conditions were always within 5% of the tension measured during the test.
- The prediction of motion envelopes was good.

Table 2–1 gives a summary of this comparison with regard to maximum top tensions of a three-riser bundle in regular wave conditions. (H = 15.7 m; T = 12.7 s).

It also has been established that in irregular wave conditions, FLEXAN-C can still be used to obtain conservative estimate of significant and maximum tensions within a tolerance of 15%.

Table 2–1 Model Test Analysis: Comparison between predicted and measured topside tension under various wave and current headings in regular wave conditions. (All figures in metric tons.)

Wave and current heading	0°	90°	180°
Max top predicted tension	39.8	36.5	37.4
Max top measured tension	40.0	35.3	37.3

Static and Dynamic Analysis. First, this analysis uses trial and error methods to optimize the configuration parameters. This is done by comparing the resulting loads and motion envelopes, in particular the bending radius obtained from the computer program FLEXAN-C.

The selected (optimized) configuration parameters for the Balmoral system are as follows:

- Riser length: 245 m (803 ft)
- Distance from center of template to the riser bases: 117 m (384 ft)
- Height of mid-water arch from seabed: 50 m (164 ft)
- Net buoyancy of mid-water arch buoys (3 riser bundles): 21.8 tons

The dynamic analysis also is used to yield data for the service life analysis, design loads (tensions and moments) for the QCDC (Quick Connect Disconnect Coupling) at the upper end, and design loads for the riser bases at the lower end.

The viewgraph gives the motion envelopes of a three riser bundles for wave headings of 0° and 180° and for two extreme positions of the FPV, i.e., near position (FPV offset towards the riser base) and far position (FPV offset away from the riser base) under 100-year storm conditions.

The dynamic analysis also assesses the stability of the mid-water arches, i.e., the risk of weathervaning due to the drag forces if the restoring moment caused by the catenary was not sufficient.

The Balmoral riser was installed successfully in the field in the autumn of 1986 as shown in Figure 2–18.

Fig. 2–18 Balmoral, Two 8-in. Water Injection Risers Installation in October 1986 (a photograph)

Chapter 3

Subsea—Templates

TEMPLATE PRELIMINARY DESIGN

From the point of view of neat geometry, system compatibility, ease of interconnection, and maintenance, it is preferable that clusters of subsea wells are arranged in a pattern rather than at random. Two methods of spacing wells in a regular pattern are the:

1. Unitized template method
2. Modular template method

Deviated drilling of well clusters from a jack-up or floating rigs, through conductor support frames or jacket structures, is common practice; this technique eliminates the possibility of misalignment between subsea template wells and platform structure. Also, underwater connections related to well completions are simplified. If a pre-drilling template is used with a jack-up rig, well-bore verticality at the mud-line will probably be better than that obtained with a floating rig—due to the use of a drilling conductor back to the fixed rig deck. Massive experience, however, particularly in the North Sea, demonstrates that a 2 or 3° tolerance can be achieved without any difficulty at all.

Unitized Template

This is a welded steel frame which incorporates a plurality of well slots (24 maximum to date), each suitable for accepting a subsea well-conductor housing. The frame also may include guideposts to position a blow-out preventer (BOP), a completion tree, interconnecting pipework, commingling headers and control lines, additional structural members to provide protection against dropped objects, and fishing gear, etc. Plan sizes of unitized template have ranged from 15 m × 15 m (49 ft × 49 ft) to around 60 m × 45 m (197 ft × 148 ft), with weights up to 2,200 tons. Figure 3–1 shows a typical design. Figure 3–2 represents a large template designed for the Balmoral subsea production system.

Modular Template

One objective of a modular template is to provide a system which can be installed through the moonpool opening of a floating drilling rig which measures 7 m × 7 m (23 ft × 23 ft). A primary base module is lowered and piled to the seabed and one or two 30-in. well-conductors may be installed also. Before any additional wells can be drilled, extra slots must be added, and special modules can be lowered on the end of the drill pipe and attached to the base module (see Fig. 3–3).

A modular template system is recommended for guideline drilling systems and for use with drilling programs where flexibility is required. The modular template system employs a template structure that is smaller than the unitized template and consists of several interlocking modules. Figure 3–3 shows a five-well modular template system.

Modular templates are generally chosen for use when the number of wells to be drilled has not been firmly established when drilling begins. The use of modular templates requires a lower capital investment to determine reservoir characteristics while providing the capability to expand the system, as desired. A modular primary base structure also can take the place of a standard permanent guide base, allowing the operator to index additional well slots, enabling production from either a tie-back or subsea production system. This is possible because the modular primary base structure is designed with enough flexibility to accept a retrievable plumbing structure for subsea production, or other peripheral structures required for a subsea system.

The modular template system gives the operator the option of selecting either a tie-back or subsea production development program after the exploratory well has been drilled. In addition to well modules and cantilever pile modules used in a tie-back system, the template base structure can accommodate modules for subsea production. A flowline module is cantilevered from the base structure for a single-well subsea production system, or a combination of well and plumbing modules is used for a multi-well system. No provision for leveling is necessary with the modular primary base structure since the conical shape of

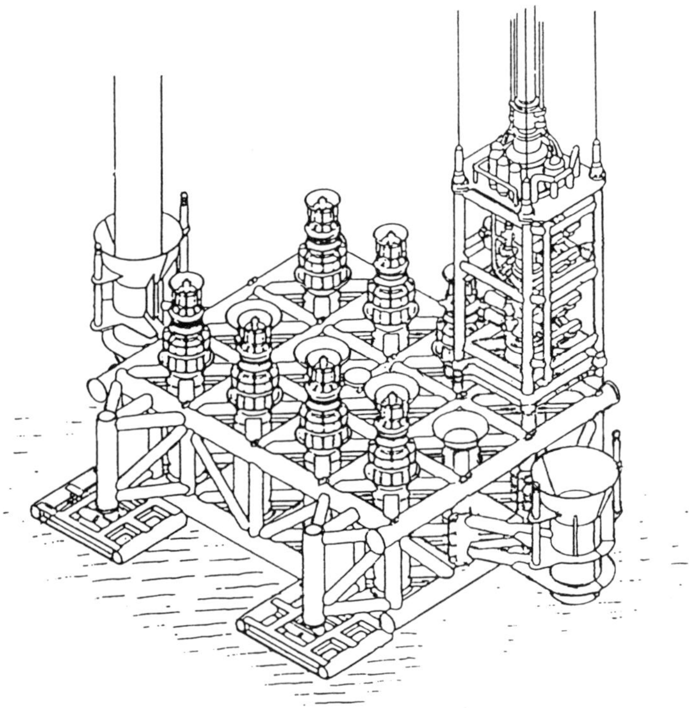

Fig. 3–1 United Subsea Template Tie-back System (Source: Cameron)

the temporary guide base landing ring and the gimbaled base of the primary base structure assist in leveling. Leveling of the modular primary base structure is a function of drilling. A vertically drilled hole and proper cementing of the 30-in. conductor are essential for successful use of the modular template system. With care, a hole can be drilled to within one-half degree of vertical. Some methods of achieving this degree of verticality include making frequent downhole surveys, using a low penetration rate, high speed of drilling, and low mud velocity.

Template Design

Pre-drilling templates for floating rigs and fixed platforms between 1978 and 1984 were used on the field developments listed in Table 3–1.

It may be said that a ''good design'' considers all of the customers' requirements (some of which may be in conflict) and reaches a compromise solution best fitted to the end use. To obtain fitness-for-purpose often requires an answer to the question ''What happens if . . . ?'' Some of the important aspects to be considered when designing for regular spacing of subsea wells are considered in the following pages.

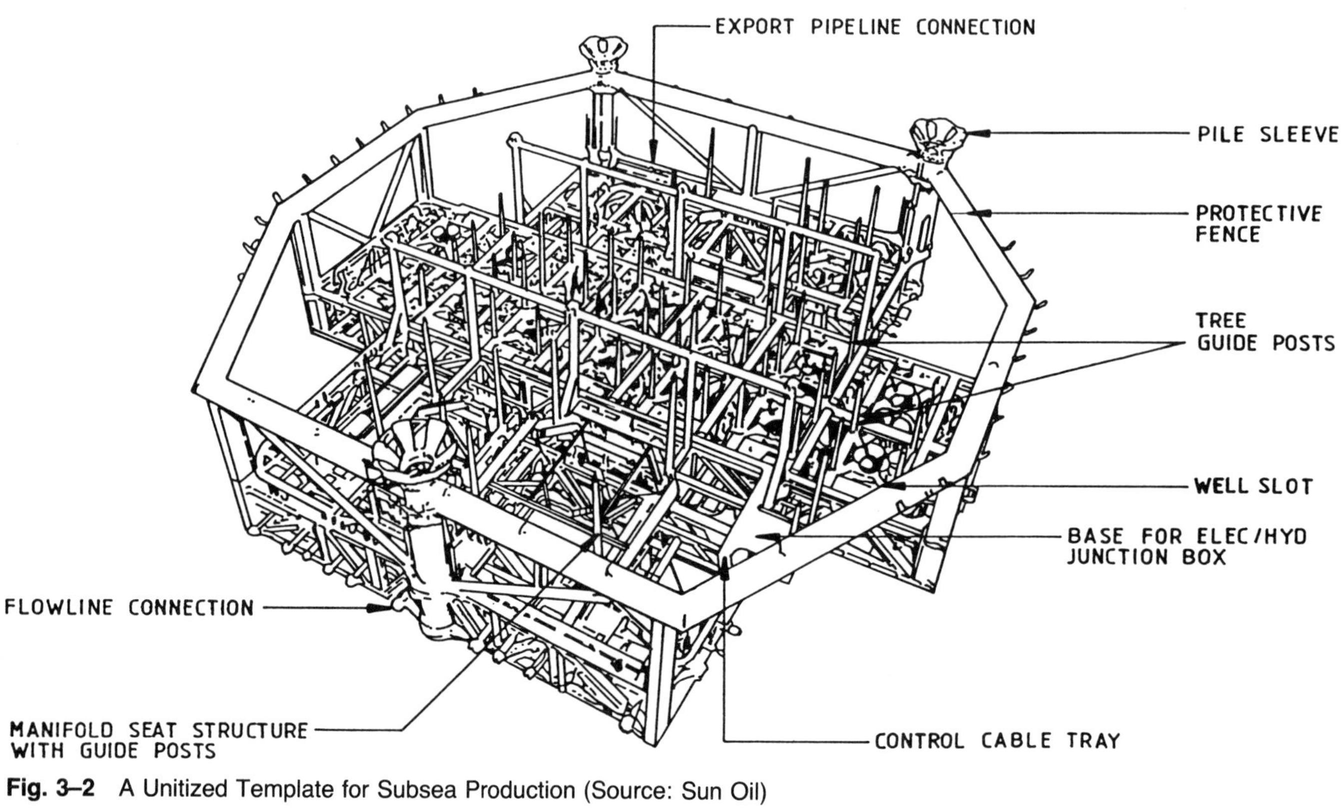

Fig. 3–2 A Unitized Template for Subsea Production (Source: Sun Oil)

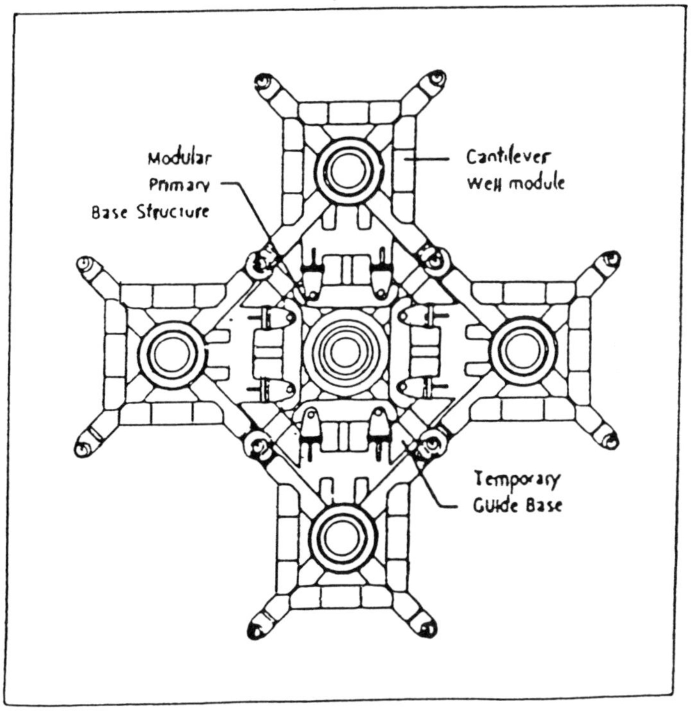

A. FIVE-WELL MODULAR TEMPLATE SYSTEM ILLUSTRATES ONE CONFIGURATION POSSIBLE USING CANTILEVER WELL MODULES AND A MODULAR PRIMARY BASE STRUCTURE.

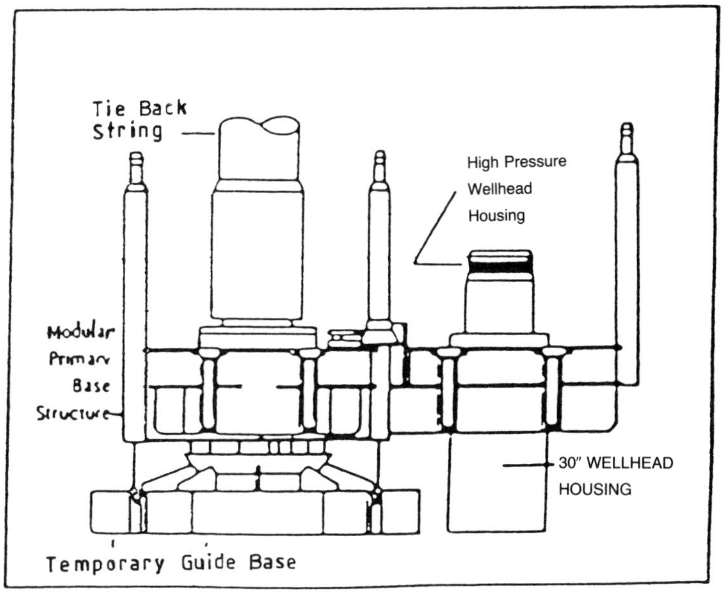

B. MODULAR PRIMARY BASE STRUCTURE WITH ONE CANTILEVER WELL MODULE ADDED.

Fig. 3–3 Modular Template

Table 3–1 Predrilling Templates by Field Development (1978–1984)

Date	Operator	Field	Water Depth (m/ft)	Type Template	No. of Slots	Type of Structure
1978	Shell	Fulmar	76/250	Modular	4	Conductor support frame
1978	Petrobas	Garoupa	130/426	Unitized	15	Piled steel jacket
1979	Phillips	Maureen	107/351	Unitized	24	Steel gravity structure
1979	Amoco	NW Hutton	146/479	Unitized	20	Piled steel jacket
1980	Mobil	Beryl B	114/374	Unitized	21	Piled steel jacket
1984	Sunoil	Balmoral	145/476	Unitized	14	Floating production vessel
1984	BP	South East Forties		Modular	16	Piled steel jacket

EVALUATION OF ALTERNATIVE DESIGNS

Factors Governing Layout

Support for 30-in. Conductor

The template structure provides space and support for the 30-in. conductor casing as well as support for the drilling assembly and trees. There should be adequate clearance for the BOP to be placed over a well for workover with trees in place on adjacent bays. A 14-ft (1 foot = 0.3048 meters) spacing between well centers has been found to be the optimum dimension.

Clearance for Drilling Workover

The provision of good clearance for drilling and workover results in a template larger than the minimum size possible. But if costs are compared, the additional template structural cost will be far outweighed by savings in drilling-rig time and the extra time required for special drilling techniques.

Manifolds and Piping

The template may accommodate manifolds and piping. This piping should not be vulnerable to damage during transportation to site, installation, or operations. Damage from dropped objects during operation can be minimized by providing the template structure with overhead protection. The piping should be designed with a minimum of valves and flanges to limit potential leakage points.

Ease of Fabrication

The template structure should be easy to fabricate and install. I-sections are easier and cheaper to assemble than tubulars, but non-buoyant templates restrict the installation options.

Protection and Loading

The tempate must provide protection for the trees, manifolds, separators, piping, etc., against possible damage from large dropped objects, dragged anchors, trawl boards, etc. The structure should be sufficiently robust to resist the loads due to the following;

- Dragged anchors and trawl boards
- Lifting (dead weight) in air
- Operational loads from risers and drilling, assembly—including workover riser
- Load from anchor pulling a satellite flowline
- Conductor expansion

Two-Tier Structure

The template structure could be either single- or two-tier. The two-tier type is preferred by reason that it:

- Provides greater rigidity under all loading conditions
- Facilitates more robust connections between the template and the supporting piles and well-bays
- Provides better protection for the template piping

Above the two-tier structure, a 4.8 to 6 m (16–20 ft) high tubular protective barrier is recommended around the perimeter of the template structure. In addition to providing protection, the buoyancy rating of the upper barrier rail gives stability to the template in the submerged mode during lowering and installation.

I-SECTIONS OR TUBULAR MEMBERS

Advantages of I-Section Designs

Structural members may be either I-sections or tubular members. Tubular sections are usually preferred because they provide buoyancy which increases the number of installation options.

If it is established prior to design and fabrication that installation is to be from a heavy-lift semi-submersible, cost savings can be made by substituting I-sections for tubular members.

Installation Buoyancy

When I-section design is selected, the top barrier rail should be kept as a tubular member for stability during installation. In addition, buoyancy tanks can be considered to support the non-buoyant I-section template. Experience proves, however, that the weight and cost of these tanks could offset the weight and cost savings from the use of I-sections.

Comparative Weight and Costs

Template structural analysis indicates that stresses are dominated by bending effects rather than by axial loads. Thus, since bending moments are mainly uniaxial, and since I-sections are more efficient in bending than circular sections, the use of I-sections leads to the lightest structural scheme.

Comparative estimates of weight for tubular and I-section designs also show the latter to be about 13 to 15% lighter. It is also important to note that the uni-fabrication costs for I-sections could be less expensive by a similar percentage.

Recently there has been a trend to incorporate some external trawl/anchor protection system in the overall template. The tubular profile offers a snag-free shape and is often adapted for the outer framework of the template with I-sections restricted to the inner area.

STRUCTURAL ANALYSIS

Computer Analysis

Computer programs can be used for the structural frame-space analysis. They carry out linear elastic analyses for a number of load cases and combine the results. For many members, the stress levels at the connections govern the design. In addition to loads in the operational condition, loads associated with installation and possible accident cases must be allowed for.

Loading Cases

The load conditions to be considered are the following:

- Operational loads with or without riser
- Lifting load particularly during the dry-lift period
- Drilling loads
- Anchor chain dragging onto template
- Anchor chain pulling satellite flowlines at export pipe

REVIEW OF ALTERNATIVE LAYOUTS

Possible Configurations

Typical template well spacing is chosen to permit the 10,000 psi wellhead system to be used without interfering with the adjacent well. The template well can be arranged in linear, multi-row linear, or circular configurations.

The objective is to find the configuration that minimizes template dimensions and weight, while providing enough space for manifolds, separators, subsea control modules, riser base, and other modular components to be installed.

Various configurations for a 10/11-slot template with three piles only are shown in Figure 3–4.

Square Configuration: (Scheme 1)

A square, arranged on two rows of wells, with the manifolds, separator, and riser-base spaces on each side of the well-bay area. It is a compact structure with a good access from a semi-sub point of view (drilling floating production platform).

Circular Configuration: (Scheme 2)

A circular/radial layout where the manifolds are placed in the central core and with a perimeter pipe-bay. Three piles are sufficient, and the structure is well balanced. It is interesting to note that a configuration like this one can be used for an articulated column base with the necessary reinforcements to hold the universal joint and additional piles.

Extended Rectangular Configuration: (Scheme 3)

An extended rectangle arranged on two rows of wells. It is a simpler structure than Schemes 1 and 2, but it limits access by the drilling semi-submersible (offset required).

Compact Rectangular Configuration: (Scheme 4)

A more compact rectangle arranged on three rows of wells. It is a heavier structure than Scheme 3, but it offers good access from a semi-sub, good space for the piping and manifolds, separators, and control modules. It also offers a lot of space for piping connections to the flowlines and export pipe, etc.

WELL-BAY DESIGN—TYPICAL

Conductor Pipe

Figure 3–5 shows the wellhead in position. Then the tree is connected through a standard 18¾-in. 5,000 psi wellhead connector. The wellhead and supporting 30-in. conductor pipe sit inside the well guide-tube on the template. The tube is usually 10 ft long to provide a guide length adequate for starting a straight well.

The well guide-tube is a 42-in. diameter cylinder with an entry funnel at the upper part and strengthened by vertical stiffeners. An internal shoulder and lock-ring groove is provided at the upper part for seating and locking in position the 30-in. conductor head. This guide-tube concept accepts a wellhead system from any manufacturer.

Guide-tube/Structure Connection

The guide-tube is connected to the main template structure by four beam sections which are strong in vertical bending to support the conductor casing. They also are strong enough to resist loads that might result from horizontal excursion of the drilling semi-submersible with tools in the hole.

Mud and Cuttings Exit

Escape slots are provided in the guide-tube for the exit of drilling mud and drill cuttings during drilling.

Guide Posts

There are four 8-in. guide posts located around each wellbay. They provide guide for the drilling BOP Christmas tree. Particular attention must be given to fabrication tolerances in construction of the well-bays to ease hook-up problems.

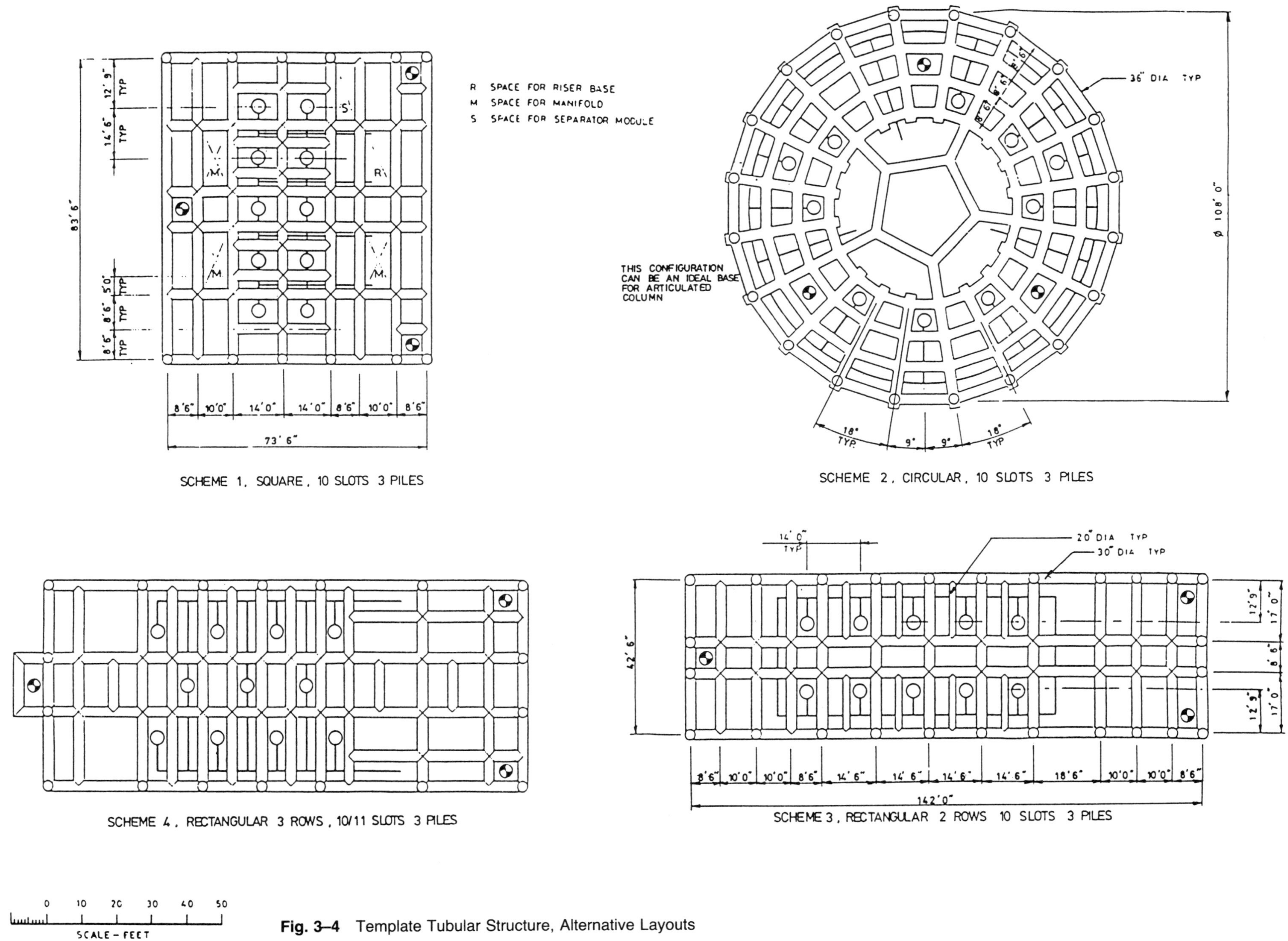

Fig. 3–4 Template Tubular Structure, Alternative Layouts

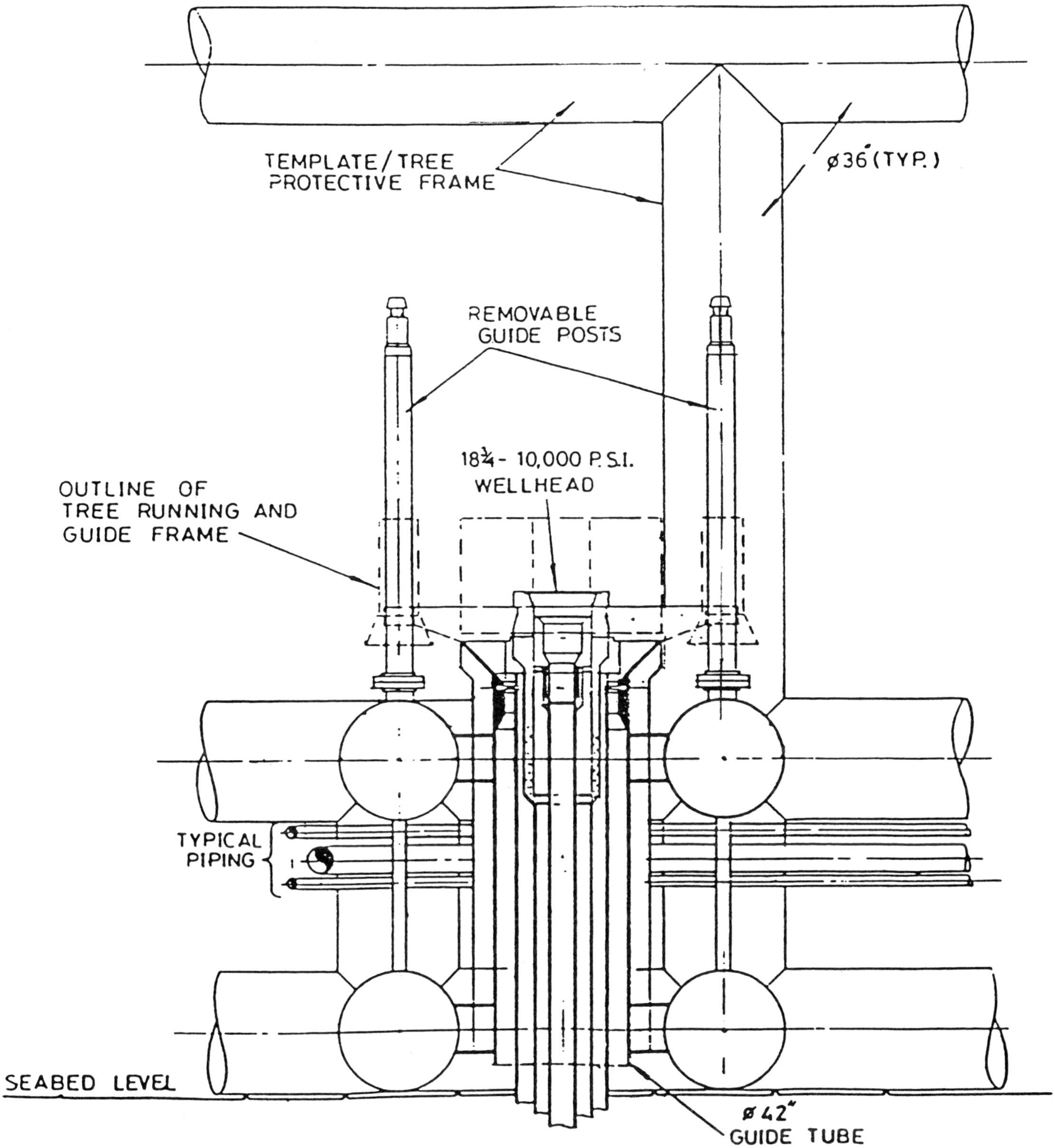

Fig. 3–5 Wellhead and Conductor Pipe Within a Tubular Structure Unitized Template

TYPICAL WELL SPACING

Factors to be considered are addressed in the following sections.

Drilling Verticality Requirements

It is critical that the verticality of a 36-in. hole drilled from a floating rig is controlled. The rig is subject to environmental forces which may cause movement from the directly overhead position, and upper soil layers may not be homogeneous. It is easy to appreciate that a boulder in clay can deflect the drill bit from a truly vertical path; 36-in. hole verticality should be within $\pm 1°$ at mudline, if the well is to be tied back to a fixed platform. Templates with pre-installed flowline pipework and vertical stab connectors also need accurate control of wellhead lateral and angular position. Drilling slots may need to incorporate provision for guide bushings and accurate surfaces to locate 30-in. housings truly vertical. Tight alignment tolerances can lead to precise template-leveling requirements, and bull's eye levels are a simple and effective method of measurement. It is important to keep tolerance of misalignment within practicable limits. High precision requirements are not compatible with conventional subsea drilling and cementing procedures.

Drill Cuttings Disposal

Cuttings disposal should be considered for templates incorporating piping, wet manifolds, and maintenance vehicle tracks. A hose-type disposal system was used for the Shell/Esso Underwater Manifold Center (UMC), but 36-in. drilling bit returns have no riser to permit remote disposal. For each 100 ft of the 36-in. hole, over 700 cu ft of soil is deposited on the surrounding template/seabed. The pattern of deposition depends on jetting-fluid flow rate and current velocity near the seabed.

Cuttings build-up has caused a problem with jacket installation over a pre-drilling template. The lowest bracing level came to rest on a pile of cuttings/cement, and a considerable jacking operation was required to bring the jacket close to vertical. Lowest jacket bracing should have 10 m clearance above the seabed. During drilling, cement built-up around the 30-in. conductor should be monitored by remotely operated vehicle (ROV) and controlled.

Accidental Loads Due to Dropped Objects

During the drilling phase, a very heavy object might be dropped, e.g., drill-collars or a BOP. The latter can weigh up to 200 tons, and there are a few press reports of them being dropped. Other BOP accidents include dropping through the moonpool and pulling off the wellhead during severe weather. The frequency of these accidents appears to be around 5% of holes drilled (from 1983 records) but it is improving.

Operators drilling through a vulnerable template tend to slacken guide wires, move the rig off to one side, and run the BOP down within 10 to 20 m of seabed at a safe distance. A twin stack drilling system reduces weight/impact force but doubles the number of "trips." The template structure of the Exxon Submerged Production System (SPS) was designed to limit damage propagation, but the designers conceded that the most exacting damage load investigated was the impact of a dropped BOP stack on the production manifold.

Effect of Broken Guidelines

Guideline drilling methods are used for 50 m (164 ft) to over 300 m (984 ft) depths, and lines can break or are dropped. Well-spacing design should consider prevention of entanglement and ease of recovery.

Interconnecting Pipework and Control Lines

The unitized templates at Enchova/Bonito, North Cormorant UMC, and Balmoral included inter-connecting pipework and some control-line. At Buchan, the template was used for well spacing only, and all flowline and control line spools were tailored to be fitted after well-completion. Similar spools could be used for wider well spacing.

If a template is damaged, integral pipe spools could be difficult to remove and replace. Open-plan spools are readily accessible. Complex manifold piping can be spaced away from the drilling/workover area to reduce the risk of damage.

Simultaneous Production and Drilling/Workover

BOP and riser operations near live/producing wells become relatively safer as well spacing increases. A 14-ft spacing is the most desirable.

Ease of Access for Intervention Systems

Cramped conditions make maintenance access more difficult. Again, work on a dead subsea tree can proceed more safely at a greater distance from live wells.

Risk of Entanglement of Broken and Intervention System Umbilicals

High density equipment arrangements are more likely to lead to entanglement of wires and umbilicals which lay across them. Discussions with diving company personnel at the design stage can lead to less hazardous operations for both manned and unmanned intervention systems.

Aspects of Blow-Out Prevention and Control

Although few in number, offshore well blow-outs are usually spectacular and attract a great deal of interest. A recent study in the southern United States showed that on a per well basis, offshore wells have a higher risk of blow-out than those onshore. The Ixtoc incident had catastrophic effects. Blow-outs occurred during "tripping" when drill-collars, positioned across the BOP rams, stopped the rams closing.

Numerous ideas have been proposed for equipment to control the flow from a subsea blow-out; most of them involve the deployment of a large inverted cone over the wild well. Some experimental work has been done to identify a basis for cone design. It is advantageous to reduce entrainment of seawater into the system by forming a seal between the cone bellmouth and the seabed. This would appear easier to accomplish (possibly by using a compliant skirt) if the well is not surrounded by a template framework.

There are practical limitations to the large-cone method, but whatever approach is tried, easy access to the well by an intervention system is desirable. For instance, at Ixtoc the BOP was still latched onto the wellhead, and divers attemped to close the rams by attaching hydraulic lines.

Conclusions

Each of the basic categories of subsea well cluster have different priorities with respect to spacing requirements. For direct vertical tie-back of each well to a fixed structure, it is essential that template spacing is similar to guide spacing within the structure. A grid pattern based on approximately 3 m × 3 m (10 ft × 10 ft) is common. Well slots need not attempt to restrain the 30-in. conductors in a vertical position, as this can be controlled by monitoring 36-in. hole verticality while drilling from the platform, but the minimum or closer well spacing is in the order of 90-in. square without removing out-riggers of the BOP stack.

When well cluster spacing is to be decided at the conceptual design stage, there are many requirements which need consideration, some of which may be in direct conflict. It should be possible to reach the best compromise in design by following a careful assessment of drilling, operational, and safety aspects so that the result is best suited to its purpose.

PACKAGE DESIGN OF MULTI-WELL TEMPLATES

Maintenance Requirements

Detailed assessment of anticipated subsea maintenance requirements, and related facets of offshore operations, has highlighted the significant economic benefits emanating from the provision of a floating production facility of at least a simple workstring derrick and, preferably, a full workover facility. To achieve an optimized floating production facility, integration studies must be undertaken in which subsea equipment design and topside maintenance support requirements are fully evaluated.

Integration of Topside and Subsea Facilities

Several studies have been conducted into the development of subsea multi-well template systems—for operation in conjunction with both floating and fixed processing facilities. These have concentrated on the absolute need to integrate the design of subsea equipment and associated maintenance intervention vehicles and tools. The necessary degree of emphasis has not always been given to the optimization of topside support facilities for subsea maintenance intervention tasks, together with the attendant space and payload constraints these impose on a floating production facility. When a dedicated field surface support facility is provided, such factors do not need to be so extensively evaluated. However, in field developments for which dedicated facilities cannot be justified, on economic or other grounds, the impact of subsea maintenance intervention requirements can have a dominant effect upon a floating production facility and, in the ultimate analysis, on field development economics generally.

Package Approach

Subsea equipment design for deep-water hydrocarbon production is dominated by the need to achieve the highest practicable reliability coupled with the ease with which subsea maintenance intervention operations can be conducted.

Attainment of maximum equipment reliability can be achieved best by strict adherence to a philosophy of simplicity in design, maximum practicable utilization of field-proven equipment, and minimizing known potential failure sources, such as seals and active components. The ability to maintain, or restore, any particular subsea component or equipment to a predetermined state of operational effectiveness is dependent on:

- Effectiveness of selected intervention systems in the performance of all necessary subsea maintenance activities
- Design and layout of subsea-installed equipment
- Time to effect routine or planned maintenance or equipment repair
- Cost of maintaining, or restoring, subsea equipment to fully operational status.

Subsea multi-well template systems have, in the past, been produced in several forms. Recent developments have, however, demonstrated that the achievement of maximum reliability and maintainability of template components and equipment can be achieved more readily by modularizing, or packaging, such equipment—each package being designed for installation and retrieval by unmanned techniques. In the event of an individual component malfunction within a package, the complete unit would be isolated, retrieved to the surface, and replaced by a new or reconditioned unit. Equipment repair philosophy therefore is that of specifically not conducting repair in-situ. This minimizes the necessity for major and complex subsea intervention operations. It also results in subsea-template intervention requirements being limited to the relatively elementary tasks of general observation, inspection and, possibly, simple manipulation.

Typical Eight-Well Template

A typical design for an eight-well, wire-line serviced template with a production riser embodying the above principles is shown in Figure 3–6. A more advanced scheme to illustrate the modular arrangement of components is shown in Figure 3–7.

Insert-type valves also could be arranged readily in modules, in which case either the complete unit or individual components within the module could be retrieved and replaced. This approach does provide an additional element of flexibility, but at the expense of increased overall module weight and size due to the need to provide access to individual components.

Package Bounderies and Drawbacks

Determination of package boundaries/interfaces is largely dependent on component functional requirements and reliability, together with overall unit size and weight. For example, components most prone to failure, and thereby necessitating most frequent replacement, are those which are operated frequently,

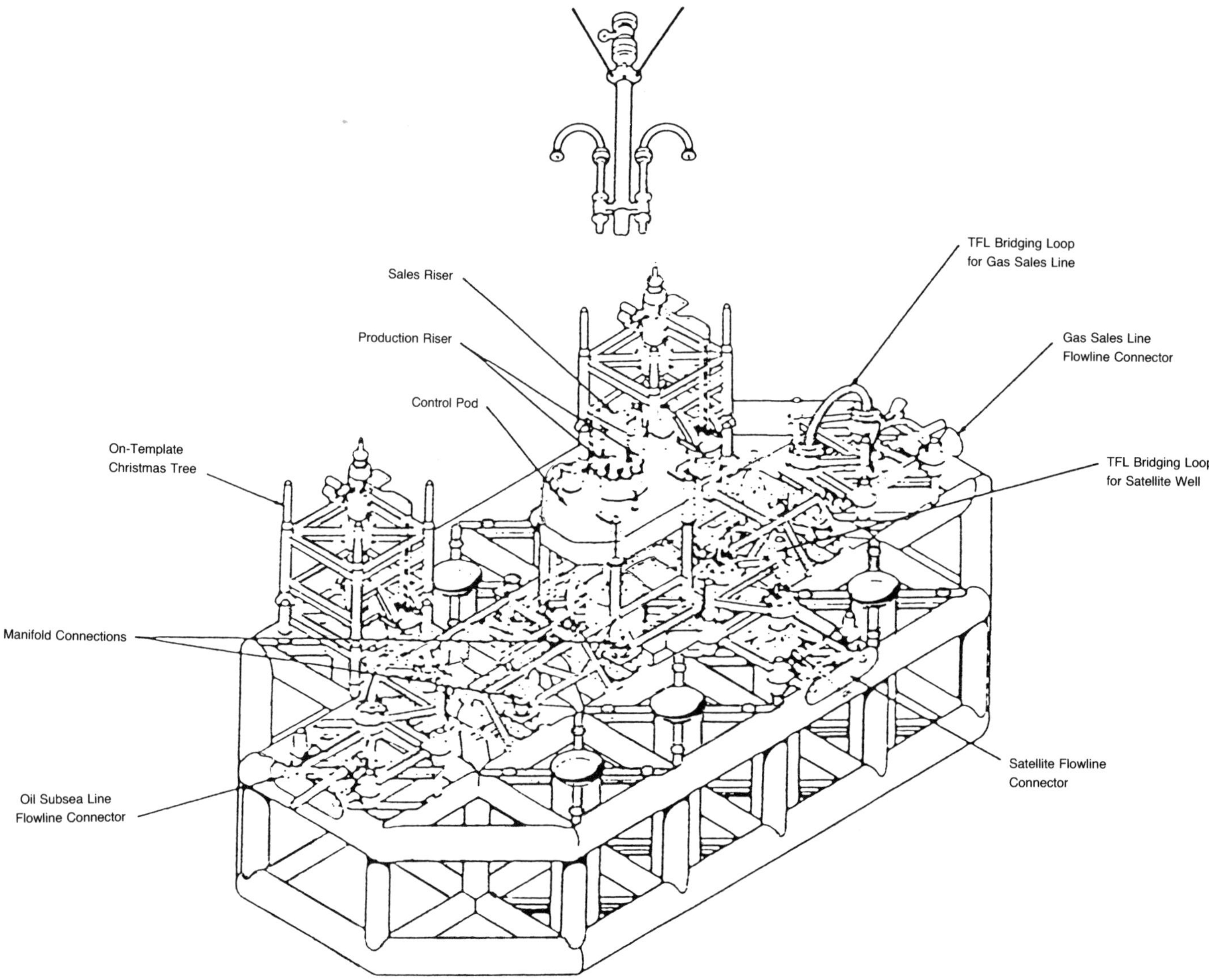

Fig. 3–6 Packaged Eight Well Template System (Source: Cameron)

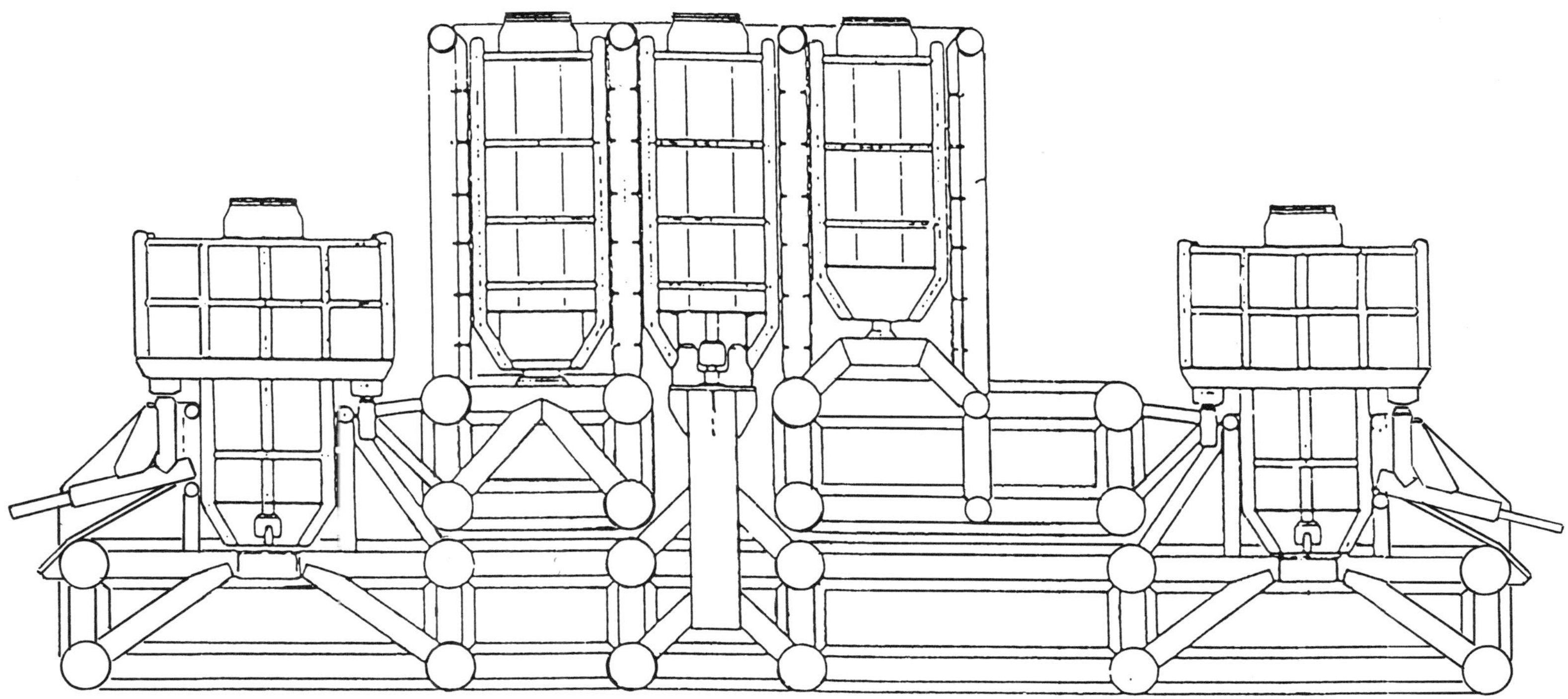

Fig. 3–7 Vertical Modules—Advanced Concept (Source: ACB)

complex in design, and not fully field proven. It is therefore highly beneficial to group such components together to facilitate rapid replacement and to minimize disruption to other template systems.

The valve package shown in Figure 3–8 is a typical example of this design approach. Here the production choke and Christmas tree production wing valves are grouped within a single unit because of the requirement for maintenance intervention on these components. Similarly, the control module is independently retrievable. It can be removed and replaced without disturbing other template-installed equipment.

Module Drawbacks

Equipment packaging therefore has some significant advantages. This design approach does, however, possess some inherent drawbacks also including the following:

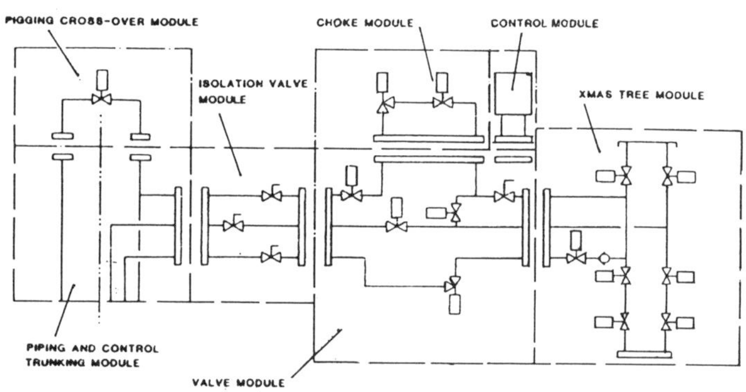

—PIPEWORK SCHEMATIC AND MODULE BOUNDARIES

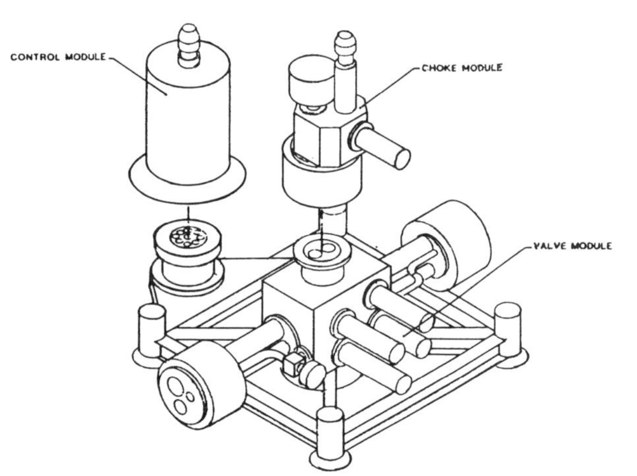

—GENERAL ARRANGEMENT OF TYPICAL VALVE MODULE

Fig. 3–8 Pipework Schematic and Module Boundaries General Arrangement of Typical Valve Module

- Overall template size can increase considerably, especially when several modules are required. This can be alleviated by stacking modules vertically, but this complicates intervention system access and component-replacement procedures.
- Adequate pipe flexibility must be provided to facilitate all necessary pipework connections at each module interface. Free pipe spans can provide the required flexibility, but these impose significant constraints in module design and layout and can restrict access for inspection or unplanned maintenance intervention activities.
- Physical size and weight of packages are too great to facilitate retrieval by remotely operated vehicles which have the capacity to retrieve individual components. Consequently, workstring or surface-cable techniques must be utilized. Units should therefore be arranged for installation and retrieval by means of guide-post arrangements. But access for inspection, or unplanned maintenance intervention activities, will be further restricted by these structural provisions.

Overall Advantages

Not withstanding the above drawbacks however, the philosophy of packaging components on any deep-water, subsea-template system is, on balance, mostly advantageous. The requirement for repair in-situ is virtually eliminated.

Maintenance Intervention Requirement

It has been stated previously that the achievement of high production levels from deep-water subsea completion equipment will be largely dependent on the extensive and effective application of remotely operated, diverless techniques to maintain equipment at the desired operational state.

Intervention requirements for a modularized subsea system will basically comprise inspection or module replacement. Installation and retrieval of modules incorporated within the template would be required. These modules include the:

- Christmas tree
- Control module
- Valve module
- Choke module
- Isolation module
- Pigging cross-over module
- Piping and control trunking module
- Insert valve
- Flowline seal-connection module
- Manifold module

Insert Type Components

In the case of packages based on insert type components, the insert valves or choke assembly could be recovered individually, or as part of the total package assembly.

Typical Dimensions

Typical overall dimensions and weights of these modules and insert components are given below.

Table 3–2 Details of Modules and Insert Components

Equipment	Overall Size (meters) (length x breadth x height) (typical)	Weight (kg)
Christmas tree	5.2 x 2.6 x 5.0	15,000
Valve module	4.0 x 3.6 x 2.3	6,000
Choke module	1.3 x 1.8 x 1.4	2,000
Isolation module	2.3 x 2.6 x 3.0	3,000
Controls module	1.0 x 1.0 x 2.0	2,000
Pigging cross-over module	4.5 x 2.0 x 2.0	7,500
Piping and control trunking module	21.0 x 2.0 x 1.5	18,000
Insert valve	0.6 x 0.6 x 1.2	450
Flowline connection module (8″ pipe)	2.4 × 2.2 × 3.6	14,693
Riser manifold	6.5 x 6.5 x 5.0	70,000
Water injection manifold	5.0 x 5.0 x 4.5	60,000

PIPING

Typical Piping Scheme

Figure 3–9 shows a type of scheme for a 10-well layout in two rows with three manifolds and the connection lines to the riser system.

Piping Design Considerations

The template wells are arranged in two staggered rows to provide essentially a well-bay linear template. The satellite-well manifold and satellite-template manifold are located at opposite ends, with the riser manifold adjacent to the satellite-template manifold. The template-well piping is located in the two edge bays along the perimeter.

The piping for the template wells is arranged off five ring headers (production 1, production 2, water injection, well testing, and annulus access) which tie in the wells to the riser manifold. Manual gate valves isolate each tree from the headers. These valves should have gate and seal assemblies which are diver replaceable, without removal of the valve body. Great flexibility in well designation is allowed for; first, all well-bays are piped to permit production from production zone 1. Additionally, four of these bays are piped so they can, alternatively, produce from production zone 2. These bays are located within the template by using available knowledge of reservoir positions. The other bays could be additionally piped for use as water injectors. An identification plate is provided at each well bay.

The ring headers are located under the upper layer of main tubulars (see Fig. 3–5) for protection against dropped objects, and longitudinal deflectors are provided for additional protection. The five headers are arranged in two rows with good clearance from the mudline to avoid cover-up by cuttings or cement and to make for easy diver inspection. In the event that a pipe has to be replaced, clamped connections are arranged in the branches to each well so that the header may be disconnected. The header itself is of welded construction.

At the wellheads, the template flowlines are fitted with male stabs for connecting the tree flowlines. Clamped connections are provided below the stabs to allow replacement of either a template flowline or male stab. The stabs are protected by a temporary cover which is supported off two of the tree guide posts.

The piping connections at the manifolds are made with diver-installed jumper spools. The connections at each end of the jumper pipe are made with 5,000 psi weld-neck flanges. The flanges are positioned so that generous space is provided for diver access when making the connection.

The jumper-pipe sections are designed to allow for some lack-of-fit between manifold and template pipe flanges. This is achieved by providing misalignment flanges at each end. Nevertheless, the pipe-flange positions on both template and manifold should be fabricated to close tolerances so that diver installation time is minimized.

The pipe supports are arranged at sufficiently close intervals to avoid possible vibration. At the same time, the supports allow limited flexing of the pipes at the jumper connections during installation of the jumpers.

At the edge of the template, the template flowlines to the satellite wells and satellite templates, if any, are provided with flanges which are braced back to the main structure. The bracing beams are capable of resisting horizontal loads up to 200 tons and are designed to prevent possible damage to the template piping and structure should an anchor snag the seabed flowlines. The flanges are positioned to give good access for diver connection. Details of the flowline connections to the template are discussed in Chapter 7. Figure 3–10 shows a simple four-well template.

Pile Connections

Local seabed soil conditions have critical influence on both the design and installation of the template structure. A soil investigation must be conducted to confirm the suitability of the driven piles. The depth of penetration of the piles will be determined by the skin friction between pile and soil which, in turn, is related to the shear strength of the soil. The size of piles, however, is governed largely by the deformation characteristics of the soil at the upper end of the piles under horizontal

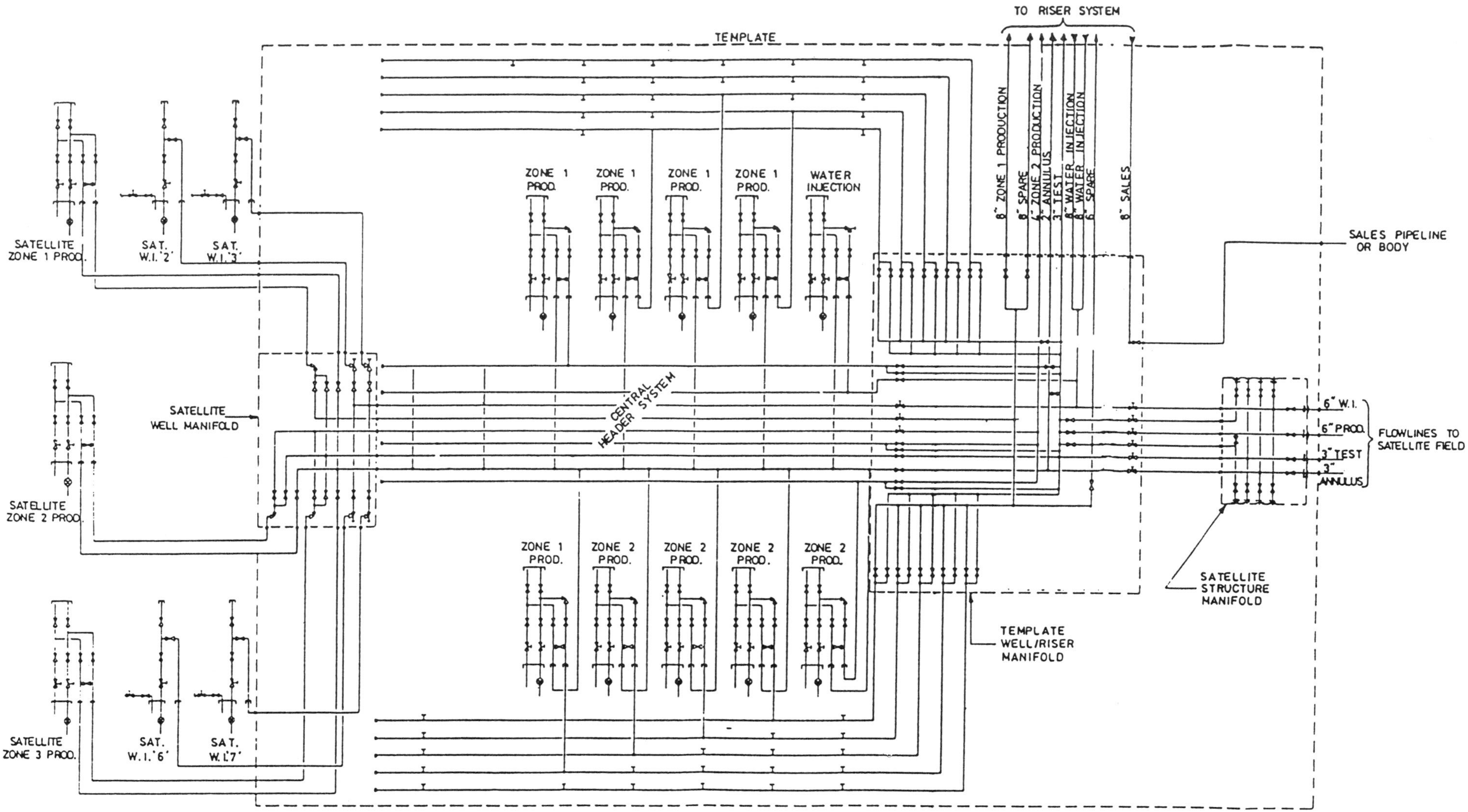

TYPICAL TEMPLATE PIPING/SYSTEM FOR SCHEMATIC
WITH:- DUAL PAY ZONES (1 AND 2)
- 7 SATELLITE WELLS (PRODUCTION FROM 2 PAY ZONES AND WATER INJECTION)
- 1 SATELLITE FIELD
- 10 WELLS ON TEMPLATE
- 3 MANIFOLDS AND A RISER SYSTEM
- 1 SALES/ EXPORT SYSTEM (PIPELINE OR BUOY)

Fig. 3–9 Typical Template Piping System Schematic

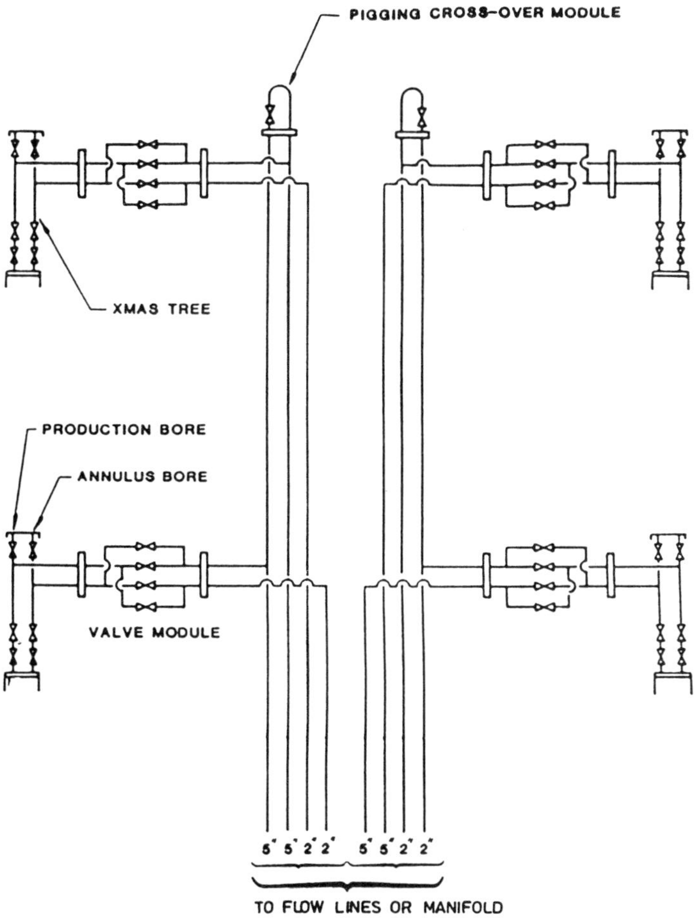

Fig. 3–10 Typical Simple Four-Well Template Piping Schematic With 2 Pigging Modules (No Manifold but Special Valve Modules)

loading. Under anchor-chain loads, the soil will deform horizontally, and severe bending movements will be caused in the piles. The soil investigation should provide data which is needed to check the suitability of the proposed piles.

Pile sleeves or guide tubes are provided within the protective fence in the corners at the riser end and the third is located outboard at the opposite end. Adequate clearance is provided for an underwater hammer. Two guideposts are provided for each pile.

The soil investigation establishes the allowable seabed bearing pressure for the detailed design of leveling pads. It also should provide an estimate of settlement from which the need for a perimeter steel skirt can be judged.

The piles are driven to a depth of 50 to 100 ft below the mudline through 48-in. diameter guide tubes.

The connection between the pile and the guide-tube is grouted. Shear lugs are provided on both pile and guide-tube to reduce the bond stress to an acceptable level. The same connection principle is used whether the piles are driven or drilled and cemented. For drilled piles, a seating shoulder is needed on the guide tube to support the pile prior to grouting. The arrangement is similar to that on the well guide-tube, which supports the conductor casing (see Fig. 3–11).

The grouted pile/template connection is preferred to a mechanical connection because it can readily absorb tolerances produced by vertical misalignment of the piles and by variations in the depth of penetration. A new mechanical connector system by BUE Hydra-Lok is now available however. This was used twice on templates during 1984, by BP and Sun Oil, with considerable success.

Conventional Pile Connections

The production of oil and gas offshore requires extensive installations on the seabed above the reservoir. These include such structures as template, jackets, floating production facilities, and satellite wellheads. Each structure requires a substantial foundation, and this is commonly provided by driving tubular piles deep into the seabed through structurally mounted pile guides and sleeves. This operation is carried out with the aid of a subsea pneumatic hammer mounted atop each pile in turn.

The grouting method requires the use of packers to isolate the cavity. The packers require pipework to inflate them, and additional pipework is required to route the grout into the cavity. Obviously, in the case of templates this pipework can be installed during fabrication. It is nevertheless a substantial captial investment, since in a diverless system each pile usually needs two grout lines, two packer lines, and the necessary packers.

Problems with Subsea Installations

For subsea installations with no fixed structure to the surface, some of the pipework is temporary and requires extensive diver involvement to make the necessary connections to an appropriate service manifold. The overall cost of such an installation is high and could delay the operation.

The HYDRA-LOK System

General Principle

A company has developed a pile/structure connection system called Hydra-Lok. The system's main advantage over other techniques is the total elimination of the high-capital-cost equipment on the structure together with the total elimination of divers during subsea installation. The process is simple and is performed by a tool deployed and operated from the surface. No expensive installation equipment remains subsea after completion, and the tool is totally reusable, making the technique extremely cost-effective.

The concept of making joints by producing a mechanical interference between two cylindrical components is extremely

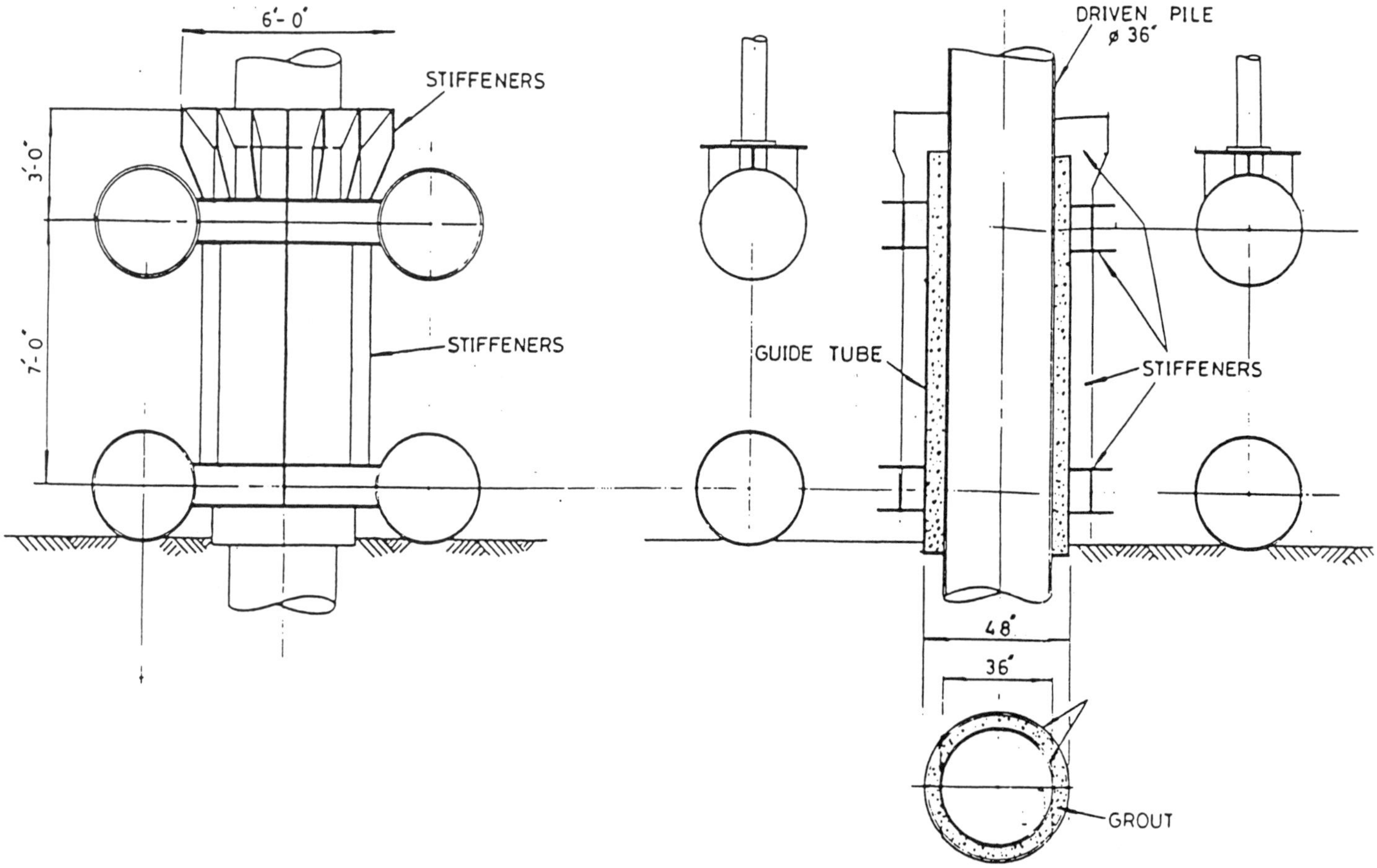

Fig. 3–11 Template Pile Grouted Connection (Typical)

old. A more directly comparable case is the expanded joints used extensively in the tube/tube-boiler connections.

Past experience with grouted connection has established a basic arrangement in the connection of pile to structures. The loads carried by the pile are transferred to the structure through a sleeve. This may be an integral part of the structure, for example the jacket leg, or it may be attached by shear plates. It is prudent to follow this practice for the Hydra-Lok connection. A typical arrangement and sequence of operation is shown in Figure 3–12. The general arrangement is very similar to the grouted case, although the overall length will be substantially reduced.

A Hydro-Lok connection consists of one or more regions where the pile has been swaged into performed grooves in the sleeve. The number of required grooves depends on the pile dimensions and loading imposed. Typically, one to four grooves might be required. This along with a 36-in. jacket pile produces a maximum connection length of 10 ft. The entire swaging process is performed in one operation, creating a rapidly formed high-strength interference fit connection.

Hydraulic Power Sypply

The hydraulic power and control circuit operates with a fresh water supply delivered from its own supply tank housed in the control cabin. This is connected to a suitable external supply on site. The circuit includes a dual pump to provide back-up if required or additional capacity. The pumps are driven by a low-pressure air supply giving a hydraulic outlet of 10,000 to 12,500 psi, respectively, at a flow of 1 gallon/minute. Because such small hydraulic deliveries are involved, ¼-in. bore high-power hoses are quite acceptable for the system and umbilical requirements. An air-heating system was introduced to avoid any possibility of freezing on the pump air outlets, a common problem with such pumps when operated over prolonged periods.

Hydraulic Power Monitoring

As a requirement of the method to monitor pressure against volume of water delivered, stroke counters are fitted to the pumps to provide an accurate indication of total delivery at any time, pressures being indicated by the respective pressure

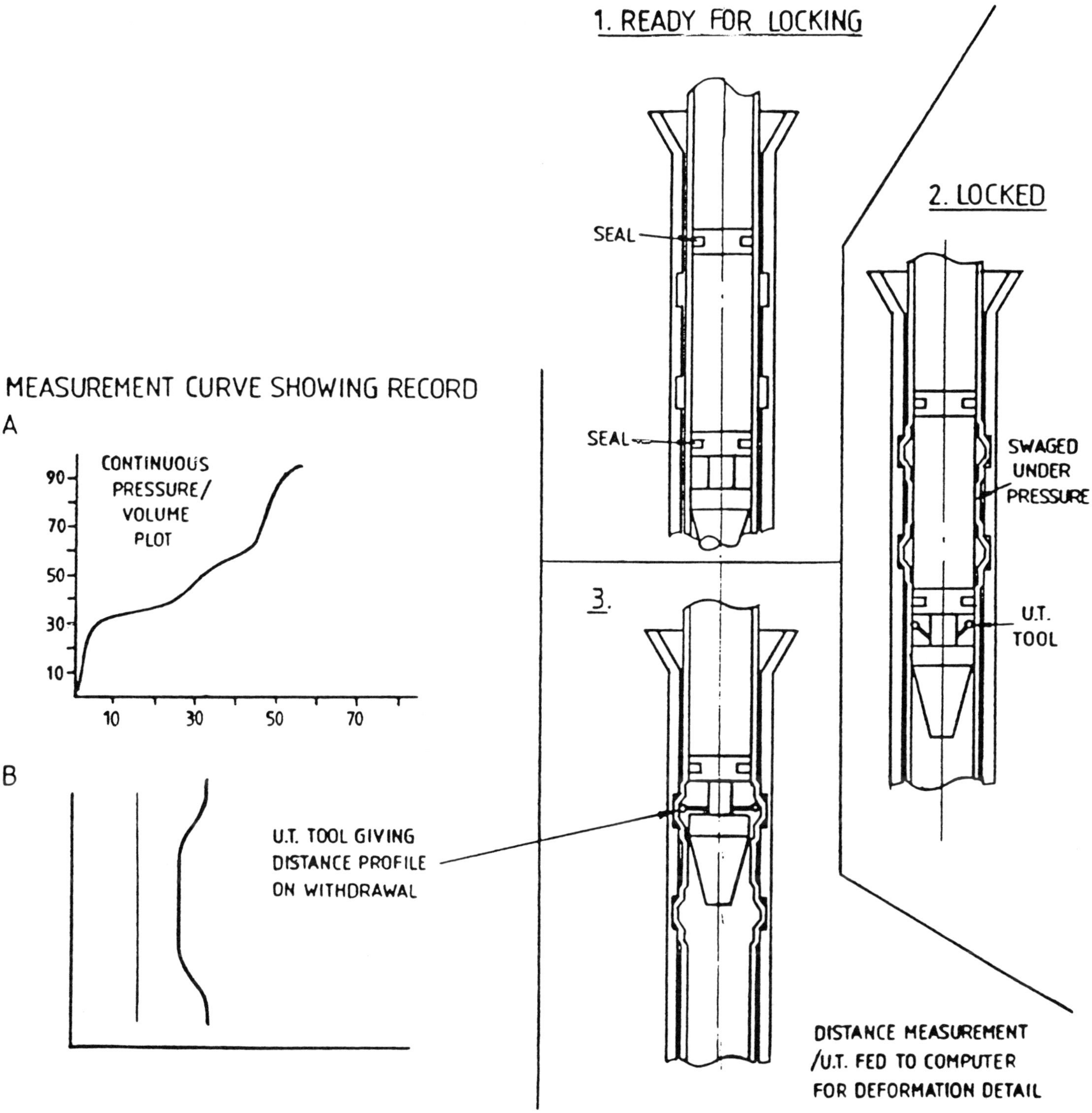

Fig. 3–12 Hydra-Lok 3 Steps Operation and Control

gauge. Relief valve and a pressure-dump facility are incorporated in the cabin control circuit also. Suitable hydraulic terminations for connection to the umbilical are provided at a supply terminal within the cabin.

TEMPLATE INSTALLATION

Range of Installation Methods

The template must be designed to be suitable for a range of installation options. Although a heavy-lift semi-submersible vessel may be used readily, the installation program will not be dependent on the availability of this type of vessel.

When fabricated, the template may be either launched and towed to site or transported to site on the deck of a barge or other vessel. Barge transportation is preferred since this is quicker and safer than towing the buoyant structure. Also, the possibility of damage to piping and other components during towing is eliminated. Transportation by barge is considered more suitable than by self-propelled shipshape vessels, because barges are less costly and more available. Furthermore, having a shallower draft, the barge has increased accessibility to the fabrication yards. Finally, if the template is to be launched by skidding off the stern of the vessel, this may be carried out only from a barge.

By the time fabrication starts, it may not have been established whether installation will be either by heavy-lift semi-submersible or by a barge with adequate crane capacity for a dry lift, namely in excess of 1,000 tons. Therefore, to allow for all these options and for launching from a barge, the template must be designed to have good positive buoyancy and stability when floating or submerged. Figure 3–13 shows the three key steps of such a delicate operation.

Barge Launching

If launched from a barge, the template then would be ballasted so that it could be lowered directly into position by a crane barge or, alternatively, keel-hauled and lowered by a semi-submersible drilling vessel. In this event, some buoyancy would be retained to minimize the required capacity of the crane employed and to ease any leveling operations that might be required. A submerged weight of 50 to 100 tons is suitable. The majority of structural tubular members will not be flooded, and these are left permanently sealed. Thus, in the structural analysis, stresses due to hydrostatic pressure are included. By flooding the lowermost layer of tubulars, the submerged stability will be maximized, and there will be no tendency for the template to tip over during lowering. Refer to Figure 3–14 for more details.

Seabed Tolerance Level

Once on the seabed, it may be necessary to level the template, although this is not considered likely. The template should be

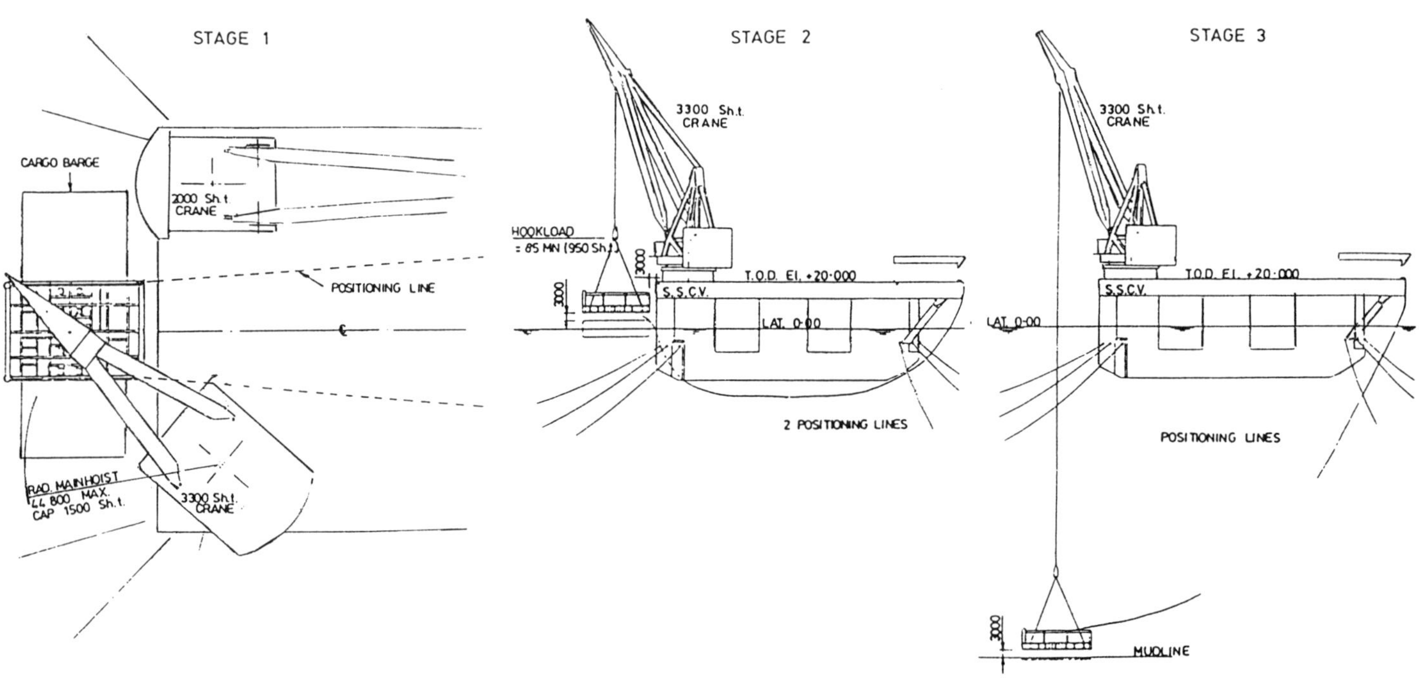

Fig. 3–13 Template Installation

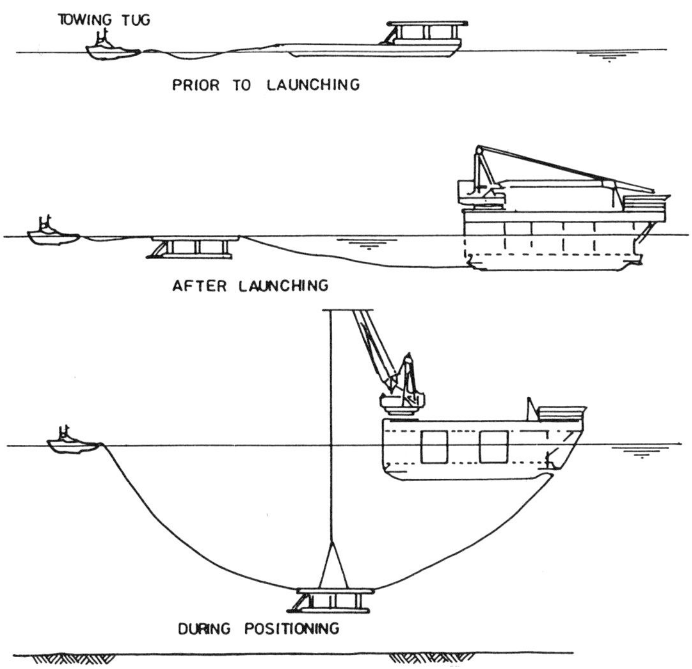

Fig. 3–14 Template Launch From a Barge

level to approximately $\pm 5°$ for drilling operations. The bathymetric surveys should indicate if the seabed is level or not and the kind of tolerance to be taken into consideration. However, local seabed irregularities may be encountered.

Seabed Leveling Techniques

If leveling is needed, this may be carried out either from leveling pads or from piles. Leveling pads are recommended since no special detail at the head of the piles is required. The soils investigation will indicate the allowable bearing pressure for leveling pad design. If the seabed soil strength is low, then the submerged weight of the template may be reduced by controlled ballasting. Also, a perimeter steel skirt could be introduced to limit settlement by preventing the soil from squeezing outwards. However, since there is a possibility that leveling pads may not be suitable because of local deep mud, some comments are made on possible methods of leveling from the piles.

Leveling from the piles may be carried out either hydraulically or mechanically. The only proven system is the Vetco hydraulic slip-leveling appartus. (Refer to Fig. 3–15 for more details.) This is, however, expensive. Alternative untried methods, based on well-proven principles, incorporating hydraulic jacks have been developed and can be used readily.

Hydraulic jacks are recommended for the leveling pads since this method is well tried and likely to require less time than the use of mechanical jacks. The jacks will be actuated by a diver from a template mounted control panel.

If the template is designed for possible installation from a work barge, then driven piles may be used. However, if it is decided during fabrication that a drilling semi-submersible will be used for installation, then drilled and cemented piles may be substituted readily.

TEMPLATE POSITIONING

Introduction

Special positioning equipment is required to achieve, for example, the level of certainty and accuracy demanded by the Conoco tension leg platform (TLP) anchors. The Sun Oil Balmoral template was positioned with the assistance of an autonomous underwater-instrument system which uses acoustics for both measurement and telemetry. The system is called "Micro-Nav."

System Description

The system comprises a microprocessor-based control and telemetry unit with internal sensors for roll, pitch, depth, and water temperature. Figure 3–16 shows a typical VDU display. Primary external sensors are four transponder/remote transducer units mounted at convenient points on the structure being positioned to form a short-baseline acoustic system. Other extenral sensors include an altimeter to accoustically measure height above sea bottom, attitude sensors, and a north-seeking gyro compass.

In operation, the system measures the precise distances between the transponder arrays on a fixed structure and those on the structure being lowered. This provides a three-dimensional picture of position, attitude, and orientation with great accuracy.

When positioning a single structure or the first of a group of structures, the system is used in conjunction with radio navigation equipment to establish absolute co-ordinates. For this type of operation, Micro-Nav is fitted to the structure being lowered and functions as an inverted short-baseline system with four external transponders ranging upwards to transducers on the deployment vessel.

For secondary positioning of additional structures, the system can be installed on both the reference structure and the one being lowered to provide high levels of accuracy. Other applications include level sensing to accuracies of 1 minute of arc, checking dimensions of complex structures, long-term monitoring of settlement, loads or corrosion potential, the control of pipeline valves, and precise spool-piece measurements. The final positioning of Hutton TLP is shown on Figure 3–18.

Close Tolerances

Conoco had specified tolerances for the final positions, orientation, and levelness of the four foundation bases after hydro-

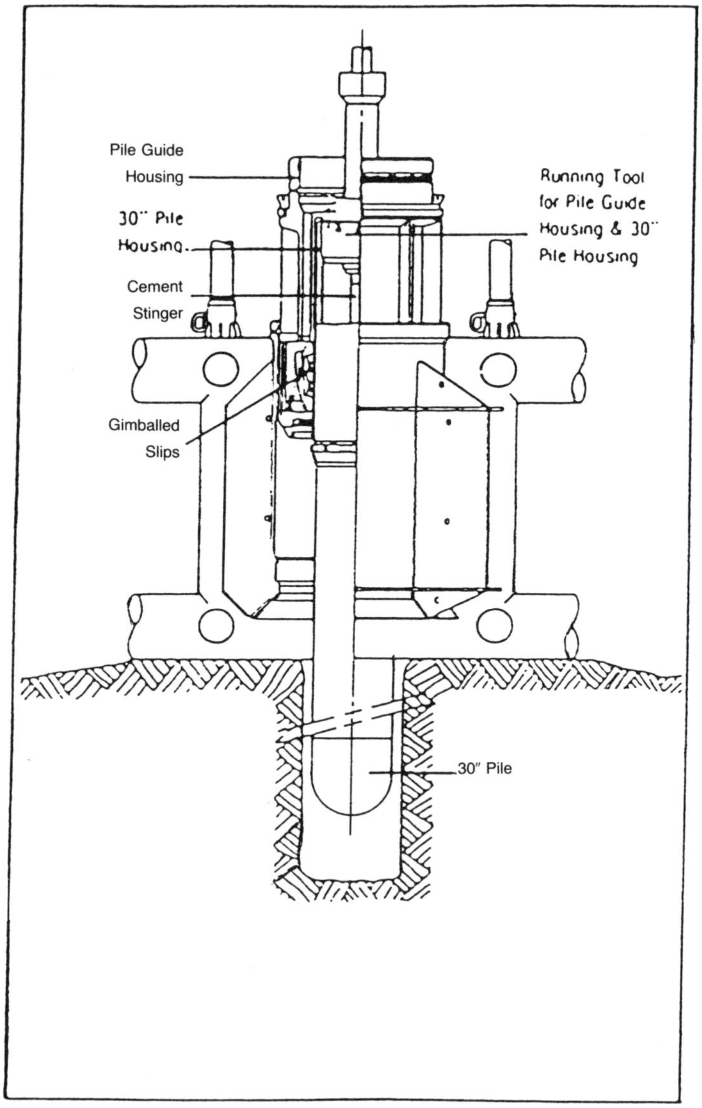

(A) PILE GUIDE HOUSING WITH PILE, PILE HOUSING, MECHANICAL STOPS AND GIMBALLED SLIPS INSTALLED IS LOWERED INTO THE LEVELING PILE RECEPTACLE IN THE TEMPLATE

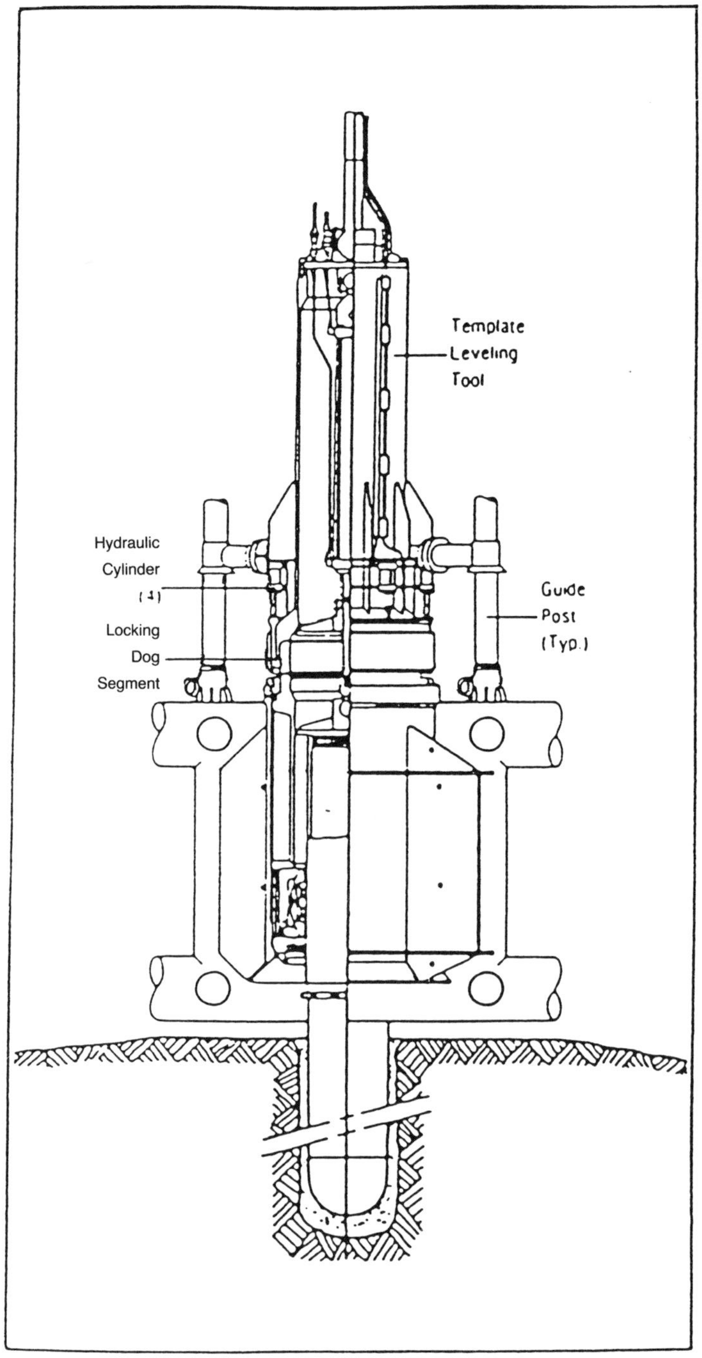

(B) TEMPLATE IS LEVELED WITH TEMPLATE LEVELING TOOL. THE 30" LEVELING PILE CAN SERVE AS AN ADDITIONAL WELL SLOT OR IT CAN BE FITTED WITH A PILE LOCKDOWN AND PROTECTIVE CAP.

Fig. 3–15 Application of Hydraulic Slip—Leveling Apparatus (Source: VETCO)

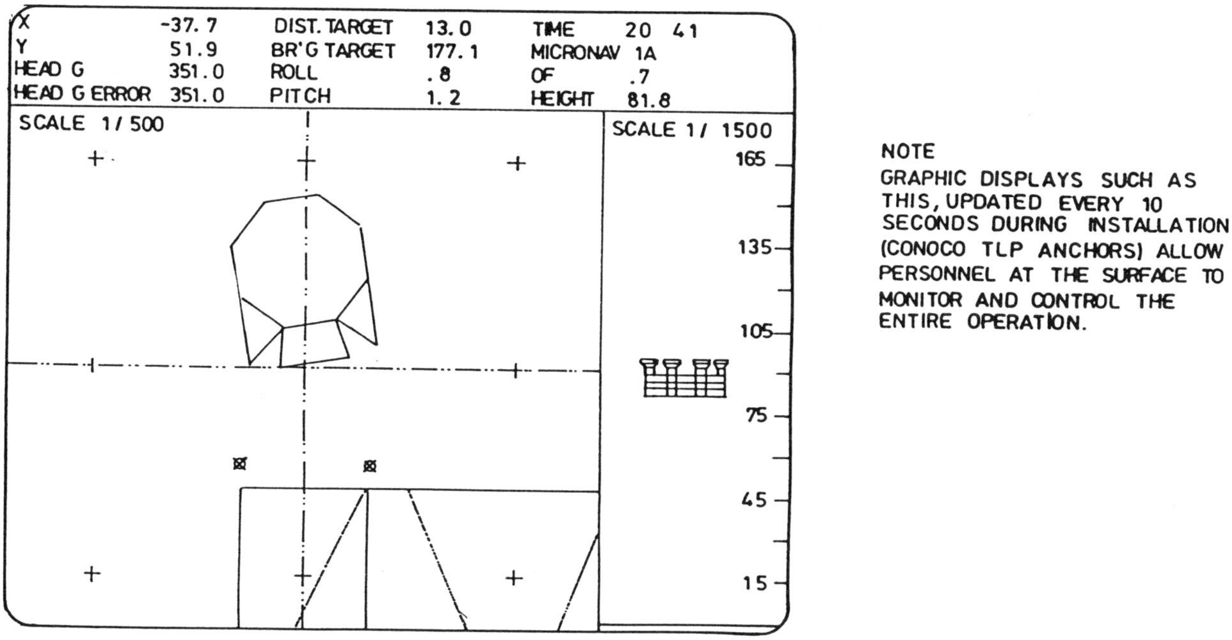

Fig. 3–16 Typical VDU/Graphic Display from Micro-uv Control System

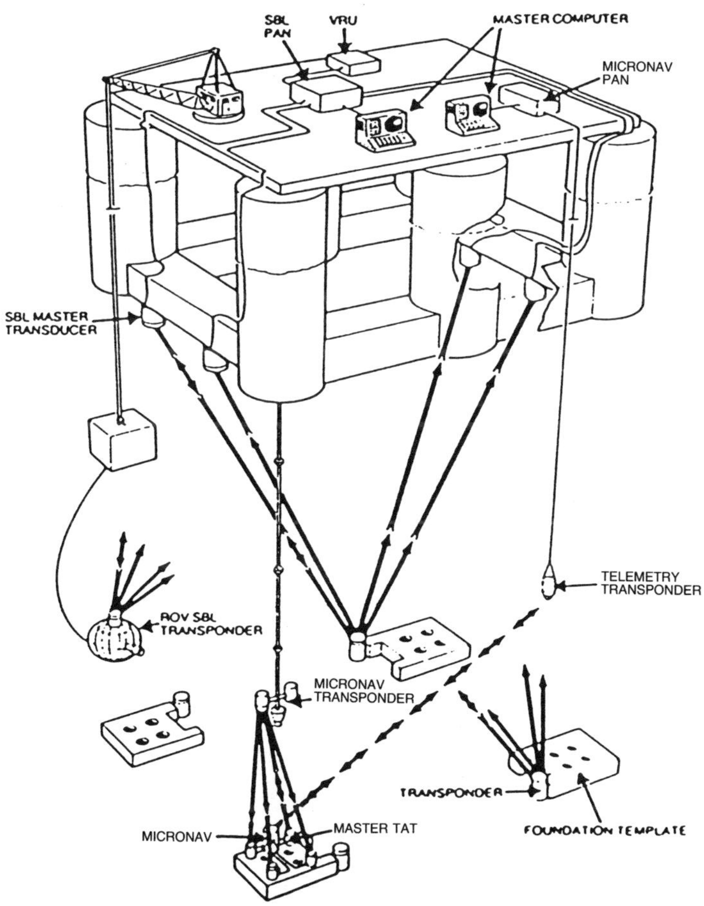

Fig. 3–17 Integrated TLP Positioning System (Conoco, Hutton)

dynamic analysis of the TLP. The most severe interpretation of their requirements dictated the following accuracies:

- Position of ±40 mm
- Orientation of ±0.7°
- Levelness of ±0.2°
- Relative height of ±50 mm.

In addition, risk-free maneuvering during development and engagement requires a data transmission rate fast enough to permit updating maximum position data at 10-second intervals.

Final dimension checks indicated that all tolerance requirements were met. The critical distance between the centers of the two lines joining the centers of diagonally opposed foundations was computed as: 14 mm, ±33 mm with 95% certainty or ±50 mm with 99.7% certainty.

Method of Achievement

To achieve this, Micro-Nav systems installed on the foundation templates being lowered were connected to four transponders positioned on the structures so as to give a clear line of sight to transponders located at the four corners of the central well template. Redundant ranges measured between the multiple transducers and transponders on the two structures, together with measurements from the attitude and depth altitude sensors on the foundation, provided input for a three-dimensional computation of the relative-position and critical-offset parameters. Final engagement of the guidance frame and foundation bases was only carried out after visual confirmation with ROV.

Typical Template Installation Project

A typical program covering engineering, procurement, sea-fastening, and offshore installation is shown in Figure 3–18.

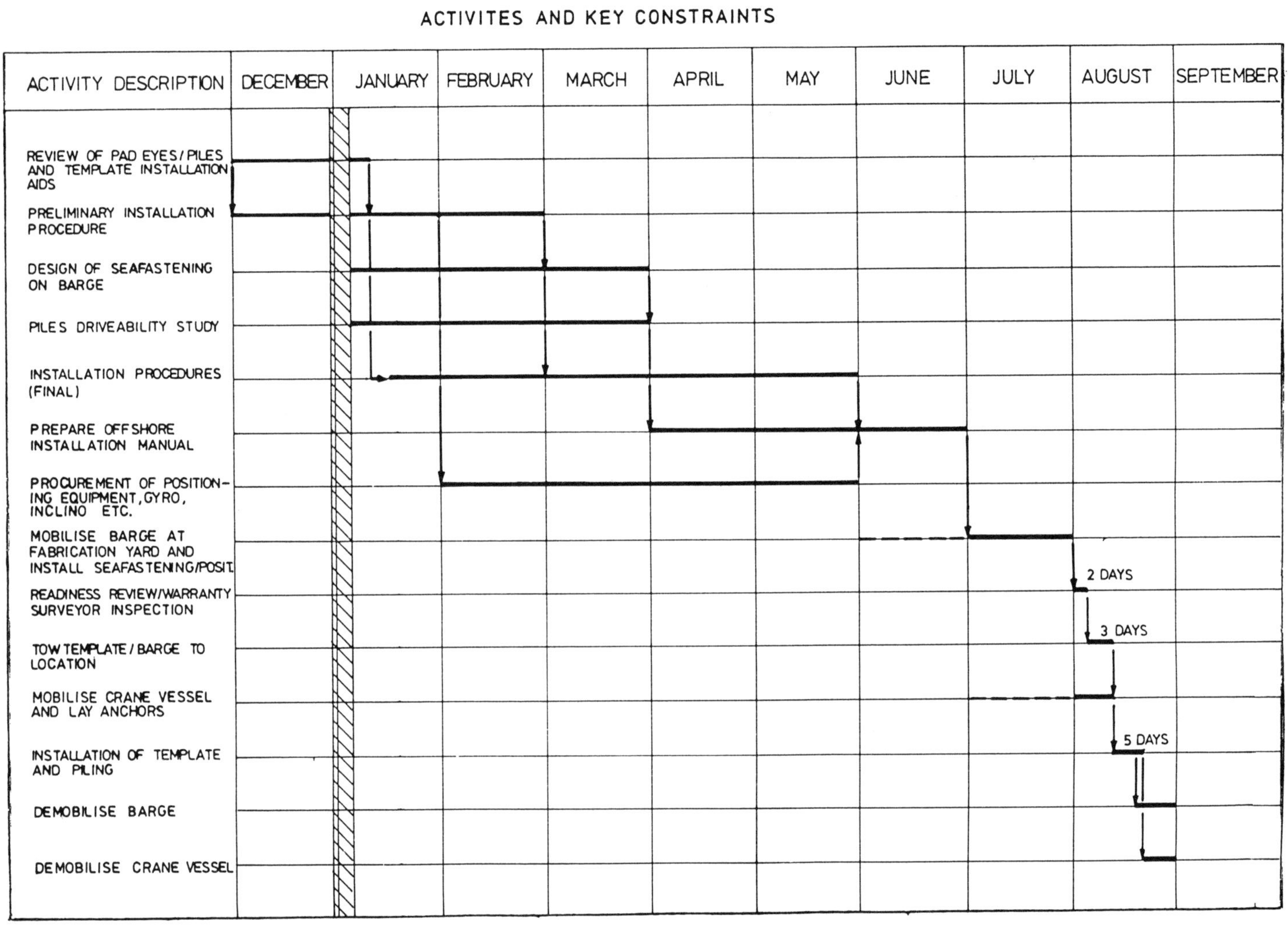

Fig. 3–18 Typical Engineering Template Installation, Procurement and Installation Schedule

Chapter 4

Co-Mingling and Manifolding

DESIGN PHILOSOPHY

Range of Manifolds

Subsea manifolds range from a simple pipeline end manifold (PLEM) to large structures such as the Shell/Esso Underwater Manifold Center (UMC). Within this range there are numerous types of manifolds. For example, it is becoming increasingly common to operate several subsea wells, both satellite and template, drilled from a single production facility. With several wells, it is desirable to reduce the number of risers required to operate these wells. This approach assists with assembly, maximizes the number of wells operated from one riser system, and reduces construction and operating costs.

Commingling. To reduce the number of risers, it is necessary to commingle the production from each of the wells subsea. In certain developments, this has been achieved without the use of either fixed or variable chokes, for example North-East Frigg. This cannot always be the option, and the use of chokes should be considered at an early stage.

Choke valves. In the past, the difficulty with using choke valves has been the need to calibrate and monitor their operation from the surface. Recent developments in chokes and their positional sensing now provides the degree of operational reliability required to safely operate multiple subsea completions through a common manifold.

To achieve subsea commingling, manifolds require numerous valves and chokes. Commingling brings together the fluid flows into a common header located within the manifold structure. This can be achieved in several ways. The selected configuration determines not only the physical size of the subsea manifold but also the quantity and size of template piping. This is due to the way in which the template and off-template wells are operated.

The design philosophy must therefore take into account piping simplicity, minimum number of valves, and system maintainability. It also should allow for the change of service of a satellite well or major repairs to a choke system without the shutting-in of the complete field. This is usually achieved with a retrievable subsea manifold.

Secondary commingling. Secondary commingling can occur within a header system on a template mounted manifold. This allows the manifold to maintain a unit size compatible with ease of handling and retrieving through a moonpool.

Additional requirements. Generally, manifolds require a common production header with each fluid flow provided with a choke. Additional requirements arise if provision is made for enhanced oil recovery (EOR) during the field's life, plus initial kicking off of the production wells and the requirements for testing and killing.

The actual grouping of valves and chokes to form a manifold has to consider not only the physical size of piping runs and valves but the functions. Specifying the functions is the most efficient approach to minimizing piping and providing for phases of field development. The functions to be undertaken depend on the specific characteristics of the field being developed.

Reference to field developments and existing installations shows the extent of the functions to be considered.

Buchan Field

The Buchan field development (see Fig. 4–1) has a stacked manifold configuration located on a nine-slot template, with four template producers and satellite wells. Subsea production is not commingled, and therefore there are 11 separate risers on a guide-arm spider. The riser blender has multiple stabs and hydraulic connectors, which has caused difficulties with their close tolerance limitations.

Valves and piping. The manifold consists of a system of valves and pipework arranged around a central structural member, the export riser. An upper frame carries the production riser connec-

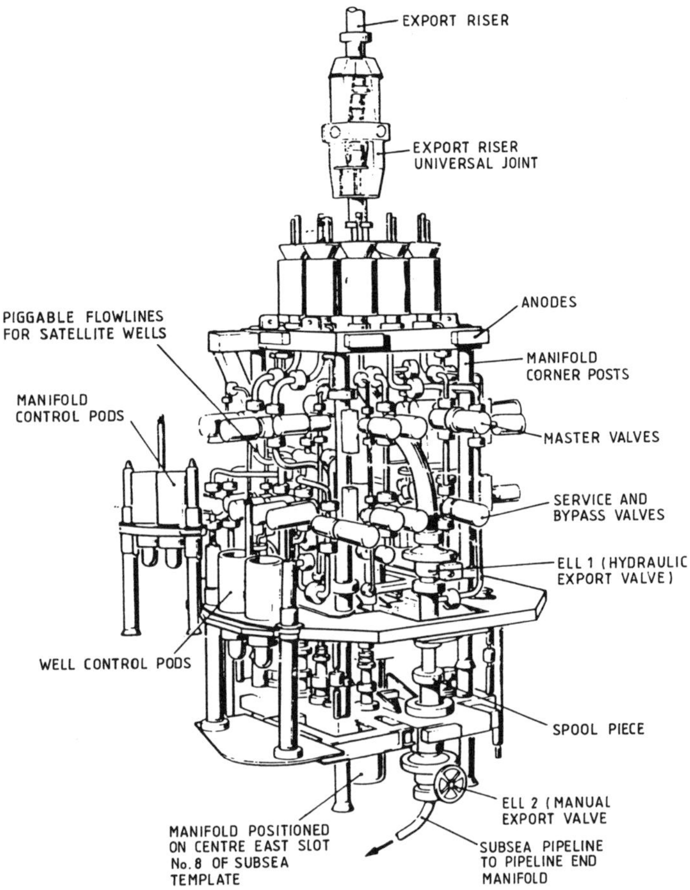

Fig. 4–1 Buchan Field Manifold, General View

tors. Another frame at the bottom carries stabs which interface with the manifold base (see Fig. 4–2). From the top of the upper frame to the seabed is approximately 10.2 m (33.4 ft). The manifold can be retrieved to the surface if servicing of the hydraulic valves is required.

Connectors. Riser connectors, for connecting the manifold flowlines to the platform, are mounted in a cluster on the top of the upper manifold frame. There is one central export connector, eight production/annulus duplex connectors, one for each well, and two service connectors.

Piggable flowlines. The production pipework, within the manifold, connects the well and gas-lift flowlines in the base to the risers. Production flowlines for the satellite wells are piggable.

Base. The base serves as the interface between the manifold and the subsea well flowlines and export system. It consists of a frame with attached piping spools to connect the manifold and template piping and is not usually retrieved.

North-East Frigg

A simple manifold system operating in the North Sea is the North-East (NE) Frigg gas development. The manifold is template mounted and uses 4½-in. horizontal flowlines from the Christmas tree to a central header. Each line is capable of being isolated remotely with 4½-in. 5,000 psi valves. At one end of the header, facilities are provided for a pig launcher (see Fig. 4–3). Chemical injection is an additional function which uses the umbilicals to transport methanol from the adjacent articulated tower to the subsea trees and manifolds.

Shell/Esso Underwater Manifold Center (UMC)

At the other end of the range, the Shell/Esso UMC is a complex system for production of the Central Cormorant field. From the UMC, the commingled flow is transported back to the South Cormorant "A" platform.

The UMC manifold system is located on both sides of a corridor provided for the remote maintenance vehicle. Fluids produced by the template and satellite wells are collected by the manifold pipework. Then pipelines from the UMC transport the fluids to facilities on the Cormorant "A" platform. Treated injection water is pumped from the platform and distributed through the manifold to injection wells.

Figure 4–4 illustrates the UMC manifold system piping. It shows main headers, inter-connecting piping, location and type of valves, pressure sensors, and chemical injection distribution points. Also shown are the connection points for wells and pipelines. Both the main headers and the two service headers are connected by a return loop to permit circulation of pigs through pipelines and headers.

UMC operations. In addition, the manifold system serves several functions, such as well testing, well killing, injection of chemicals for hydrate control, plus routing of through-flowline (TFL) tools used for well servicing the line cleaning.

To fulfill these functions, the manifold has two main 8-in. I.D. and two 3-in. I.D. service headers and a network of 1½-in. chemical injection piping. One of the 8-in. headers handles production fluids and the other distributes injection water. The two 3-in. headers double for well testing and TFL well servicing. These service headers have 5-ft radius bends to be compatible with the TFL tool strings. Well testing is accomplished through the service pipelines in conjunction with an 8-in. pipeline between the UMC and the platform.

Manifold Valves

Design Specifications

One of the main design requirements of the subsea system is to locate all remotely controlled valves within retrievable mani-

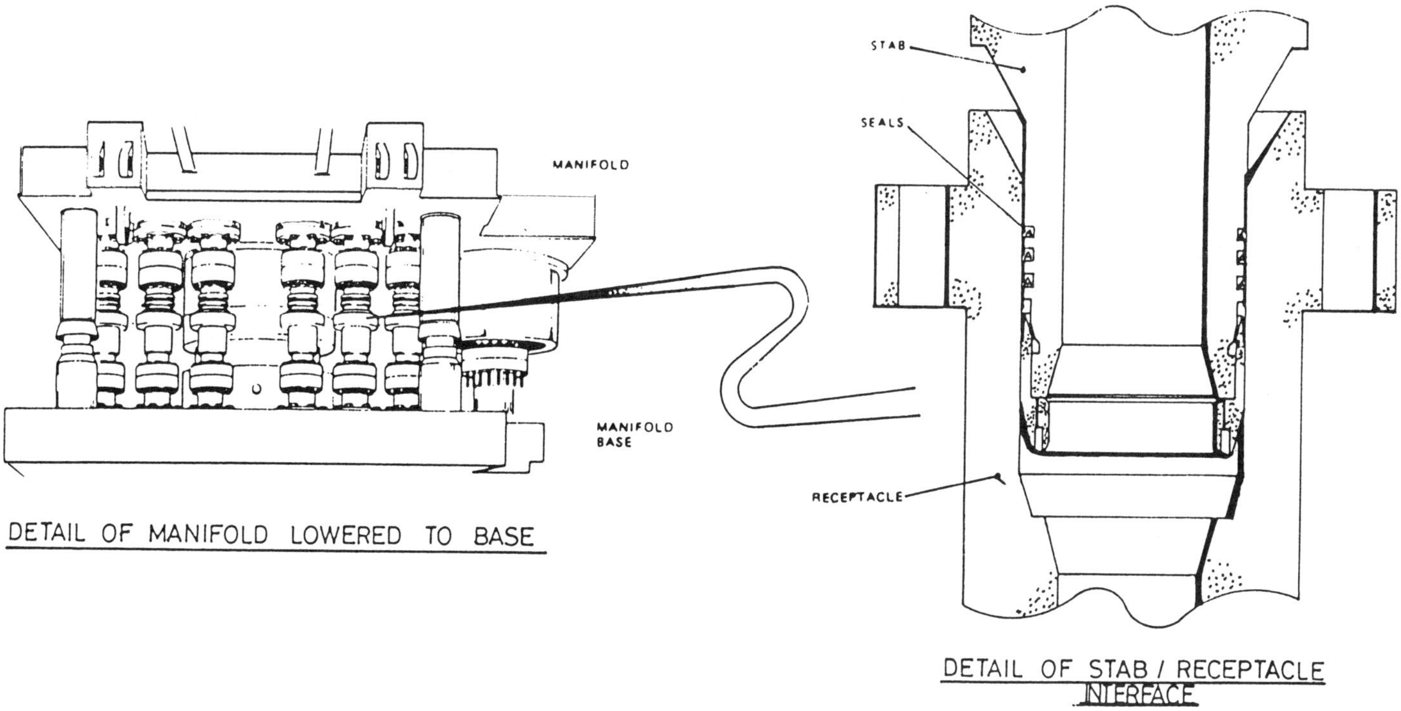

Fig. 4–2 Detail of Stab/Receptacle Interface

fold frames or modules. The selection and location of the valves on the manifolds is of prime importance. Design specifications of valve type, size, pressure rating, and actuator function are required prior to detailing manifold layouts.

Valve Type and Size

Options of either gate valves or ball valves may be considered for use in subsea applications. Gate valves are preferred owing to a long history of use in subsea BOP stacks, trees, and manifolds. Both the valve and the valve actuators have been through extensive development with the addition of proven field use and design improvements. They have been proven as relatively reliable devices. Ball valves also are proven items, however, their subsea actuators are not as well developed and reliable as gate valve actuators. Ball valve actuators with "failsafe closed" spring return cartridges also are larger than corresponding gate valve actuators and require more space in a manifold.

Gate valves are available in sizes up to 6⅜-in. bore. Ball valves are available in sizes up to 30-in. bore. Therefore, gate valves are preferred for all the smaller sizes in which they are available, and ball valves are preferred for sizes 10 in. and larger. Where 8-in. lines are required, 6⅜-in. gate valves can be utilized if the corresponding pressure drop across the valve is judged not too excessive. In all other cases, the valve size should be the same as the nominal pipe size used.

Pressure Ratings

Subsea gate valves should be supplied to API 6A specifications and can be obtained in 3,000 psi, 5,000 psi, or 10,000 psi working pressure designs. As piping system test pressures are at one-and-one-half times working-pressure, care must be taken when selecting the appropriate pressure rating of the valve as the gates are only tested at working pressure. However, the body can be tested at higher pressures.

As an example, for a water-injection system, with injection lines at 2,300 psi, the test pressure required for the piping would be 3,450 psi. A 3,000 psi valve would not be adequate, if the valve is needed to isolate the test pressure, although the working pressure of the line is below the 3,000 psi rating. In this example a 5,000 psi-rated valve would be required.

Flanging

Further care should be taken to ensure that adjacent piping flanges are compatible with the valve flanges. This is particularly important when dealing with API 6A type flanges. There are numerous occasions where this particular point has been overlooked. This omission often results in valves being installed with steps between the pipework and the bore of the valve. Within a large system, this causes unacceptable pressure losses and turbulence. In this situation, pressure sensors and other instrumentation in close proximity to the installed valve can suffer from noise and distorted signals.

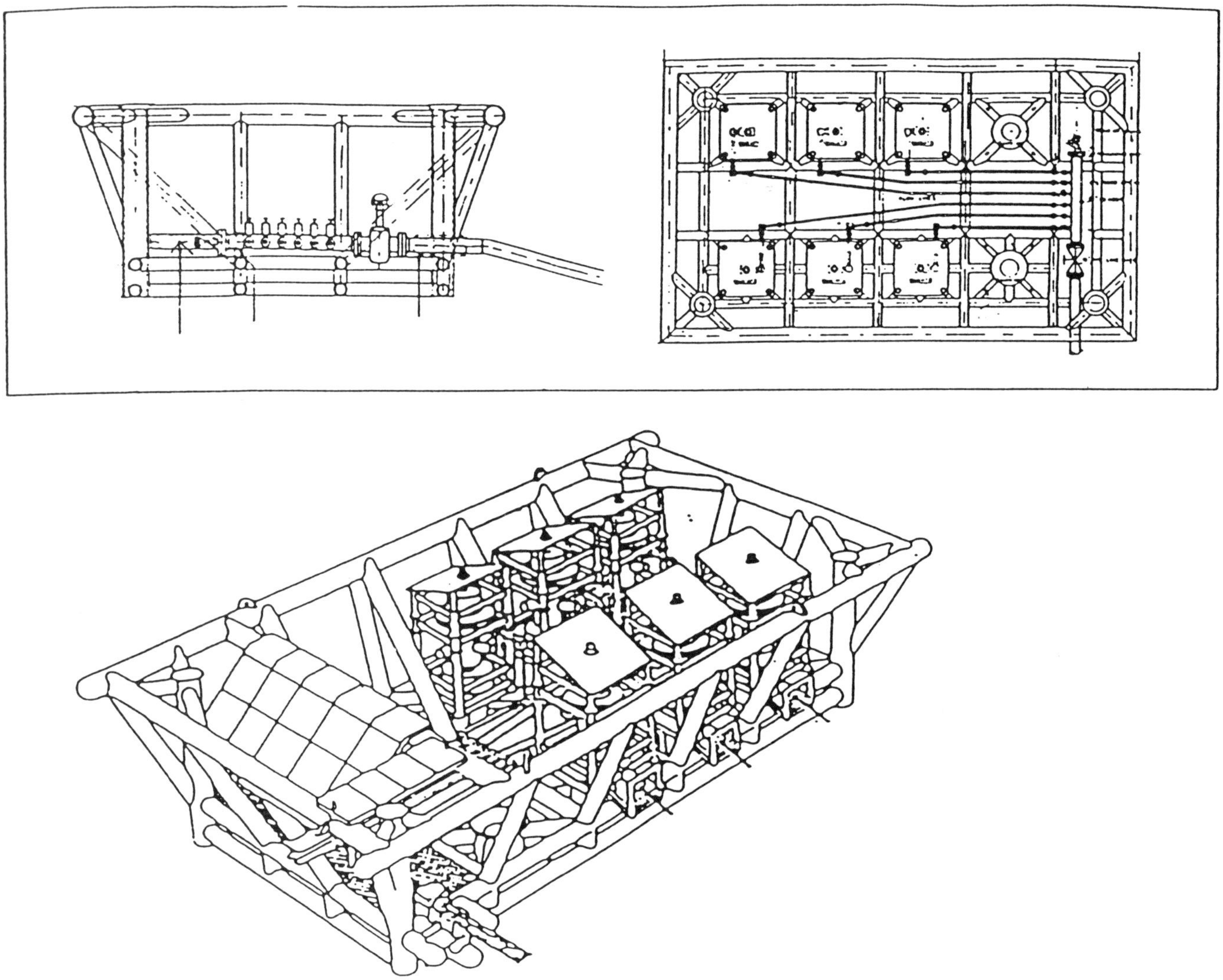

Fig. 4–3 North-East Frigg Template

Actuators

For safety, the actuators used on subsea valves normally close against full-line bore pressure in the event that the control system hydraulic power is interrupted. It is essential that the actuator is pressure-balanced on one side of the system to seawater pressure in order to avoid hydraulic locking.

There are two ways to achieve this pressure balance. Figure 4–5 shows an actuator with an integral bladder which alleviates the need for a separate accumulator system. This can be an advantage where space limits the use of separate accumulator banks. This is particularly relevant on manifolds with large numbers of valves and in certain cases large sizes. A standard type of actuator is shown in Figure 4–6 illustrating the use of an external accumulation system to prevent the ingress of seawater to the spring cartridge after actuation of the valves.

Operating Fluids

The systems on a number of previous installations used an open cannister type of accumulator filled with SAE 30 grade oil. This should be avoided for two proven reasons. First, the oil migrates slowly to the seawater eventually exposing the spring cartridge to a corrosive environment. Second, a large number of control systems use aqueous control fluids. There is, therefore, a risk of a contaminating gel forming at the actuator piston seals due to the incompatibility of the control fluid and the mineral oil.

Operating Pressures

Hydraulic valve actuators are normally supplied for use with 1,500 psi hydraulic control pressure. However, designs are avail-

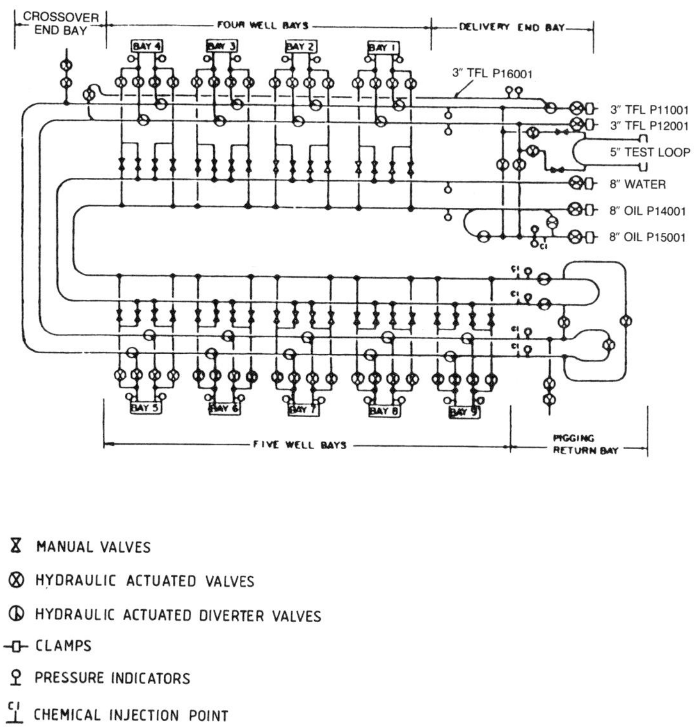

Fig. 4–4 Central Cormorant UMC System Piping

able for units rated for 3,000 psi. The decision to adopt either system depends on water depth, operating times, and the control system used. A typical example of the psi operating range for a 2¹⁄₁₆-in. subsea valve is given in Figure 4–7.

Valve Trim

Subsea valve trims are established to comply with National Association of Corrosion Engineers (NACE) MR-01–75 guidelines and must have traceable materials and documentation. This is the accepted practice within the industry, and a range of trims are provided by all the major subsea valve manufacturers. A comprehensive list of trim designations is given in Figure 4–8. The presence of hydrogen sulphide and carbon dioxide dictates the trim used, and as the general guideline, the service can be broken into three principal groups.

1. Standard service: Primarily for non-corrosive liquid, gas, or crude.
2. Sour oil and gas: For use where hydrogen sulphide is encountered in the well production fluids. This type of trim also may be used where carbon dioxide concentrations are *less* than 1%. Carbon dioxide partial pressure is the working pressure at the wellhead multiplied by the percent in decimals of the carbon dioxide gas content.
3. Sour and/or corrosive oil and gas service: Primarily used where high levels of hydrogen sulphide and carbon dioxide exist in the well production fluid.

High or Low Temperatures

Care must be taken when service temperatures are above 121°C or below −29°C or when chlorides are present in the production fluids. In these instances, consultation with the manufacturers should be made at an early stage of design.

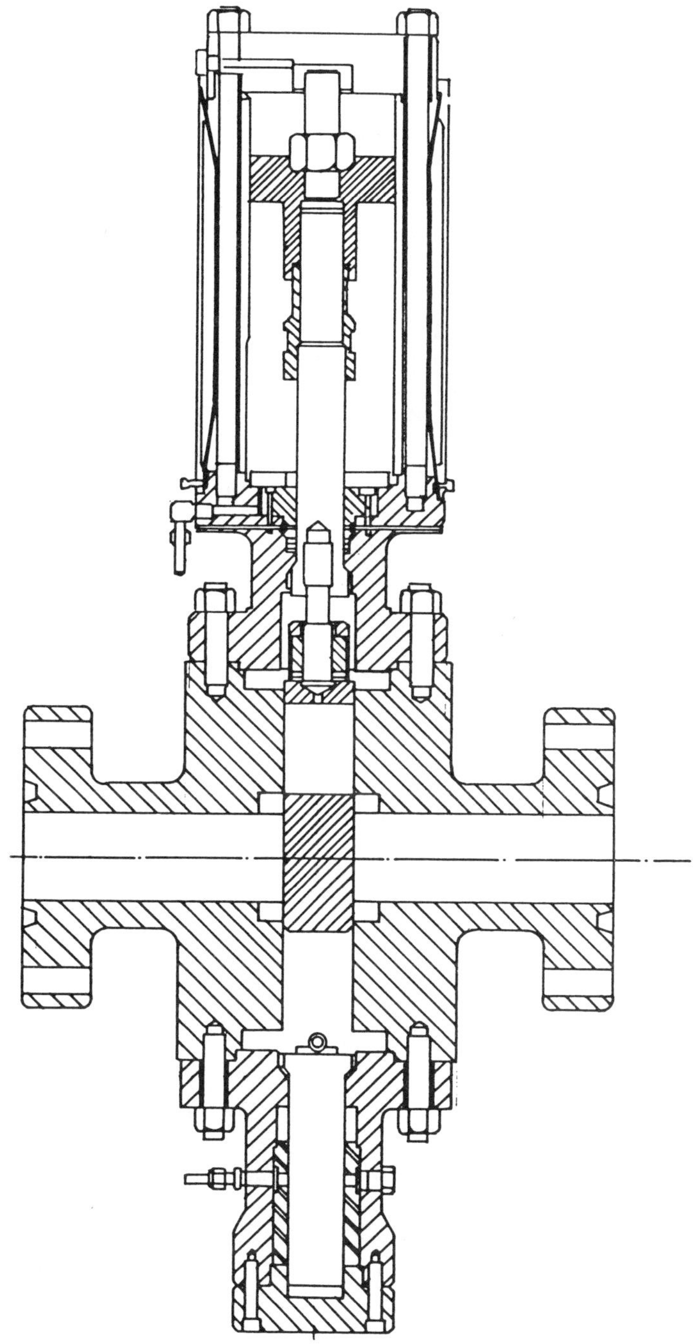

Fig. 4–5 Gate Valve with Integral Bladder Accumulator

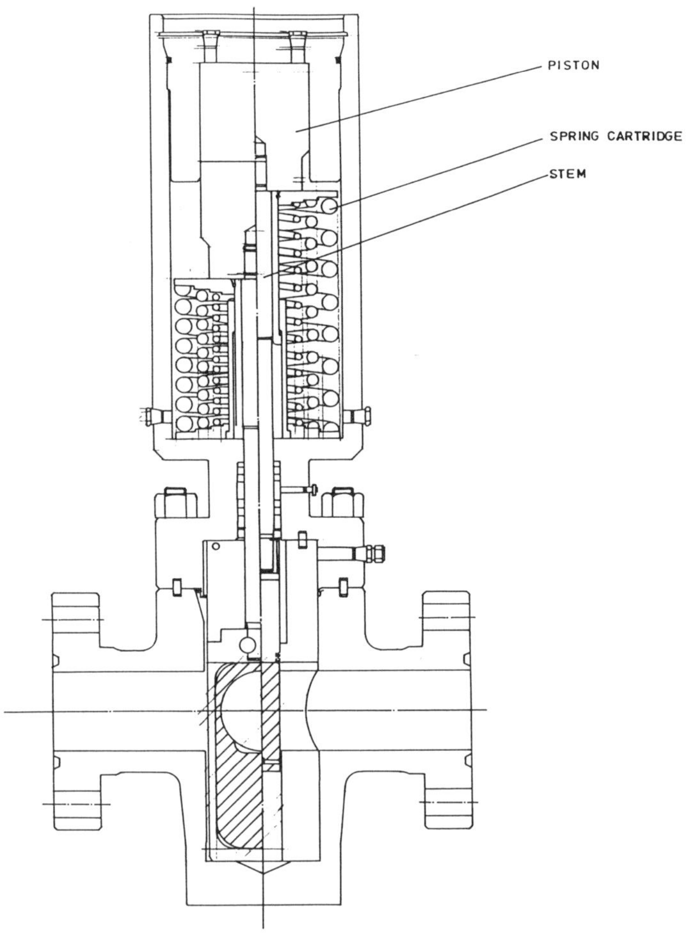

Fig. 4–6 Standard Spring Cartridge Using External Accumulation

Materials

Materials used to provide the required trim vary between manufacturers. However, as a general guideline, the list of materials shown in Figure 4–9 is indicative of the most commonly used materials.

Gate Assemblies

Gate assemblies dictate the features and life for any one make of valve, and they also vary from one manufacturer to another. The gate and seat are the most critical components of the valve. Gates provide a metal-to-metal seal against the seats and are finished to a high center line average (CLA) value, usually by lapping. This provides a finish in the region of two light bands (23.2 × 10.6 inches) when measured with an optical flat or interferometer. Increasing the number of light bands decreases the finish and is detrimental to the efficient sealing of the gate.

It is becoming increasingly attractive to operators to be able to change out the subsea gate and seal assemblies subsea. To meet this requirement, the gate and its carrier should be a cartridge type construction (see Fig. 4–10A). There are diverless designs available for horizontal or vertical retrieval also.

Unfortunately, this is not common to all valves. The majority of valves have the gate seats that are removed from the valve body using special extractor tools. A suitable method is to provide a seat carrier, as in Figure 4–10B, providing a means to remove the gate and seats in one piece leaving the body in line.

Whatever type of valve an operator chooses, the valve-gate assembly for manifold valves must seal bi-directionally. A means also should be made to combat wash-out which may occur when the valve is opened initially or just prior to closing.

Valve Features

A problem occurs when closing manifold valves in a close-packed configuration—that is, the pressure locked in the pipe system altering the volume within the valve body. This situation is most unsatisfactory but can be overcome by providing a balanced stem. A balanced stem maintains a constant volume within the valve body which prevents the lock occurring. Other means may be used to overcome this problem, but preference should be given to the balance stem design.

Mechanical Overrides

All the leading manufacturers provide the option of mechanical overrides and visual position indicators. However, manual overrides should be unnecessary owing to the current high reliability of fail-safe-closed actuators, but visual indicators are still recommended.

Electrical Position Indicators

Electrical valve position indicators have a history of failure and should be critically assessed before the decision is made to include this form of instrumentation into the control system.

Manifold Valves Summary

The foregoing is a brief description of the basic design and an indication of the philosophy prior to selection of a valve for a manifold system. The subject and equipment variations are extensive, and only the base cases have been reviewed. Proven designs must be used. Awareness must be the order of the day. Even a minor seal change, made as a product improvement, can turn a field-proven valve into what must be considered a prototype for field application use.

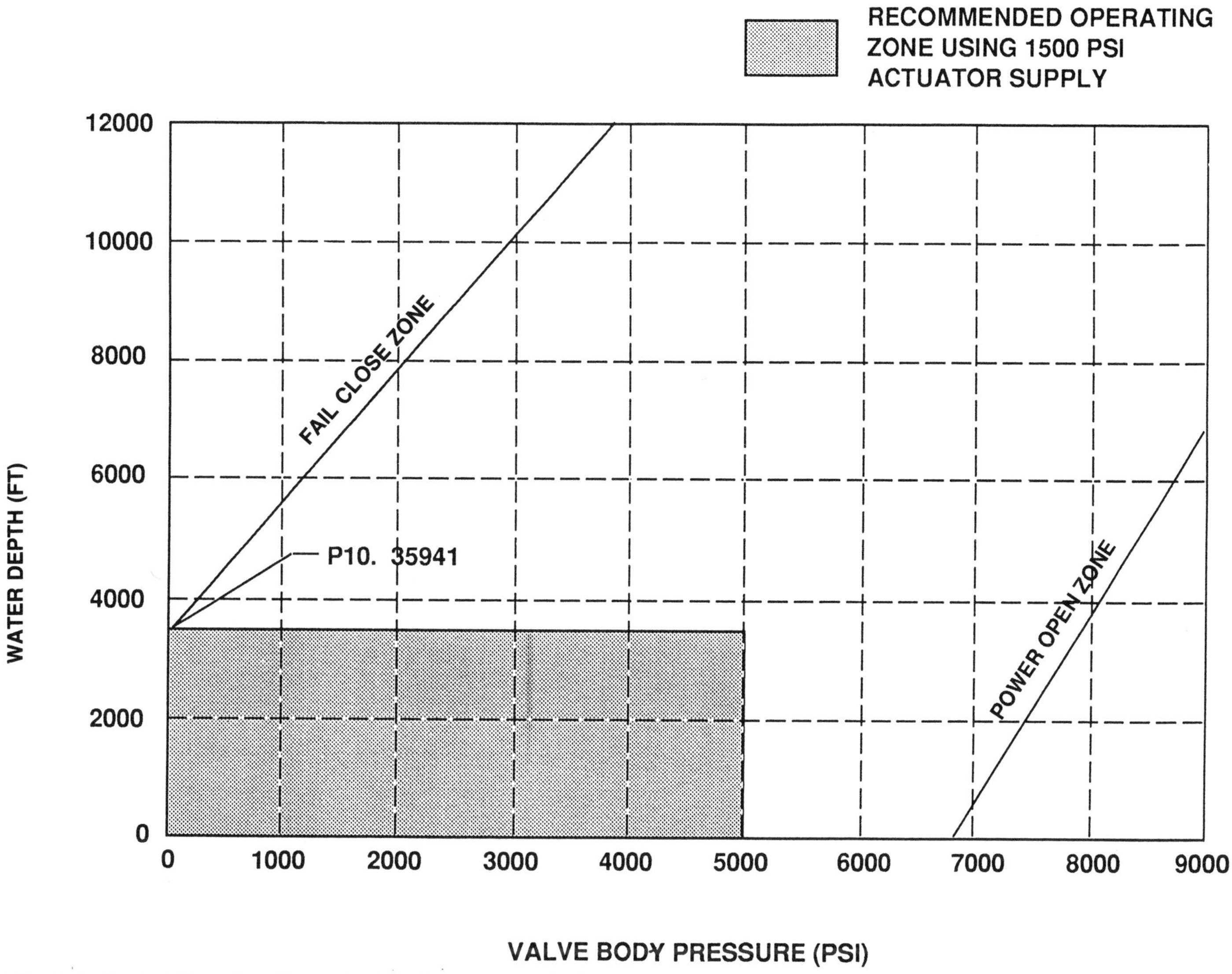

Fig. 4–7 Typical Operating Range for 2 1/16-in. 5,000 PSI Subsea Gate Valve with Hydraulic Actuator

Pipework Materials and Fitments

Pipework Connections

When laying-out piping, consideration must be given to the type of piping connection. Intermediate piping connections are needed for two reasons: valve replacement and to minimize assembly problems.

The main types of piping connections available are: clamp hubs, swivel-ring flanges, and the standard weld-neck flange.

Clamp Hubs

Clamp hubs are often used for subsea connections because they are small and have a reduced number of bolts to handle, install, and tension. They can, however, be difficult to disconnect on occasions. Since the clamp hubs are pulled together by the tensioning of the bolts (wedging the clamp halves on the tapered hub profile), a residual stress and friction resists removal of the clamp halves, even with the bolts removed. Divers often have to hammer or use a special tool to strip the clamp halves off the hubs, particularly if they have been subsea for a considerable length of time.

Swivel-Ring Flanges

Swivel-ring flanges have the advantage of being able to provide rotational alignment between bolt holes. However, since the initial fit-up and assembly will be done on land, this extra facility is generally not required. There will always be general applica-

TRIM	NATIONAL	WKM	OCT	CAMERON	GRAY	McEVOY	VETCO	FIP	WILLIS
General Oil-Field Service Oil & Gas	N-1	T-21	R	REGULAR	A	RM-12	TR-10	F-21	STANDARD
General Oilfield Service - Mild Corrosion - Some CO_2 S.S. Trim	N-2	T-22	S	S.S. TRIM	B	RM-21	TR-12	F-22	STYLE 1
Extreme CO_2 No H_2S Full Stainless	N-3	T-23	F	FULL	C	RM-23	TR-13	F-23	STYLE 45
Sour Gas & Oil H_2S more than CO_2	N-4	T-24		SUPER TRIM	D	RM-32	TR-14	F-24	STYLE 8
Extreme CO_2 Some H_2S Full S.S. Controlled Hardness	N-6	S-24	FB	FULL SS SUPER TRIM	E	RM-24			STYLE 9
Extreme Sour Service - CO_2 & N_2S Conforms to NACE IFI	N-6	T-26	FB	FULL SS SUPER TRIM	E	RM-33	TR-16	F-26	
Water Flood Service Untreated or Saline Water	N-7	T-27	WF	NICKEL PLATING		RM-22		F-27	
High Temperature Service above 350°		T-30				RM-99		F-30	
Low Temp Sour Gas & Oil Resistance to H_2S at -50°		T-37							
Low Temperature Non Corrosive Lading -75°		T-38		MODEL 75		RM-13			
Low Temp Sour Gas & Oil - 75° No CO_2		T-39							

Fig. 4–8 Valve Trim Designations

tions for this type of flange, but this particular type of fitting is not recommended for subsea.

Weld-Neck Flanges

Three types of weld-neck flanges are available and accepted for subsea systems: API types-6B and -6BX and ANSI B16.5. Type-6BX flanges are incompatible dimensionally with standard line pipe weld fitting. In many cases, they require that a reducer/transition fitting be used with every flange. This requires additional space which is not a desirable feature.

API Types

The API type-6B and ANSI B16.5 flanges are dimensionally identical and match standard pipe fittings. They are available in a range of pressure ratings. The API type-6B flanges are manufactured from API type-4 materials, with a minimum yield of 45,000 psi and a higher pressure rating than the equivalent ANSI flange. External loading from vibration, temperature changes, and lifting requires a connection to resist internal pressure as well as external loads. This together with the objective of using API components where possible, favors the use of API type-6B weld-neck flanges for manifolds, template piping, and on subsea trees.

Gaskets

API type-6B flanges are normally supplied with RX gaskets. This type of gasket, when made pressure-tight, means the flange faces do not make contact with each other. Any moment loading on the connection is therefore transmitted through the gasket and bolts. This load reduces the sealing ability of the connection, which is undesirable, and the stand-off gap in the subsea environment will promote corrosion and marine growth. For these reasons, BX gaskets are preferred and the flanges supplied with full face contour and flat faces for better resistance to moment loading and corrosion. Experience in the field has shown that after prolonged periods of subsea use, full face contoured flanges are relatively free from corrosion on contact surfaces. Divers,

GENERAL TRIM DESCRIPTION		N-1 REGULAR TRIM	N-2 STAINLESS STEEL TRIM	N-3 FULL STAINLESS STEEL	N-4 SOUR OIL & GAS (H_2S)	N-6 SOUR OIL & GAS (CO_2 & H_2S)
BODY		LOW ALLOY API TYPE 2	LOW ALLOY API TYPE 2	STAINLESS API TYPE 2	LOW ALLOY API TYPE 2 HRC 22 MAX	STAINLESS API TYPE 2 HRC 22 MAX
BONNET		AISI 4140	AISI 4140	AISI 410SS	AISI 4140 HRC 22 MAX	AISI 410SS HRC 22 MAX
STEM		AISI 4140 MOLYDISULFIDE COATING	17-4 PH	17–4 PH	17–4 PH	17–4 PH
GATE NUT		AISI 4140	AISI 410SS	AISI 410SS	AISI 410SS HRC 22 MAX	AISI 410SS HRC 22 MAX
BONNET GASKET		AISI 1018	AISI 304SS	AISI 304SS	AISI 304SS	AISI 304SS
PACKING HEADER		AISI 1018	AISI 302SS	AISI 302SS	AISI 302SS	AISI 302SS
BONNET FASTENERS		ASTM A-193 B7	ASTM A-193 B7	ASTM A-193 B7	ASTM A-193 B7M HRC 22 MAX	ASTM A-193 B7M HRC 22 MAX
	SEAT	AISI 4140	17–4 PH	17–4 PH	17–4 PH	17–4 PH
SEAT ASSY	HUB SEAL	NITRILE	NITRILE	NITRILE	VITON	VITON
	FACE SEAL	CTFE	CTFE	CTFE	CTFE	CTFE
GATE	GATE	AISI 1045	AISI 410SS	AISI 410SS	AISI 410SS HRC 22 MAX	AISI 410SS HRC 22 MAX
	COATING	SURFACE HARDEN	HARDFACED	HARDFACED	HARDFACED	HARDFACED

Fig. 4–9 Material Trim Designation

however, can have some difficulty in holding the BX gasket in position while tightening the flange. Also, plans must be made to assure that no pressure exists in the pipework where divers could be asked to break any flange open subsea.

Manifold Piping

Two of the most commonly used types of pipe are the ASTM A106 grade B and the API 5L grade B. A prerequisite for manifold piping should be the use of seamless pipe only. While ASTM A106 is provided in seamless, only the API 5L can be supplied in electric-resistance welded and submerged-arc welded forms.

A typical list of the main requirements for manifold piping could read as follows:

- Conform to B31.3 code.
- Covered by API specifications.
- Hardness less than HRC 22 per NACE MR-01–75.
- Ductile to resist fatigue and brittle failures.
- Carbon equivalent less than 0.45% so that materials will be easily weldable and minimize fabrication problems.
- Standard material grades to maximize availability.
- Seamless construction.
- Layout should consider the need for pigging (minimum radius ends, barred branch connections).
- Produced sand accumulation and pipe erosion to be considered.

Certain complications will arise when the above requirements are modified to provide for the use of higher strength pipe in the API 5LX strength ranges. Hardness in the weld-heat affected

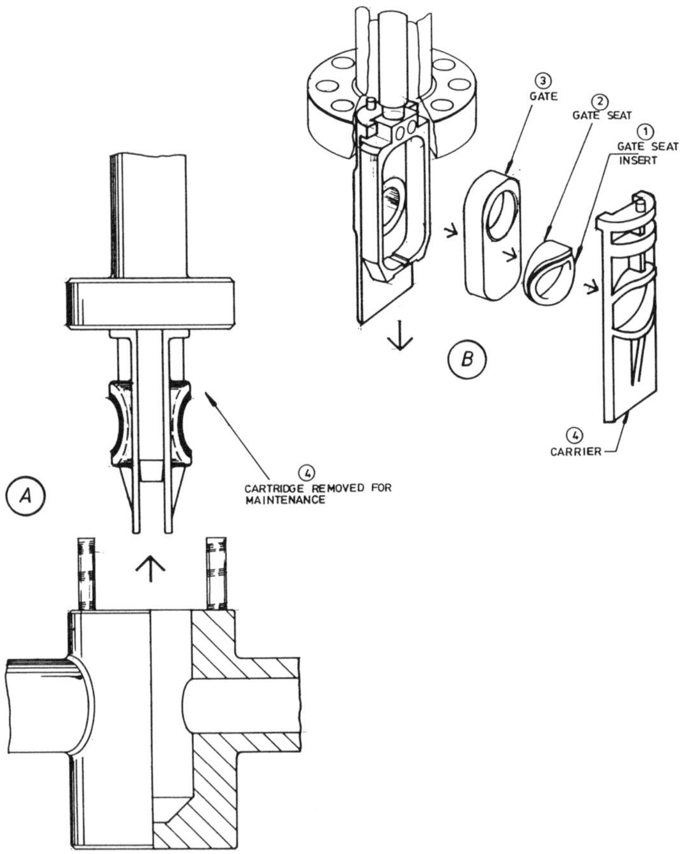

Fig. 4–10 Cartridge Type Gate Valve Assembly (Source: Vetco)

zone will require close monitoring if NACE MR-01–75 requirements are to be met.

Handling Hydrogen Sulfide H_2S

Even though the initial basis of design may show no hydrogen sulphide content, it is often cost-effective to consider the possibility of future contamination. If this is the case, attention to every detail should be made to design in the ASTM A106 grade B and API 5L grade B range. This decision can be particularly important to avoid special welding problems associated with using the API 5LX range.

Pipe Fittings

When using the higher strength grades, pipe fittings must be matched to the strength requirements of the pipe. As the strength range of standard fittings is limited, this can mean that special non-standard fittings and formed elbows have to be accepted.

Manifold Design

Subsea manifolds present additional problems not normally met on surface plants. The first and most important is the need to maintain the geometric tolerances at the make-up interfaces to the template or manifold base. This need further underlines the requirement for face-to-face flange types, with pipework anchor points, at the interface between the manifold and external pipework. In addition, these assist the breaking-out of spools and their repair during service.

A second area requiring some thought is the pipe-support design. Conventional pipework supports can lead to rapid destruction of the pipe due to fretted corrosion caused by the tendency of the pipe to vibrate within a loose-fitting saddle. To overcome this particular problem, seal welds should be used on each of the support saddles, and the pipe runs should be provided with adequate ''U'' or other loops to provide flexability and facilitate movement without exceeding the recommended pipe-stress limits.

Subsea Connections

Tolerances

Subsea connections may be either remote or diver make-up. In either case, problems occur if the limitations of manufacturing tolerances are not fully understood. This lack of understanding can lead to unrealistic tolerance limits which make connections impossible. When it is appreciated that structures, fabrication, and mechanical components all have their own tolerance limitations, the practical connections can be made.

Connection Types

The type of connection selected must be consistent with overall design and operational policies, for example restrictions on diver and surface-vessel utilization. Connections also must provide a reliable seal, compensation for small misalignments, resist handling damage, allow reasonable tolerance limitations for the fabricators, and minimize installation and retrieval costs. The three field-proven methods are:

1. Diver-installed spools
2. Male/female couplings
3. Small-bore hydraulic connectors

When the manifolds are landed, the final alignment tolerances and adjustments can cause some severe problems. To aid the final design choice, a decision matrix should be compiled.

A field of 137m (450 ft) typical water depth is well within the range of saturation diving, so that diver-assisted connection systems can be used. The North Sea offers a competitive enough market in such diving systems for diver-assisted installation, maintenance, and replacement operations to be viable considerations.

Some of the tasks performed by divers would include:

- Engagement and disengagement of the manifold lock down mechanism if required
- Connection of hydraulic jumper hose bundles between the control pods and the manifold junction plates
- Connection of instrument signal lines between the control pods and the manifolds
- Manual opening of header isolation valves

Divers also are required to make or break the manifold piping connections to the template/manifold base.

Diver-Installed Jumper Spools

Typically, a jumper spool can be diver-installed between a flange or hub and an upward facing template flange (or hub), and it may use one of the several manually operated end-connections available on the market today. These include:

- Grayloc or Cameron clamps
- Swivel ring flanges
- Misalignment flange unions
- API or ANSI bolted flanges

Connector Comparison

All of the above listed end-connections are available and can meet the design requirements of pressure rating, design life, and reliability with metal-to-metal type seals. In order to select a preferred type of connection, a ranking matrix could be used (see Fig. 4–11). Weight factors are designated for each criterion on the basis of importance and desired impact to the remainder of the system.

The clamp and misalignment flanges are ranked highest, followed by the swivel-ring flange. Of these three, the misalignment flange is probably the easiest to install since it can accommodate a wide range of positional tolerances. The possibility of using short lengths of flexible hose for jumper spools was omitted from the decision matrix due to the excessive cost of end-connections, shorter expected design life, and the large minimum bend radius.

Male/female Stab Couplings

Male/female stab couplings have been operated successfully subsea, for a number of years, as choke and kill line connections in drilling risers and BOP connections. In such applications, the connections are made and broken frequently, with periodic retrieval for inspection and maintenance. Generally, they are not in use very long. Production manifolds on developments such as BP Buchan field and Petrobras Enchova field have made use of such couplings.

Buchan manifold. The Buchan manifold, with 19 stabs, has been operating since 1985. There were some initial leaks on the stab connection, but these were mainly due to damage during installation. These stab assemblies have a metal radial lip seal backed up by two elastomeric seals. This type of connection requires great care and expense during fabrication to ensure that they are dimensionally correct to within ±0.015 in. or less. It is imperative to use a high-precision jig to fulfil this requirement. Flowline pressure and thermal expansion create pin and box separation forces.

Rigid structural members are needed to prevent distortion initiating seal failure and leaks. This method is better suited for use with only a few connections per system, which must be broken periodically for inspection in diver-accessible water depths.

Enchova field. The Enchova field manifold is designed for 13 stab-type connections and is installed off the east coast of Brazil in about 207 m (680 ft) of water. Diver access to the manifold is possible but costly. The main justification for stabs in this system is to minimize subsea diver work. Therefore, a design

Type Connection	WF-3 Strength Ruggedness	WF-3 Sealing Reliability	WF-3 Installation Time/cost	WF-2 Misalignment Compensation	WF-1 Hardware Cost	Total Score
Bolted flanges	3	3	1	1	3	26
Clamps	3	3	2	2	3	31
Swivel ring flanges	3	3	2	2	2	30
Misalignment flanges	3	2	3	3	1	31

Weight Factors (WF)

3—most important
2—next most important
1—least important

Ranking Criteria

3—best
2—in-between
1—worse

Fig. 4–11 Decision Matrix for Choosing Diver Connection Alternatives Against Function Criteria that ensures remote disconnect and quick retrieval for surface repairs was preferred.

Hydraulic Connectors

Hydraulic connectors for individual flowlines or small groups of flowlines combined in one larger connector have already been successfully used on subsea production systems. Satellite flowline pull-in systems invariably use some type of hydraulically actuated connectors for remote diverless operation. Horizontal hydraulic flowline connectors have been made for deepwater applications also.

Hydraulic connectors as shown in Figure 4–12 are very similar in application to the male/female stab. Seal reliability is improved by using an AX-type metal gasket in some designs. Because the connector is locked to the flowline mandrel, it resists separation forces better and reduces any side loads on the seal area.

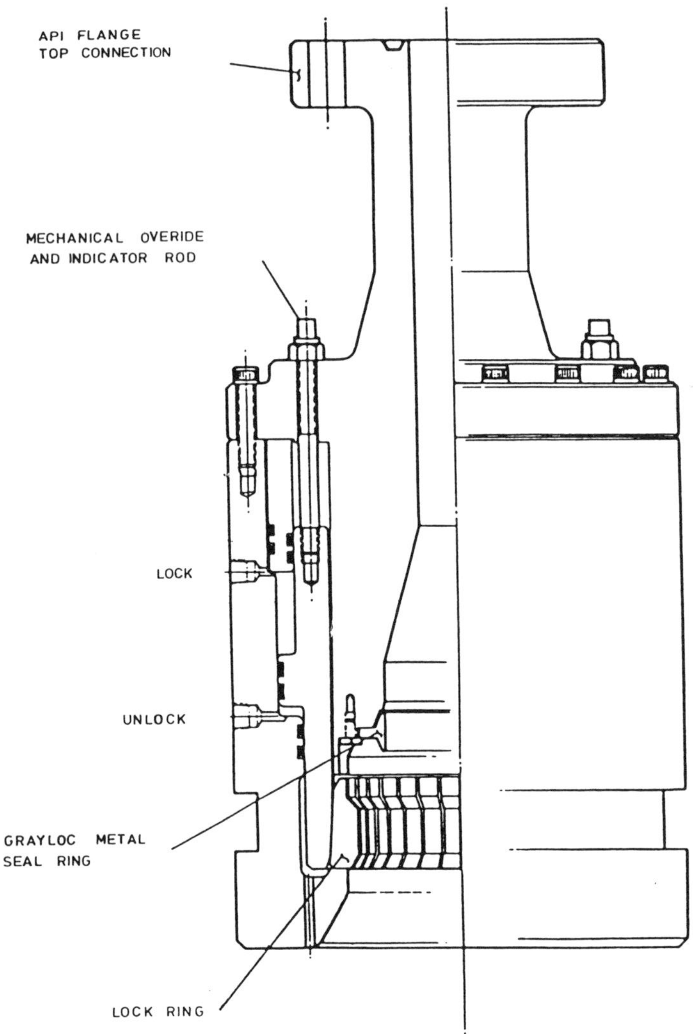

Fig. 4–12 Hydraulic Connector (Source: Vetco)

Hydraulic connectors have the capability to pull the mandrel slightly into alignment and preload the seal connection as it locks. This equipment is expensive and is best suited for remote diverless applications or where periodic or frequent retrieval is necessary.

Connection Method Comparison

The connection methods described in the previous paragraphs are evaluated in Figure 4–13. The following factors are considered:

- Sealing reliability
- Ruggedness or resistance to damage
- Ease of recovery
- Installation cost
- Template piping interface fabrication difficulty
- Resistance to operating loads
- Initial equipment costs.

Each criterion is assigned a factor according to its degree of estimated importance. Then each connection method is ranked against the other methods on a scale from one to three within each criterion.

By using this decision technique, the following three methods (in descending order of preference) can be chosen for diver assisted operations:

1. Jumper spools
2. Hydraulic connectors
3. Male/female stabs

The Advantages of Jumper Spools

The jumper-spool alternative is rated top in all categories except for installation cost and ease of recovery. This alternative is strictly a diver make-and-break connection but is inherently safe and simple and the best choice if frequent removal is not anticipated. It offers the simplest fabrication and lowest initial cost for hardware and fabrication.

Hydraulic connectors score only slightly less on the rating table. However, the level of fabrication difficulty, with close tolerances and the initial cost, are actually much greater than with spools. This method and the clearly undesirable stab methods also have limited flexibility for changing template piping requirements.

The favored connection based on the above parameters, which may change depending on the policies adopted, indicates that diver-installed jumper spools utilizing misalignment flanges give an adequate solution.

Jumper Spool Design

The problem with jumper-spool designs has been their comparative inflexibility. Figure 4–14 shows a widely used form of a

Connection Method	WF-3 Sealing Reliability	WF-3 Ruggedness	WF-1 Ease of Recovery	WF-3 Installation Cost	WF-3 Interface Fabrication Cost/Difficulty	WF-2 Resistance to Operating Loads	WF-1 Initial Equipment Cost	Total Score
Jumper spools	3	3	1	1	3	3	3	40
Hyd. connectors	2	3	2	3	2	2	1	37
Male/Female stabs	1	2	2	3	1	1	2	27

Criteria weight factors (WF)	Ranking Criteria
3—most important	3—best
2—moderately important	2—acceptable
1—least important	1—worse

Fig. 4–13 Decision Matrix for Choosing Connection Alternatives Weighed Against Function Criteria

90° bend spool in which only rotational and parallelism errors are catered for. Because the diagonal distance between the centroid of the balls cannot be altered, the desired elevation and tolerance is difficult to achieve during installation.

A solution is to use the configuration given in Figure 4–15. This arrangement uses a slip-type joint to allow the length to be adjusted, but this introduces another leak path.

Should the requirement arise for multiple stabs and hydraulic connectors, it is of prime importance that the male and female components are mounted on jig-bored plates to provide the necessary close tolerance fit-up.

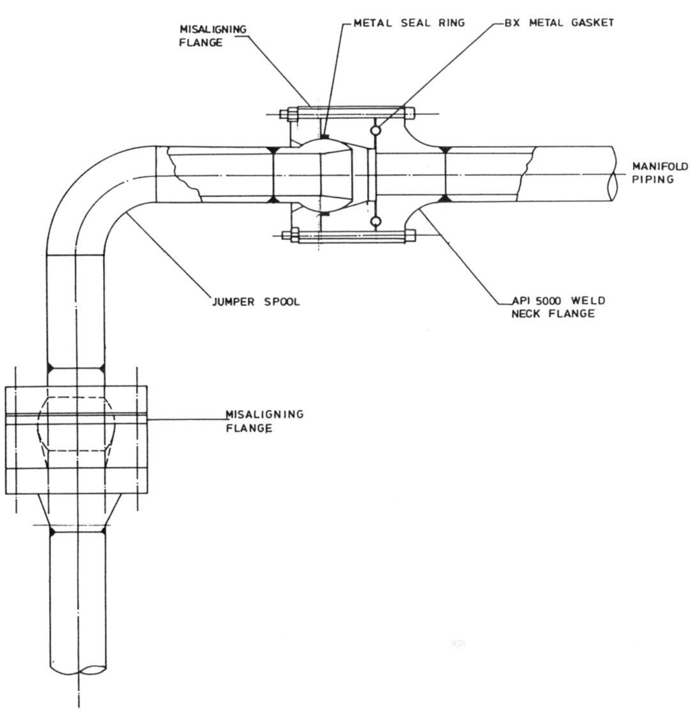

Fig. 4–14 A Typical 90° Bend, Diver-Installed, Jumper Spool

Leak Paths

Experience has shown that small vibrations and movements in subsea piping can result in the loosening of fasteners, as used in various connections. In many cases, this is a result of using torque wrenches for torquing up the flange, resulting in excessive bolt loadings and uneven tightening of the studs. The need however is for a continual program of bolt-torque checks throughout the life of a system.

A solution to this problem is to use stud tensioners during construction and service. To make use of stud tensioners, it is necessary to provide adequate clearance between flanges and adjacent fittings. In addition, care must be taken to ensure that the surfaces beneath the nuts are free from paint, etc., and there are other safeguards.

Reducing the number of seals in a leak path. Leak paths must be minimized. The aim should be to obtain an optimum balance between the size and number of modules and the number of seals to be made up subsea during installation or maintenance/repair operations. Also, seal surfaces should not be left unprotected for extended periods. Generally, sealing arrangements should have metallic primary seals provided on all fluid connections. Where appropriate, a low-pressure sealing capability against the outside environment is required also.

Leakage monitoring. At an early stage, consideration must be given to the provision of an oil- or gas-leakage monitoring system coupled to an automatic alarm or alarms in the surface-production control room.

Unitization

A Need for Simplicity

Diver intervention is a necessary part of the philosophy for the development of medium-depth marginal fields. Also, it is

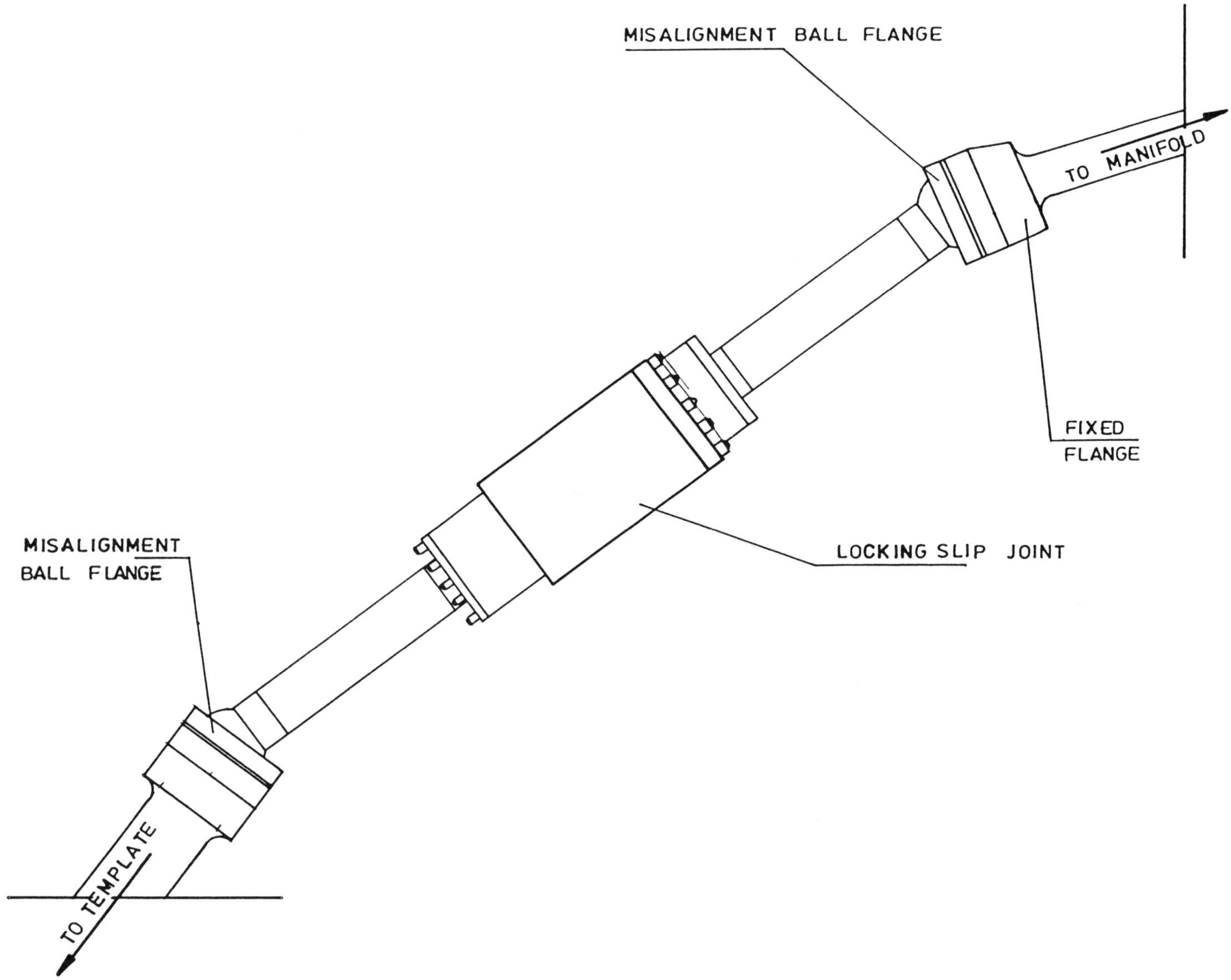

Fig. 4–15 Adjustable Jumper Spool

necessary that any tasks be inherently simple. In addition, consideration must be given to the system unitization in an attempt to reduce the impact on production shutdown during anticipated or unscheduled underwater maintenance and repair.

The modular approach must therefore be applied, with the emphasis placed on simplicity of construction and ease of handling.

Module Sizing

Module size should account for an ease and speed of installation and maintenance, with minimum sensitivity to weather conditions. Size also relates to the facilities available for the running and retrieval of each module, from a diving support vessel or a floating production platform.

Module Failure Considerations

There is a need for the system to be designed so that components with a relatively high probability of failure do not, in any failure mode, cause shutdown of more than one well or area at a time. That is, breakdown of one module should not shut in the total production of the field. In this respect, production considerations will often overide certain optimum size decisions. Additionally, module size should respect the requirements for minimizing diver intervention. This should be considered where a significant reduction in complexity or cost is evident.

Equipment configuration design should allow for safe access, either by diver or submersibles. An ability to troubleshoot an equipment malfunction from the production platform in preparation for maintenance is, however, preferable.

Typical Unitization

Where manifolds are configured by task separation only, the result is usually a simple guide post and frame construction, with modular inter-connected diver-installed jumping spools. The more complex a system becomes, the greater is the need for smaller modules. Consider the hypothetical scheme shown in Figure 4–16. This system provides for round-trip pigging, TFL lines, and data-linked connections. This system is serviced by two flowline bundles. Each of these contains four TFL/gas lift lines with four TFL diverters. All active manifold components for two wells are being arranged in three valve modules as described below.

1. *Valve module no. 1:* Connects directly to the tree and piping module. It comprises the production, TFL, and annulus isolation valves, as well as crossover and chemical injection valves. All of these components can be packaged in a module suitable for moonpool retrieval. One such module is required for each well.
2. *Valve module no. 2:* Connects to the piping module and comprises the pigging loop and its isolation valve. It also can be packaged in a relatively simple moonpool retrievable module. One such module is required for each flowline bundle.
3. *Valve module no. 3:* Connects directly to the flowline bundle and piping module and comprises two TFL diverters. TFL diverters do not lend themselves to efficient packaging. Therefore, they result in a relatively large, yet still moonpool-retrievable, module. One such module would be required for each flowline bundle.

Although the schematic in Figure 4–16 illustrates an achievable layout, there are certain points both for and against this particular method of unitization.

Advantages of proposed unitization. There are two advantages:

1. All active components are contained in moonpool-retrievable modules.
2. The introduction of two flowline bundles permits the piping modules to be split into two smaller independent modules, facilitating easier handling during installation and retrieval.

Disadvantages of proposed unitization. The main disadvantages are:

1. Division of the piping module into two limits the number of flowline bundle contingency crossover possibilities.
2. Three different types of valve module results in:
   a. A large number of remote subsea connections
   b. Low space efficiency
   c. Large amount of vertical access required
   d. Large number of guide posts required for installation and retrieval
   e. Large number of running and connection tools
   f. Generally more complex maintenance operation procedures

To avoid the possibility of damaging the flowline bundle, the TFL-diverter module should not be connected directly to the flowline bundle. This suggests that a spool piece or other provision be added to this concept, thereby increasing the size and complexity.

Variations to a Proposed Unitized System

The above scheme can be refined by placing the pigging loops and TFL diverters within a module adjacent to each tree. This approach, shown in Figure 4–17, allows all the manifold-active components to be contained in a single-valve module of approximately the same envelope dimensions as a Christmas tree. Thereby many of the disadvantages associated with having several different types of module in different locations on a template are minimized. The disadvantages of a single-valve module arise from the marginal increase in modular size over the first scheme and the introduction of two 6-in. valves required to place the pigging loop within each module.

Manifolding Systems Without Pigging

Many manifolding systems do not incorporate the requirements for either pigging or TFL systems. With such cases, the manifold configuration allows for the use of less complex module units. In many instances, the modules can be reduced to a simple frame containing one pipe-run with associated chokes or valves and a control system connection. Generally, the running/retrieval of modules, or larger units, should be achieved by the use of a drill string with suitable standard connectors, or other types of running tools.

PRODUCTION CONSIDERATIONS

Multiple Pay Zones

Separation of Reservoirs

Many developments are undertaken where the development block is divided between various operators. Each block may contain several accumulations or potential accumulations of oil or gas.

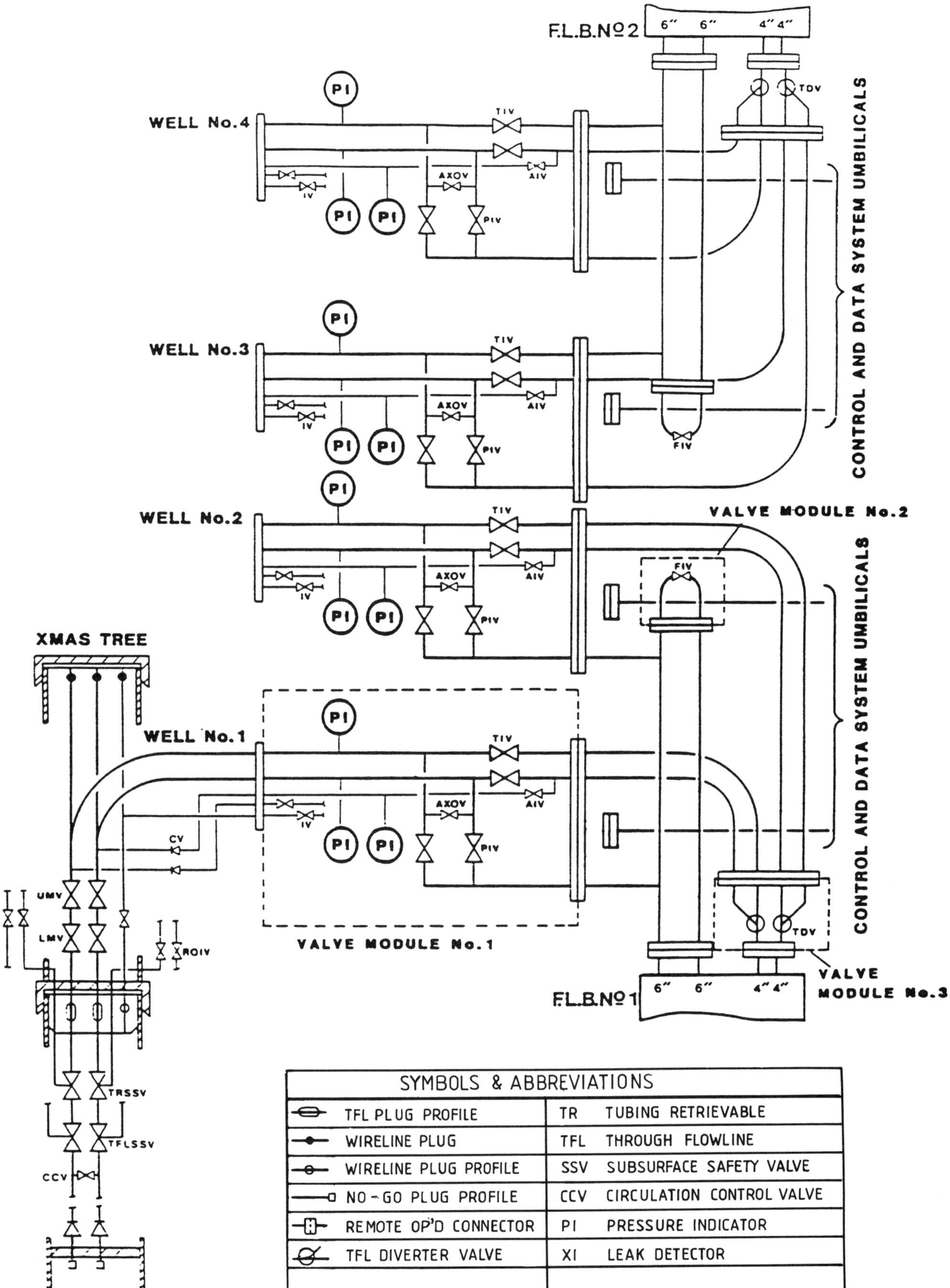

SYMBOLS & ABBREVIATIONS	
TFL PLUG PROFILE	TR TUBING RETRIEVABLE
WIRELINE PLUG	TFL THROUGH FLOWLINE
WIRELINE PLUG PROFILE	SSV SUBSURFACE SAFETY VALVE
NO-GO PLUG PROFILE	CCV CIRCULATION CONTROL VALVE
REMOTE OP'D CONNECTOR	PI PRESSURE INDICATOR
TFL DIVERTER VALVE	XI LEAK DETECTOR

Fig. 4–16 Typical Unitization Scheme, Pipework Layout Involving Three Modules Per Well.

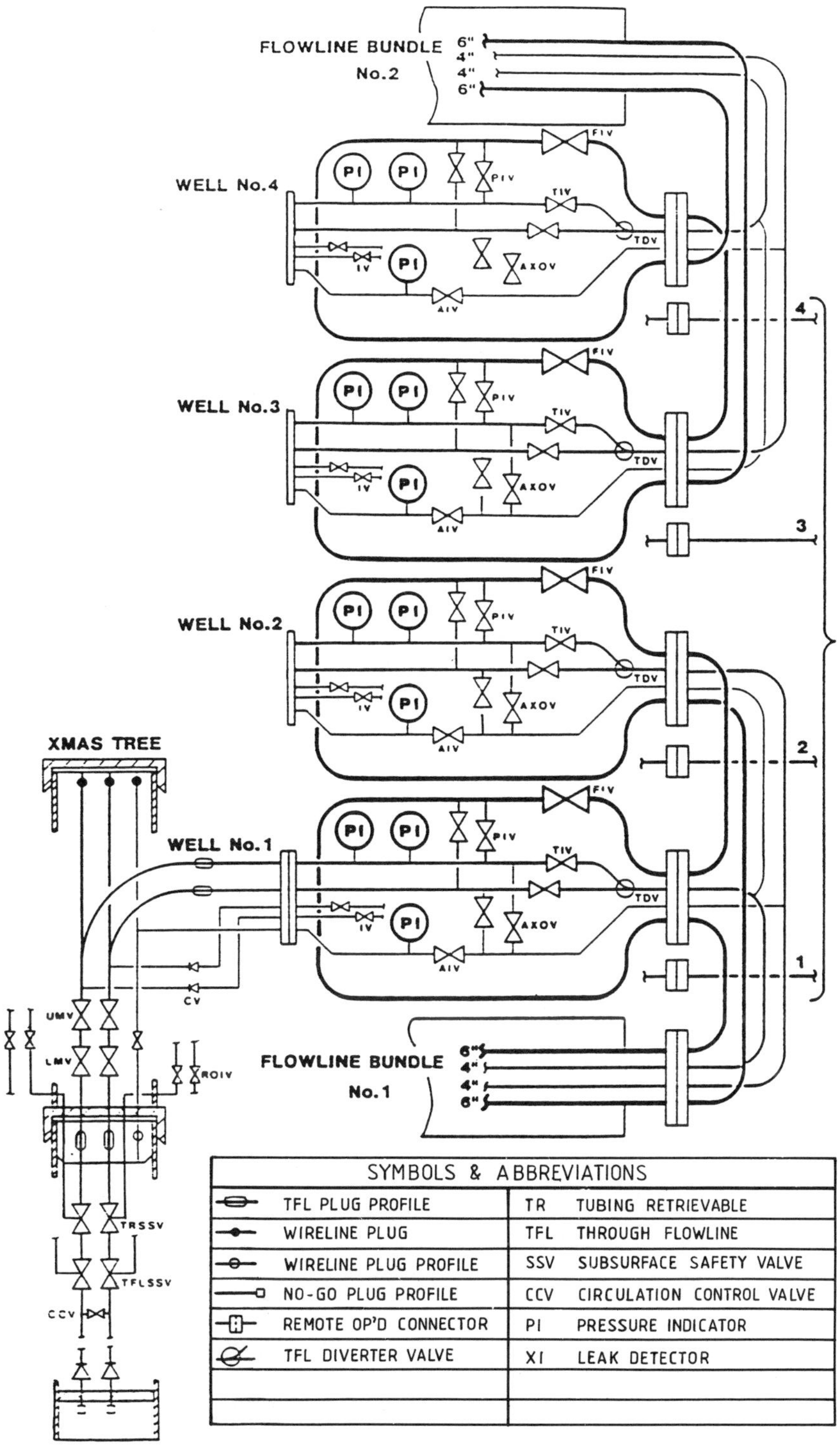

SYMBOLS & ABBREVIATIONS	
TFL PLUG PROFILE	TR TUBING RETRIEVABLE
WIRELINE PLUG	TFL THROUGH FLOWLINE
WIRELINE PLUG PROFILE	SSV SUBSURFACE SAFETY VALVE
NO-GO PLUG PROFILE	CCV CIRCULATION CONTROL VALVE
REMOTE OP'D CONNECTOR	PI PRESSURE INDICATOR
TFL DIVERTER VALVE	XI LEAK DETECTOR

Fig. 4–17 Typical Unitization Scheme, Pipework Layout Involving One Module Per Well

These reservoirs are usually made up of more than one pay zone. When the complete system is in operation all production, regardless from which reservoir it originated, would be directed to a single manifold except when the reservoir characteristics are sufficiently unknown to delay the initial design. Production from separate reservoirs and zones would be kept separate from each other until they are commingled on the surface, but all producing wells within each reservoir and zone will have their flows commingled subsea before entering the risers.

To summarize, oil from different reservoirs or geological

strata should not be mixed subsea, and each pay zone should be produced into separate header systems having separate flowlines and risers. Likewise, production from separate operators' outlying fields may have their own individual pipelines or risers. The total flow from any one of these locations can be routed to the surface test separators to obtain data for royalty-payment calculations as well as field management.

Fiscal Arrangements

An analysis of the fiscal arrangements at the study stage can indicate to what extent separation of pay zones and other reservoirs is required. Rationalization can produce simplified designs provided that agreement can be reached between the various partners and the recognized government departments. The final fiscal decisions reached will indicate the amount of operational flexibility that is to be designed into the manifolding and/or template arrangements.

Each pay zone also may provide such diverse pressure levels that, even though fiscal problems are resolved, engineering design dictates that the producing systems are mechanically separated.

Well Recompletion

A field's production plan additionally can require that a well initially completed into one zone is recompleted into a second zone at a later date. This arrangement can be accommodated by providing dual manifold headers with suitable crossover facilities.

The schematic (see Fig. 4–18) shows only one approach to crossing over pay zones. The chosen layout depends on the specific development under consideration, fiscal parameters, and the production fluid characteristics. A simple means of crossing over from one pay zone to another is shown in Figure 4–18.

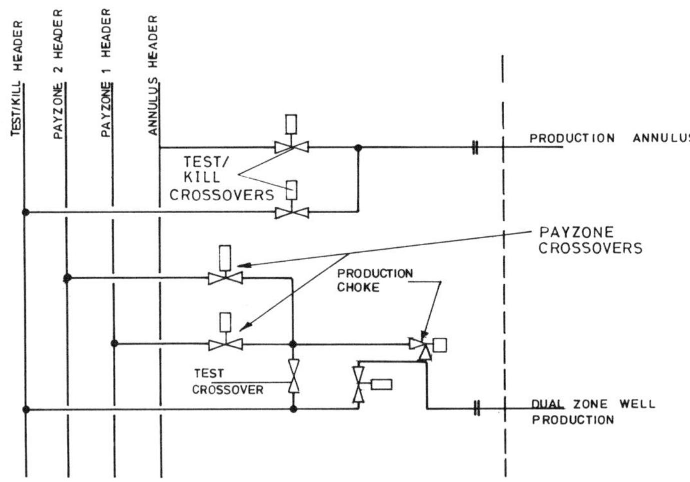

Fig. 4–18 Satellite Production Payzone Header Crossover Arrangement

Production Enhancement

Production enhancement can take many forms, from steam injection to downhole pumps. The two most common methods are gas lifting and water injection and these only will be considered here.

Gas Lifting

By adding gas lift, it is possible to increase the rate of reservoir depletion and also marginally increase the recoverable reserves. These two benefits are obtained at the expense of:

- Greater capital cost
- Greater operating cost
- Increased topsides weight
- Increased topsides space requirement.

Any decision related to gas lifting depends on a full economic analysis of its effects. Data should be provided to demonstrate the greatest production rates possible, together with an overall system-design scheme and an estimate of increased costs when gas lift is added.

Initial reservoir studies indicate the amount of gas that needs to be injected for gas lift purposes. The upper limit is determined by technical or economic breakpoints occurring within a specified range.

It is not always possible to make firm recommendations in the absence of an economic appraisal. However, it should be noted that to add gas lift at a later stage in a field's life can be very expensive. Thus it is prudent to allow for the possibility of gas lift in the initial design of subsea manifolds and equipment. This involves additional initial expenditure, modifications to the well template manifold piping and valves, extra template piping, and careful design of the tree for choke switching. Although this expenditure would be wasted if gas lift were not required, it would at least be available if, for reasons initially unknown, future economic operation required its use.

Operational aspects of gas lift. The operation of several gas-injection wells from a recirculating compressor system can present some problems. The simplest solution is to operate each well by using the gas injection choke opening—thereby varying the gas to gas lift ratio and, hence, the well production rate. The production choke is either removed or completely left open. Switching to gas lift is achieved best by configuring the manifold pipework to allow minimum downtime at changeover.

Figure 4–19 illustrates a typical gas lift crossover arrangement in which overall control remains rather unstable because movement of the choke on one well affects every other well in the system. If choke movements are too abrupt, it is possible to disturb the system to an unacceptable level because of an interaction between various components. The principal casualty in this process would be the centrifugal gas compressor and its control and surge protection equipment.

These difficulties can be minimized by correct operation. Oper-

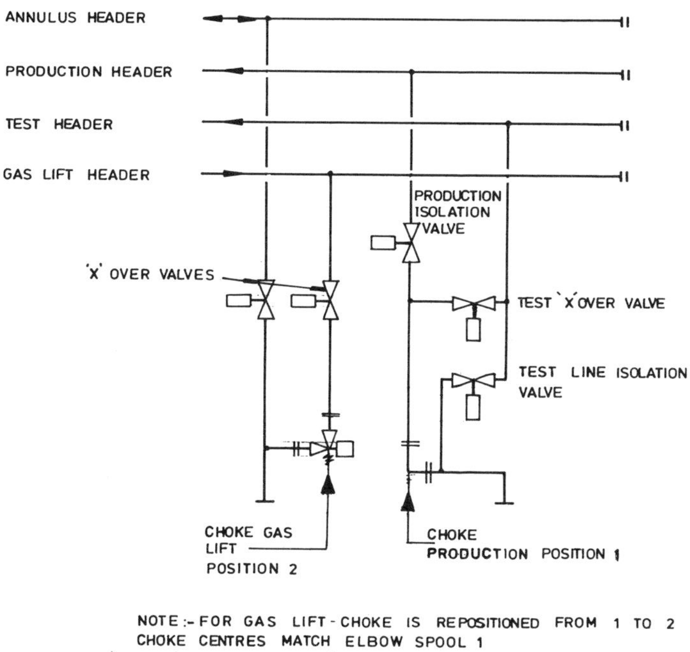

Fig. 4–19 Typical Gas Lift Crossover Arrangement

ational stability can be further improved by the availability of an excess of associated gas which is normally vented from the first-stage oil and gas separator.

Increasing Production with Gas Lift. To increase the production rate from a particular well, the gas lift ratio can be increased by the following procedure:

1. Increase the gas discharge rate from the centrifugal compressor. This causes a sudden drop in pressure that is rapidly corrected by a reduction in vent rate from the oil and gas separator.
2. After the system reaches equilibrium, the overall production rate increases slightly because all wells will have a slightly greater gas-lift ratio. The gas ring main pressure will be increased slightly.
3. Gradually open the gas lift choke on the selected well, allowing the system to stabilize after each adjustment.
4. The overall system will stabilize to a situation in which the extra gas is directed to the desired well, with the ring main pressure at its original level and the compressor is operating at the increased flow rate. Additionally, the suction pressure is at its original level with all excess gas vented under pressure control from the oil and gas separator.

Decreasing Production with Gas Lift. If it is desired to reduce the gas-lift ratio to one particular well, then this is done by:

1. Slightly decreasing the choke setting.
2. Slightly decreasing the compressor flow rate.
3. Repeating the procedure until the desired flow rate is obtained with ring main and compressor suction pressures at original valves.

By adopting the above method, transients can be minimized.

The above approach will require additions to the manifold pipework valves and control system. These can be summarized as:

1. Addition of a gas-lift header
2. Changeover of chokes from production to gas lift
3. Piping configuration changes to accommodate easy valve and choke changeout
4. Choke position indicators
5. Differential pressure tappings across chokes
6. Pressuring sensing arrangements in the header and branch system
7. Crossover valves arrangements between annulus and gas lift branches

Water Injection

As with manifolds, the number of view connections should be minimized when considering the use of water injection. This is achieved by controlling subsea flow by using a header, branch, and choke layout.

Data on the total field injected-water requirements are usually available at an early stage of the development. This, together with the field injection well count, shows whether a separate water-injection manifold system is needed or not.

Water flow measurement. The total flow of injected water can be measured at the topside facility, and from time to time individual wells may be calibrated by using the total-flow meter. As more wells are brought on stream, it will be found necessary to monitor individual well flow at the outlet branches of the manifold. This may be achieved by the use of ultrasonic flow meters. At present, the cost of this type of meter is very high, and it is not yet considered fully field proven. To circumvent this, the use of differential pressure tappings across the subsea water injection choke can be considered. This requires that preliminary tests are carried out to provide test data suitable for the production of a series of flow curves. The most commonly used curve being: flow rate, q against P, where P is the pressure differential across the choke.

Typically,

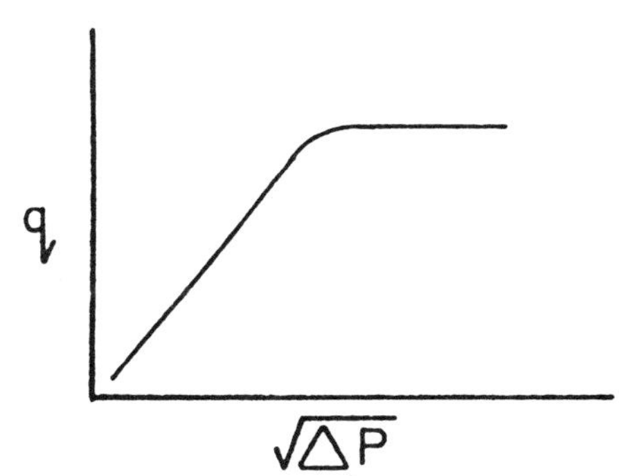

Fig. 4–20 Typical Choke Characteristic Curve

In addition, this information is invaluable in the production of a field-water injection model.

A curve of this form allows a quick determination of the coefficient of velocity (CV) relation between q and P. These curves, along with the relevant test data, may be used to establish the degree of accuracy expected from the differential pressure (DP) cell readout, to provide a coefficient for the subsea control system computer, and to supply accurate input to any field water-injection system model.

Reservoir management. Facilities to provide flow measurement, from whatever source, are essential. These are used both to obtain information during routine operations and to allocate and measure injection flow to individual wells for reservoir management.

Water conditioning. Water conditioning, to alleviate problems associated with corrosion and bacterial growth, is carried out at the surface facility and not via a separate line within the water injection manifold.

Water velocity and pressure. Owing to the higher velocities and pressures associated with injection, particular attention must be paid to calculation of the system pressure drops. Therefore, it is important that chokes are kept fully open for maximum flow and only throttled if flow exceeds the design rate. There are two main reasons why the flow may exceed the design rate: 1) the injectivity ratio may be less than assumed at the design and 2) because the flow resistance developed in the reservoir differs from that expected. These are usually clues in correct prediction of the reservoir characteristics.

Manual chokes. Manual chokes are adequate for this application because it is assumed that changes to their setting will be infrequent. Manual chokes should be diver retrievable, using a form of handling frame. An alternative remotely operated choke is recommended since the increased cost cannot be justified for a water injection system.

Manifold header. Preferably, the manifold header should be symmetrical and equipped with branches. This is accomplished by using block valves and manual diver-adjusted chokes downstream to divide the water flow proportionally between the individual wells. Since clean, filtered, and deaerated water passes through the chokes, wear to the choke orifice is minimal, and periodic adjustment probably will not be necessary. Once the system is started and operating smoothly, flow will have to be adjusted only occasionally and not on an urgent basis.

Service Facilities

The subsea manifold and template design must provide facilities to service riser flowlines, provide annulus monitoring, perform flow tests, calibrate, chokes kill wells, and export the production.

The template riser or bulk flowline manifold is the main subsea flow control unit. It gathers oil from the reservoir via other manifolds or template and routes it to the appropriate flowline/riser. Likewise, it functions as crossover for the service lines (by providing a service header with crossover paths).

The annulus-monitor line is routed into a main template header and provides for monitoring of individual wells. It should be recognized that only one well at a time may be monitored. To provide the facility to monitor a number of wells simultaneously would be very expensive and extremely difficult to achieve, without introducing unacceptable complications with the design.

Well tests can be approached in a similar manner with the line sharing its duty for well-kill. The provision of this service again calls for the use of a header system within the template or manifold main structure.

The test line also can be used for the purpose of choke calibration, in conjunction with the topside test circuit.

Further facilities may be included for chemical injection by additions to the pipework layout. It must be remembered that throughout the manifold's design phase, change occurs as additional service functions are identified.

TYPICAL MANIFOLD DESIGN

In the previous sections, the approach has been to develop a scenario for a subsea system with particular emphasis on a field development at medium depth 120–150 m (400–500 ft) for a subsea manifold/template combination having satellite and template connections. To warrant such a system, the size of the field precludes the use of a single retrievable manifold system, and three main task separations will be required.

1. Template/riser production and control
2. Satellite production/water injection
3. Outlying reservoirs

Template Well Manifold

The template well riser manifold (see Fig. 4–21) is the main subsea flow control unit. Figure 4–22 shows the design of this manifold schematically. It gathers oil from reservoir A template and satellite wells in addition to oil from outlying reservoirs. Oil passes through the manifold pipework and control valves and then is routed to the appropriate riser line. Also, injection water passing through the manifold is distributed to the satellite well, manifold, and the satellite reservoir manifold. The manifold illustrated is connected to nine template flowlines and also to the lower production riser assembly, in this instance using a rigid riser system. Only minor modifications would be required for conversion to a flexible riser layout if connections were made at the top.

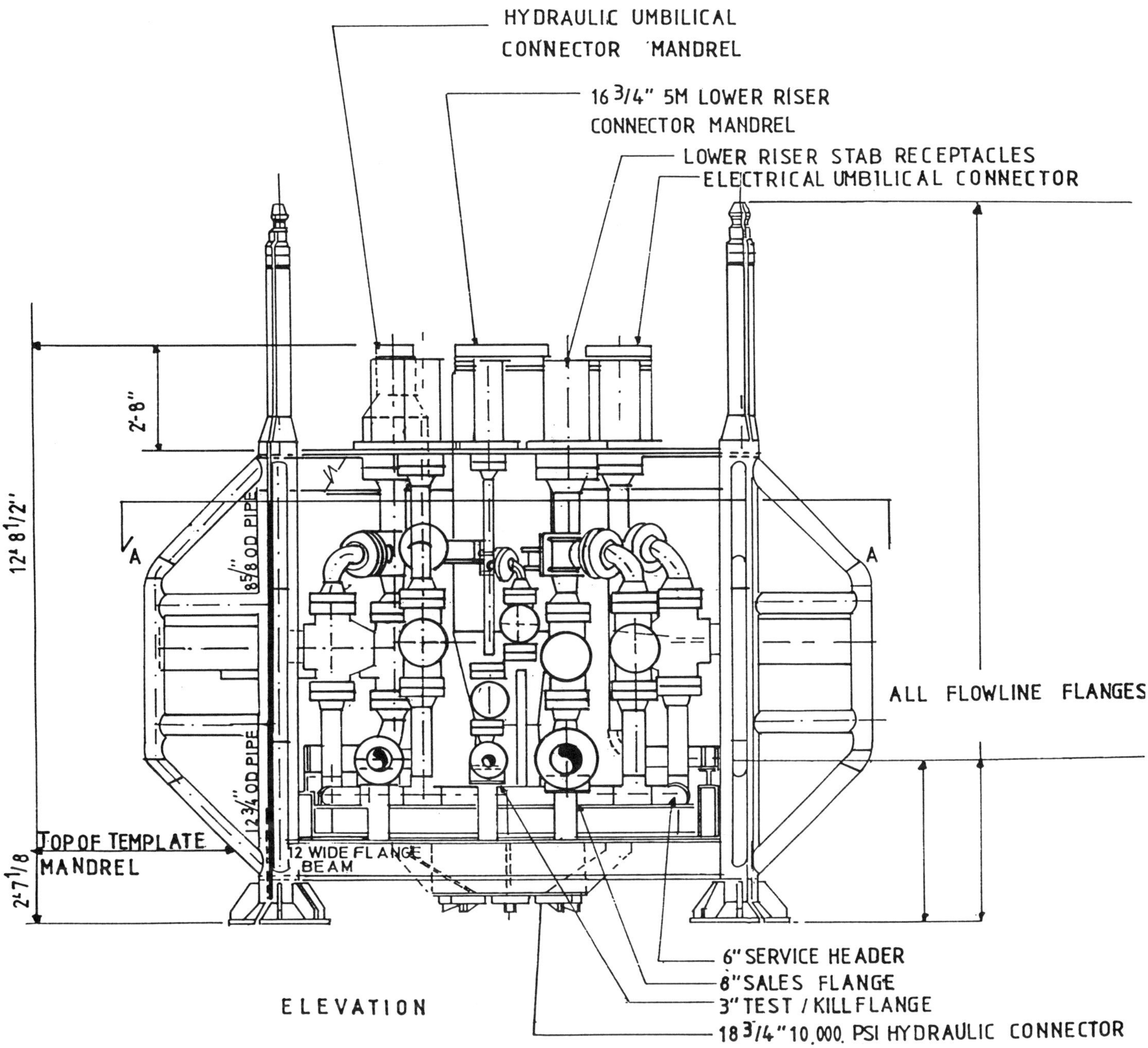

Fig. 4–21 Template Well/Riser Manifold—General Arrangement

A major requirement for this manifold is that oil produced from each reservoir is not to be commingled subsea. Likewise, the two pay zones producing oil from one reservoir will not be mixed subsea. This does not mean that each well has a separate flowline to the manifold and a dedicated riser to the surface. All reservoir A pay zone 2 wells are choked subsea and commingled into a separate template piping header. This pay zone 2 oil header is connected to the template-well manifold, and the oil is routed through it to a production riser. The pay zone 1 oil produced from the remote reservoirs B or C will be commingled similarly at their respective template and sent through a pipeline to the satellite reservoir manifold and on to the main template well manifold. Oil from each reservoir will be piped to a designated riser and will not be mixed subsea.

A 6-in. service header with a series of crossover valves is included in the design to allow each of the riser flowlines to be flushed clean before disconnection and retrieval of the riser. This service header also may be used to divert reservoir A

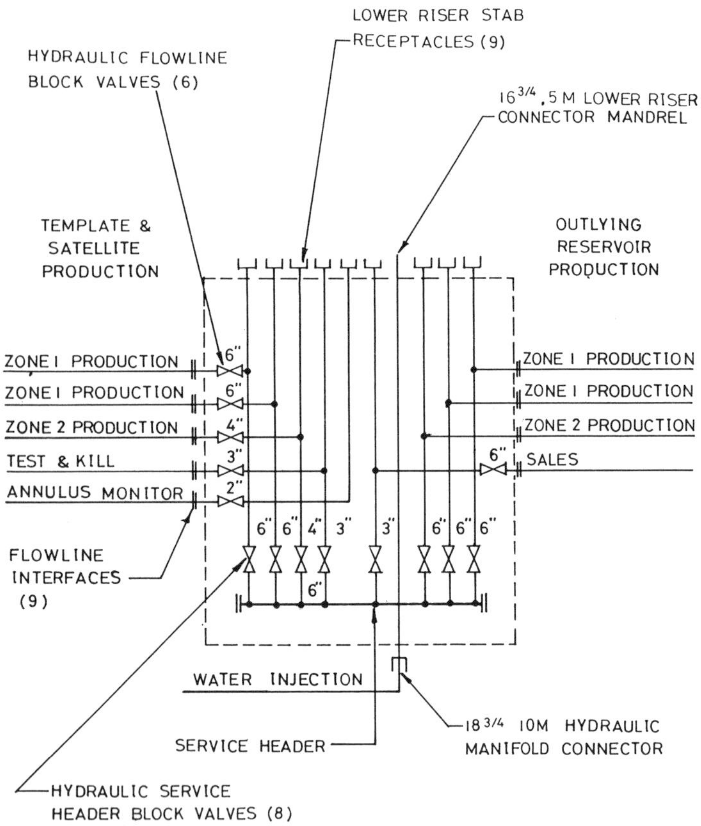

Fig. 4–22 Template Well Manifold, Flow Schematic

pay zone 1 production into any of the three risers normally used for reservoir B or C production, should these lines not be in use.

The template-well manifold must transmit the riser loads into the template and provide the capability for the lower riser assembly to be disconnected quickly in an emergency. The load bearing connection between the lower riser assembly manifold is a standard 5,000 psi hydraulic connector and mandrel, with the mandrel located on the manifold framework. This load bearing connector also provides a flow path for the injection water. A 20-in. outer diameter pipe extends through the manifold and is connected to a standard 10,000 psi hydraulic connector. This connector is the load-transmitting connection between the manifold and the template. The template water injection header is located under the manifold and is provided with a mandrel to interface with the manifold connector and has structural framing to transmit the loads into the template.

The nine riser line interfaces to the manifold are non-load bearing hydraulic retractable male/female stabs. The stab receptacles will be located on the manifold frame in a radial pattern around the connector mandrel.

In an emergency, the wells can be shut in at the trees, manifold block valves closed, riser flowlines flushed (if time permits), flowline stabs retracted, and the lower riser connector released in just a matter of a few minutes.

The sales pipeline also is connected to the template well manifold to permit the flow of processed oil from the surface into the export pipeline. A hydraulic gate valve is used to close off the pipeline should the riser be disconnected or a break in the pipeline occur. This template manifold would have an overall dimension of approximately 6 m × 6 m × 6 m (20 ft × 20 ft × 20 ft) and weigh approximately 80 tons.

Satellite Well Manifold

The satellite well manifold (see Fig. 4–23) controls the reservoir production satellite wells. The main purpose for this manifold is to provide a space for chokes, isolation valves, and for several wells to be connected to the template piping at one point to save space. This manifold must interface with the flowlines out to the satellite wells, as well as the template piping headers. These satellite wells include four water injectors, three producers, and slots for produced fluids from two pay zones. Figure 4–24 shows the functional logic of the manifold.

Both pay zone 1 and 2 oil lines are connected to the appropriate template header and have identical valve and piping logic. The normal flow path for the produced well fluids is through the manifold inlet flange, hydraulic choke, hydraulic production valve, and piping to the template header. A valve and piping is provided to permit by-passing of the choke into either the production line or test header. Therefore, either calibrating the choke position or conducting full production is provided for. The entire manifold would weigh approximately 40 tons with overall dimensions of approximately 3.3 m × 3.3 m × 8.2 m (11 ft × 11 ft × 27 ft). This size and weight permits the unit to be installed through a 12 ft square workover moonpool, using a workover derrick and riser. A small, light duty, debris cover then can be run down over the manifold after it has been landed and connected.

Satellite-Reservoir Manifold

The satellite-reservoir manifold (see Fig. 4–25) controls the flow of fluids to and from the outlying reservoir wells. Reservoirs B, C, and D each have a pipeline bundle connected to the satellite-reservoir manifold on the template, which is located over reservoir A. Each pipeline bundle contains a 6-in. pay zone 1 oil line, a 6-in. water injection line, a 3-in. production test or well-killing line and a 2-in. annulus pressure monitoring line. The manifold design is shown schematically in Figure 4–24 together with the headers and isolation valves necessary to produce and perform other routine field operations.

The three oil production lines, from the satellite reservoir manifold, are connected directly to the template well manifold via short template piping sections and are routed into dedicated risers to the surface. The test header, annulus monitor header, and the water injection header are each connected to their respective template piping header.

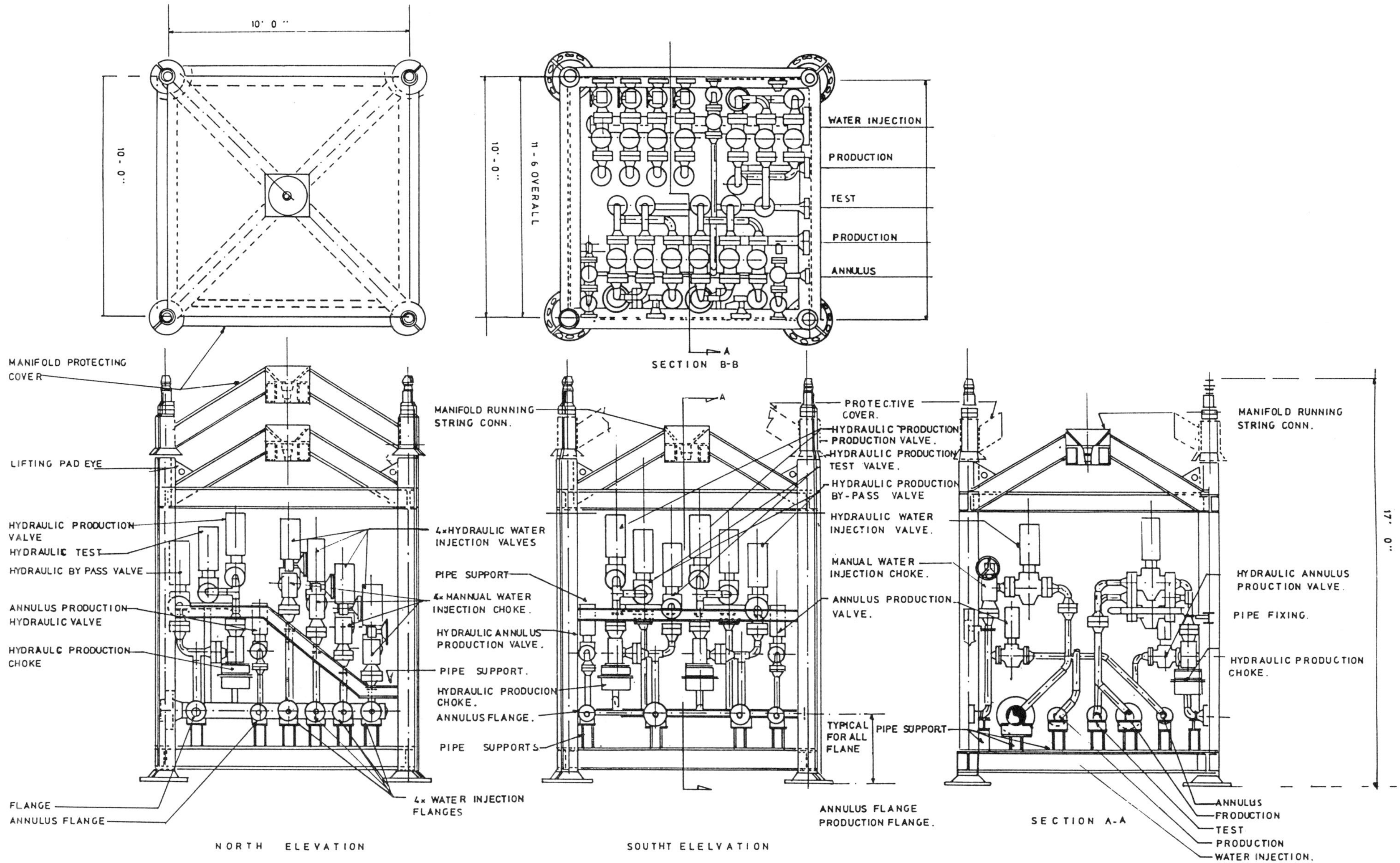

Fig. 4–23 Satellite Well Manifold Production/Water Injection

The physical location of the manifold should be adjacent to the template well/riser manifold, its position chosen to interface with the three pipelines to the template and with the main template-well manifold. The manifold would be 3.6 m × 5 m × 4.5 m (12 ft × 16 ft 6 in. × 15 ft) in overall size and approximately 37.5 tons weight and have a similar protective frame and debris cap to the manifold. Construction and installation/handling designs are identical.

If, at some later date, it does not become economic to produce these outlying structures, its presence would have very little

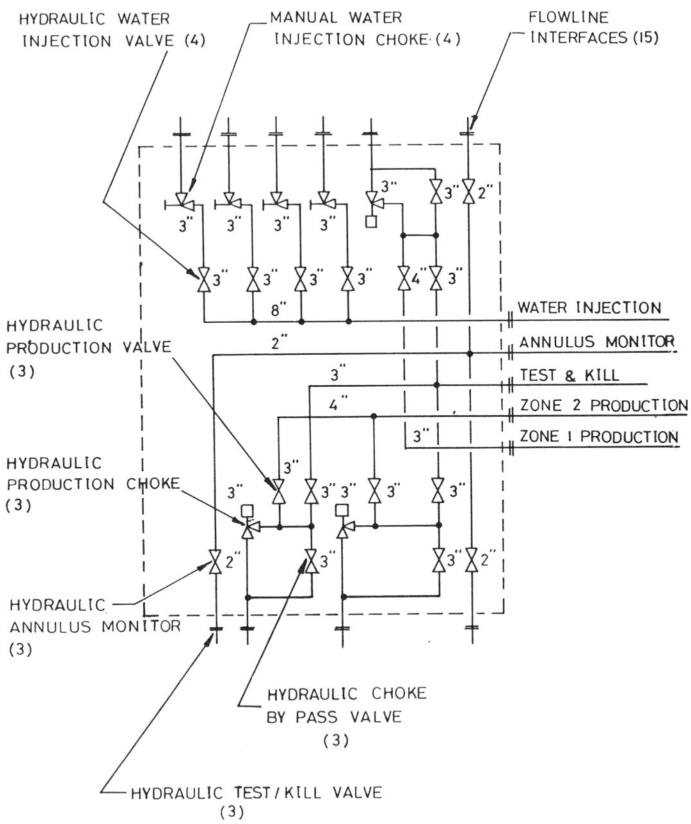

Fig. 4–24 Satellite Well, Production/Water Injection Manifold Schematic

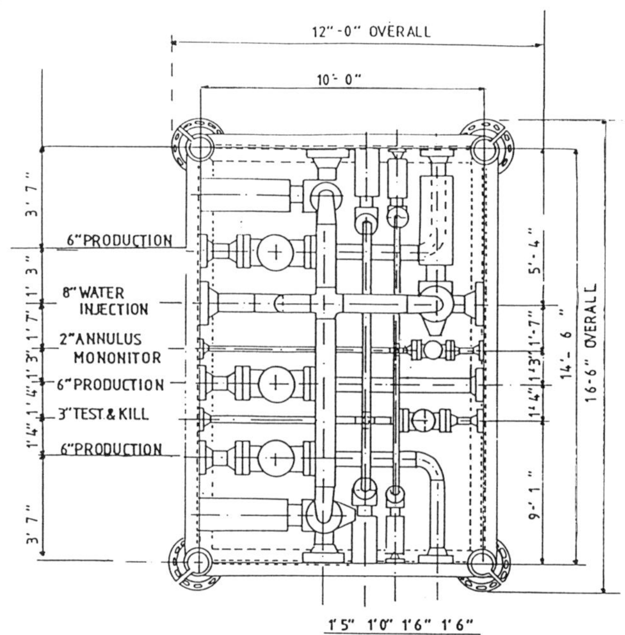

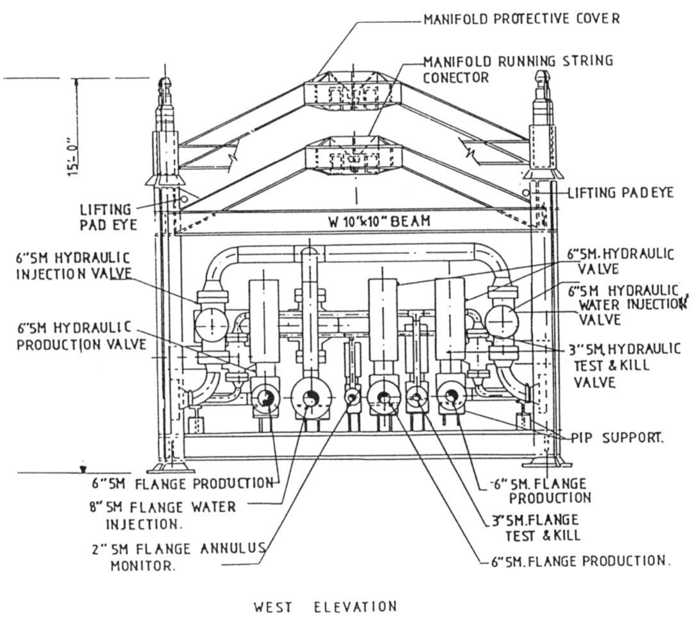

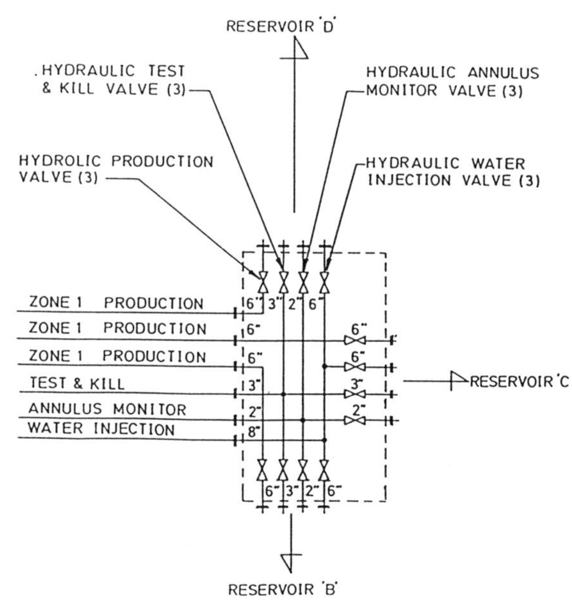

Fig. 4–25 Outlying Reservoirs Manifold Schematic

impact on the system. The satellite-reservoir manifold would not need to be fabricated until the outlying reservoirs were ready for development, and the template header flanges to this manifold could be blind flanged and capped off. Likewise, the pipelines do not need to be laid. The spare riser lines then could be used as alternate production paths for oil from the main reservoir A.

INSTALLATION AND MAINTENANCE

Running and Retrieval

The running and retrieval of manifolds and modules should be achieved in the most simple and efficient way using standard tool preparations, generally the J-slot type over the manifold or module center of gravity. The running tool may be similar in appearance to a BOP stack handling tool. Alternatively, the use of a hydraulic connector (see Fig. 4–26) may be considered, if a more automatic approach to latching/delatching to the manifold is considered.

Guidance to the support structure should be by guide-wire line, with either the wire guidance being provided by the running tool or module mounted sleeves—the basic design for the wire equipment being similar to that for Christmas trees. The manifolds may be lowered to the subsea structure using either drill string, if a derrick is available, or wire rope-line from a heave compensated crane or A-frame. The exact method can be firmed up only when the installation philosophies have indicated which vessel types are to be used. Using drill-pipe successive joints would be added to the string, while hanging off the running string in the rotary table with slips. Alternative lifting arrangements should always be provided to allow for sudden changes in installation policies.

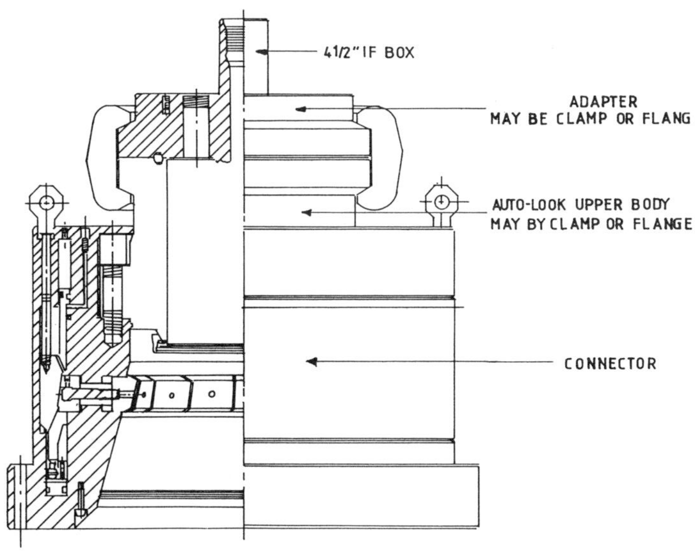

Fig. 4–26 Hydraulic Connector for Automatic Running/Retrieval

Actual running operations should be limited to conditions that would give heave of less than 1 m to the installation vessel. If not, then further precautions must be taken to provide soft landing features to the manifold. Retrieval may be carried out in more severe weather as long as it is fast enough to avoid slamming onto the manifold base structures.

Once the manifold has been placed on the seabed, an intensive work period commences for the diving team, which may occupy them full time from initial installation to final hook-up. If it ever becomes necessary to pull the manifolds to the surface for repairs, the following sequence of events would have to occur:

1. Shut in wells
2. Circulate and flush lines
3. Disconnect piping and control lines
4. Establish guide lines
5. Lower running apparatus to manifold
6. Connect and secure to manifold
7. Recover manifold
8. Locate temporary covers on subsea openings to pipework and control lines

Make and Break Times

Although diver-assisted connections are an accepted means of achieving manifold to pipework hook-up, it must be stressed that every effort should be made to use the most efficient form of connection, and so reduce make and break times.

The type of flange connection used dictates the make-up time of any rigid or flexible jumper spool. It should be remembered that many types exist that use ring type joints other than the BX, R types. It is these newer styles that generally use less studs per flange than the standard API/ANSI flange. This reduction in the number of studs greatly reduces the make-up time.

Other types of subsea connections require that their make-up include signal and power couplings, hydraulic stabs, and pin connectors. It would be futile to try and accurately predict the make-up times for subsea connections, due to the many variables involved and the environment being worked in. However, some estimates may be useful as a baseline from which to work. Times obviously vary with the size of the connection, but this is not the only factor involved; in many cases the complexity of the coupling may be dominant.

An indicative list of make and break times is laid out in Fig. 4–27. These must be treated with caution and installation schedules verified beforehand with the selected diving contractor.

Maintenance Requirements

A marginal-field subsea system should incorporate high-reliability components requiring little scheduled maintenance during the life of the field.

Type	Make up Time (hrs)	Comments
Pipe Connections		
API flange	2–4	Make up time related to size Rotational alignment problems
Clamb hub	3	Alignment problems
Swivel ring flange	2	Small misalignments only
Misalignment flange	1·5	Up to 12½° misalignment
Stab connection	0·5	Diver lockdown assumed
Controls Connections		
Pin connectors	1	Some types may require mini-habitat
Couplers signal	0·5	May have secondary fixing
Couplers power	0·75	May have secondary fixing
Hydraulic Connections		
Compression fittings	0·25–0·5	Varies with type and size
Single hyd stab	0·1 –0·2	Mainly preparation time
Multi stab plates	0·25	Uneven pressure forces may delay breaking

Fig. 4–27 Diver-Assisted Connections—Typical Make Up Times

Allowances for exceptional maintenance caused by accidents, environmental damage, or defective components should be made at the design stage. Where possible, individual components should be isolated to permit their shut-in and repair, without necessitating a complete system shut-in.

The following general principles should be adopted in the maintenance plan:

- All components must be functionally tested at the time of installation.
- Periodic pressure and/or functional testing must be carried out. In general, test periods should not be at more than one year.
- Visual inspection should be carried out at a frequency to be defined within the maintenance plan. In general, one annual visual inspection of all components would be required, more frequent inspection being performed at the operator's discretion.
- Special inspection and non-destructive examination requirements must be defined at the detail design of the systems.

It is important that an ultrasonic baseline survey, at strategic points on the manifold pipework, is carried out prior to installation. This survey is required to be able to interpret results from the subsea inspections. Such inspections are carried out to monitor the effects of erosion and corrosion during the life of the equipment. It is the designer's function to prepare suitable drawings which highlight areas most at risk.

Template piping should be constructed for access by divers. Piping connections should be flanged to permit diver make-up using torque wrenches or bolt-tensioning devices.

It is good practice for divers to periodically blast the worst marine growth and sediment from the manifold. A more thorough cleaning of specific areas is, however, called for prior to a detailed inspection or non-destructive examination.

Such inspections/examinations must be conducted on a planned periodic basis and include:

- General visual inspection of piping, connections, and structure
- Ultrasonic inspection of selected components
- Check of cathodic protection system and an anode inspection

Wherever possible, manifolds should be maintained in-situ and only recovered for major overhaul. This maintenance, carried out by divers, should (typically) include:

- General pipework and control plumbing repairs
- Sealout injection on certain types of valves
- Function checks (performed by divers and/or topside)
- Visual checks on chokes, valves, and actuators

Although initially chosen to minimize replacement during normal wear, manifold mounted chokes, valves, and associated hydraulic actuator should be replaced, and necessary repairs carried out topside.

Unitization allows removal and reinstallation times to be minimized.

INSTRUMENTATION

Current Trends

Increasing use is being made of pressure transmitters and chokes within manifold systems. There are an important part of emergency shutdown systems and production control. With the increasing use of microprocessors and satellite communications, the information gathered is becoming more essential. Information traditionally analyzed on the production facility may now be transmitted to the shore base within minutes, allowing decisions to be made with the full back up of the operator's specialist base staff.

The compilation of a history file and alarm listings requires increasingly detailed instrumentation.

Required Instrumentation

Instrumentation within a manifold must be located to provide pressure monitoring, both high and low level, within each header and branch into the header. Tappings, to mount transducers, will therefore be required at the inlet from each well to monitor line pressure. Tappings also will be required within each header to provide high- and low-pressure alarms to the production facility.

Flow measurements at each choke valve are required to allow the correct choke setting for commingling purposes. In addition, these provide an indication of each well's condition. They can give advance warning of well depletion or damage when interpreted along with downhole pressure. These flow measurements are obtained by using differential pressure transducers mounted across the production chokes. Additional pressure transmitters and differential transducers are required within the water-injection manifold circuit to provide injectivity rates to each water injection well. The total field injection rate is obtained from the topside metering system.

Additional instrumentation is required within the choke to monitor the choke position. These LVDT devices require subsea electronic conditioning of their signals before entering the control system telemetry link. Other positional devices may be used for valve position indication within the manifold pipework system if, for instance, commanded status systems do not provide the operator with a high enough confidence level.

Leak detection requirements could result in the use of hydrocarbon sniffers or density sensitive devices. The devices are normally located in a purpose-designed catchment area within a manifold or template cover/canopy.

Instrumentation at the riser manifold is required within each riser line for alarm purposes to assist in the detection of riser damage. Crossover headers, for flushing purposes, also must contain pressure monitoring devices to allow information on the pressure integrity of the crossover isolation valves to be monitored. This information is vital in any preflushing operations.

Future Developments

The use of differential pressure transducers as a means of measuring flow rates, while giving an acceptable accuracy for normal requirements, presents certain problems. These relate to the need to use calibration data and empirical coefficients within the microprocessor software—to predict flow rates through the production or water injection chokes. Future development and refinement of existing measuring devices (of the ultrasonic type) will eliminate many of these problems and provide reassurance that flow rates are accurately measured.

Present devices have not yet reached a stage of development where reliability and accuracy gains are guaranteed. The prohibitive cost of these devices does not present a cost effective alternative to DP cells. This situation is rapidly changing, and it is anticipated that in the near future ultrasonic flow meters will be the accepted method of providing accurate, reliable flow measurement for subsea manifolding purposes.

Other future development areas include the acoustic transmission of data and the further application of fiber optic links. It should be remembered however that, for marginal field purposes, these developments are presently a long way from acceptance and could impose severe reliability constraints on the subsea system.

Chapter 5

Subsea Chokes

INTRODUCTION

Increasingly, marginal-field development engineers are looking at subsea solutions to decrease the costs associated with fixed platforms. To reduce riser requirements, it is necessary to commingle the production subsea. Commingling is achieved by choking individual flows from each well into a subsea header system. The choking method used may involve either simple bean inserts or variable chokes.

The use of TFL systems has shown that bean change-out is a workable solution. However, the method that has gained increasing acceptance is the remotely actuated variable choke. These chokes come in three main formats: needle and bean, rotating disc, and orifice sleeve. Surface installations have used these chokes in the past. Their application subsea, however, has resulted in the need for an improvement in wear resistance of the choke internals and in the reliability of the actuators.

Past low-reliability reports have made the subsea designer wary of placing a high dependency on remotely actuated chokes. Product improvements and extensive testing have improved this situation, and two recent developments for Texaco and Sun Oil (on the Highlander and Balmoral fields, respectively) have made extensive use of remotely operated chokes.

Also, subsea chokes may be required to fulfill gas-lift and water-injection roles, in addition to normal production choking. It is the designer's task to select the choke that best suits the field development program. In particular, there may be a future requirement to change the position of the choke in the pipework, or its trim when switching from normal to gas-lift production.

Although water-injection chokes may be actuated remotely, it can be expected that manual chokes will be used for manifold or template installations because, generally, the setting of water-injection chokes is infrequent and the total water-injection rate is controlled from the topside facilities. This, however, does not mean that less attention be paid to the calibration and setting of the chokes. Flow-simulation studies, based on the resetting of individual chokes, may be used to give a good indication of the effect on flow rate to individual wells. They also may be used to predict the need for future water-injection and well-count changes.

TYPES OF CHOKE

Refer to Figure 5–1.

Rotating Disc

The rotating disc type of choke throttles flow through two circular holes in discs placed face-to-face. The extent of the choking that occurs is set by the amount of overlap between that one orifice and the other.

The discs are mounted in the choke body with one disc fixed and the other allowed to rotate through 360°, suggesting that flow may be choked from 0–100%. However, it is generally recommended that operation is at a choked rate of 20% or greater to avoid aggressive turbulance and possible high erosion effects. This type of choke may be found on many surface installations and is in use on the Texaco Highlander project.

Orificed Sleeve

This choke is based on a fixed and a moving sleeve. As the moving sleeve is pulled back, flow is channeled through orifices which become exposed in the seat sleeve. The rate of impingement of the flow, between the two sleeves, controls the choking rate. Also, flow is restricted by a reduction in the number of sleeve orifices exposed as one sleeve is moved over the other. Currently, this choke type is being used by Sun Oil in the Balmoral field.

Figure 5–2 shows the 1° movement mechanism used in the actuator.

Needle and Bean

The needle and bean type of choke is extremely common in surface applications, particularly on choke and kill manifolds. Choking is achieved by reducing the annulus between a tapered spear and a removable bean insert. The major risk with this

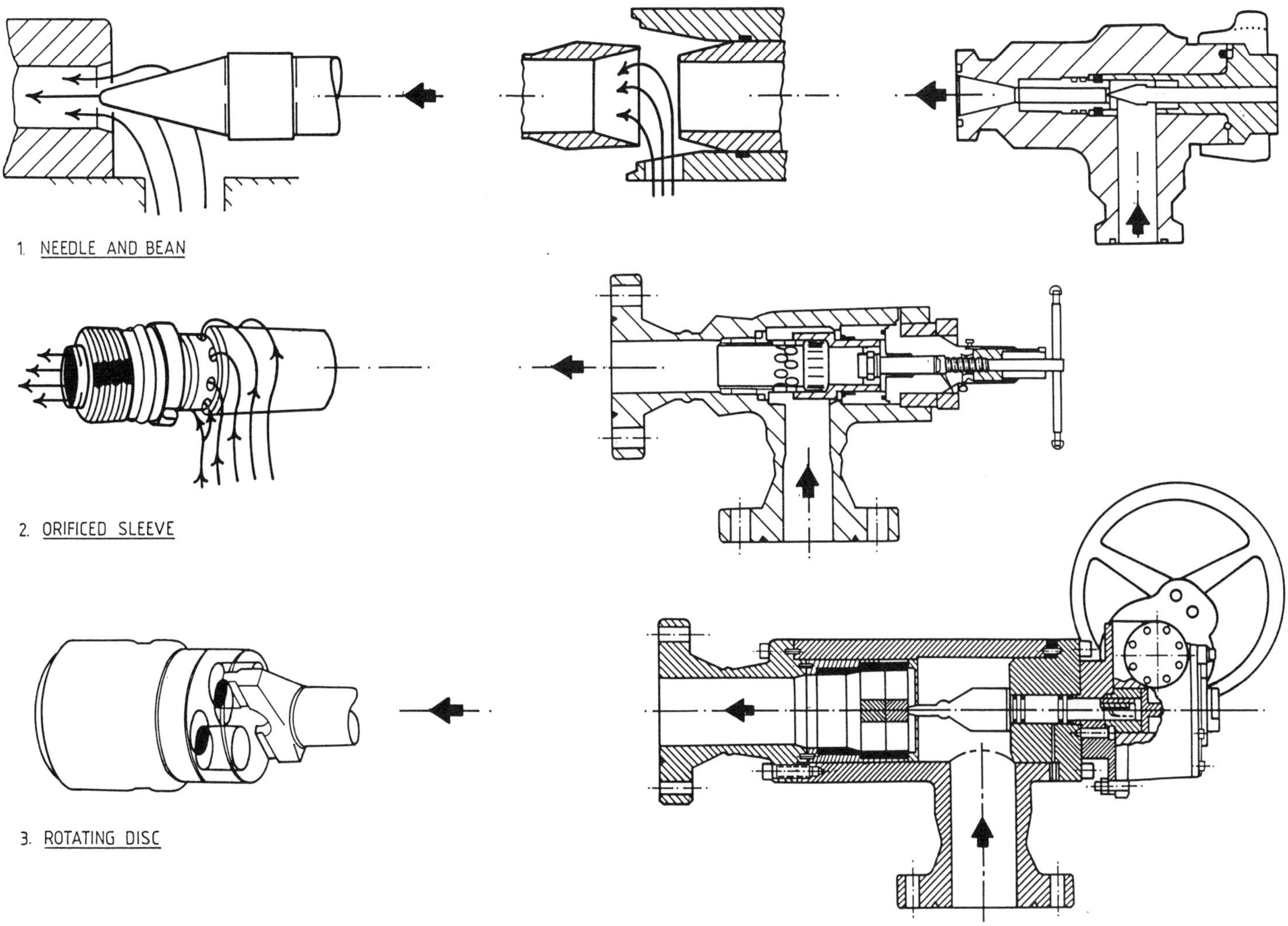

Fig. 5–1 Choke Types

arrangement is with the unpredictable wear patterns on the spear and bean.

Summary

In general, it may be said that the needle/bean combination allows the choke to maintain a uniform flow, whereas the orificed sleeve and rotating-disc arrangement can cause irregularities in the flow pattern. Neither system, however, would be able to stabilize flow within the confines of the choke body. Wear rates at the point of choking are generally better for the disc and sleeve types.

CHOKE LOCATION

The location of the choke within the subsea system must be carefully considered so as to avoid potential problems. The choke can be the most unreliable component in a subsea system, and therefore its location should facilitate maintenance.

Remotely actuated production chokes may be used for satellite trees; the choke may be located on the tree or within a module at a central manifold base or template. If the choke is located on the satellite tree (see Figs. 5–3 and 5–4) and the pressure drop over the choke is high, icing up (hydrate formations) may occur. Other potential problems arise in the size of flowline and/or the need for insulation.

For subsea maintenance, the choke should therefore be integrated into proposed pipework in a manner that allows actuator or trim change out subsea. Instrumentation requirements also dictate the chokes position within the pipework—as pressure tappings are often required to be placed at least two to four times the line diameter away from the choke inlet—or any outlets on a straight pipe run.

Fluid power is applied to the net area of the piston (1). This moves the drive pawl (2). As the drive pawl moves forward the ratchet gear (3) is engaged. The ratchet gear is keyed to the pinion shaft (4). As the ratchet gear rotates the pinion gear, the ring gear (5) is rotated 1°

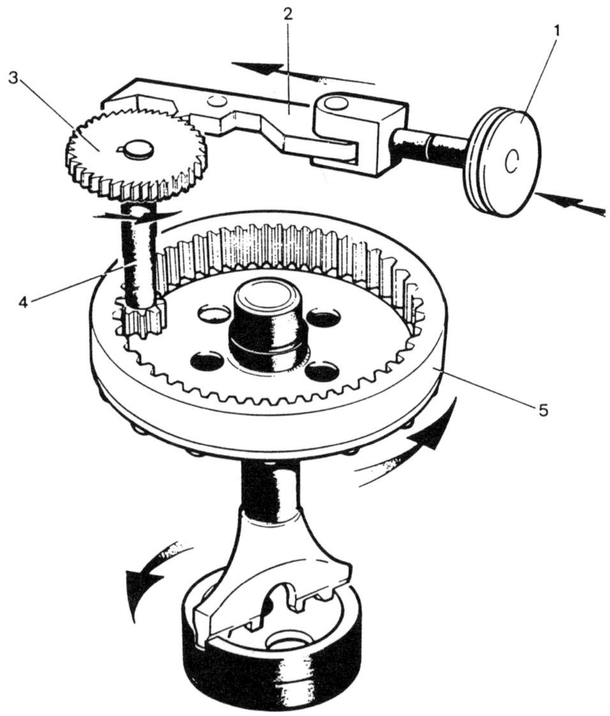

Fig. 5–2 One Degree Movement Mechanism

GENERAL REQUIREMENTS

The general requirements for production chokes vary for each specific field application. However, any chokes used should generally conform to the following basic parameters:

- The choke must be capable of controlling crude or gas and have characteristics as defined in the operator's basis of design.
- Choking should be possible from full-bore to shut-in.
- Setting should be remote and by way of the selected control system.
- A visual indication of choke position should be observable by an ROV.
- Position-indication feedback to the surface master station should be provided, compatible with the data-acquisition system.
- A manual override should be provided.
- The actuator should be resistant to corrosion problems connected with subsea service.
- The choke position should not drift when the hydraulic supply is interrupted.
- The choke trim and body should be such that erosion resistance is optimized.
- Subsea replacement of the choke actuator should be possible.

Additions to the above list are necessary if the choke's application is totally remote or if produced fluids pose any potential problem.

The use of a manual water-injection or production choke increases the need for a more efficient visual-indicator system and a low-torque requirement for manual settings. However, the general requirements will not vary greatly from the above list.

Physical Sizing

The use of actuators on chokes has increased their physical size to such an extent that problems may arise in location of the choke within the manifold pipework or on the Christmas tree. This aspect requires a methodical approach to piping layout to provide suitable diver access. Additionally, the flange-to-flange sizes of the chokes should be specified to match standard flange and fitting dimensions. This enables the operator to change out the choke with standard piping spool pieces if and when required. The total length of a choke may exceed 48 in., with the actuator having a diameter in excess of 24 in. Choke weights vary between 500 and 850 lb (1 pound = 0.4536 kg), the actuator accounting for 40% of this value.

CONTROL-SYSTEM INTERFACES

Any remotely actuated production choke requires three main interfaces with the control system.

Control Lines

A hydraulic-connection interface is required for controlling the choke. The number of lines varies from vendor to vendor but is likely to be a maximum of three. The control-line interface (with the choke) must be compatible with the chosen choke and/or the actuator replacement technique.

Monitor-Line Interface

An interface is required in the signal line to establish communication between the remote-position indicator sensor and the data-acquisition lines. If electric, an inductive interface is required. If hydraulic, a quick-disconnect interface must be provided so that the component can be replaced independently. The most common control-system type in use with subsea chokes is the electro-hydraulic multiplexed system, and in this case, the connection method will be the inductive coupler.

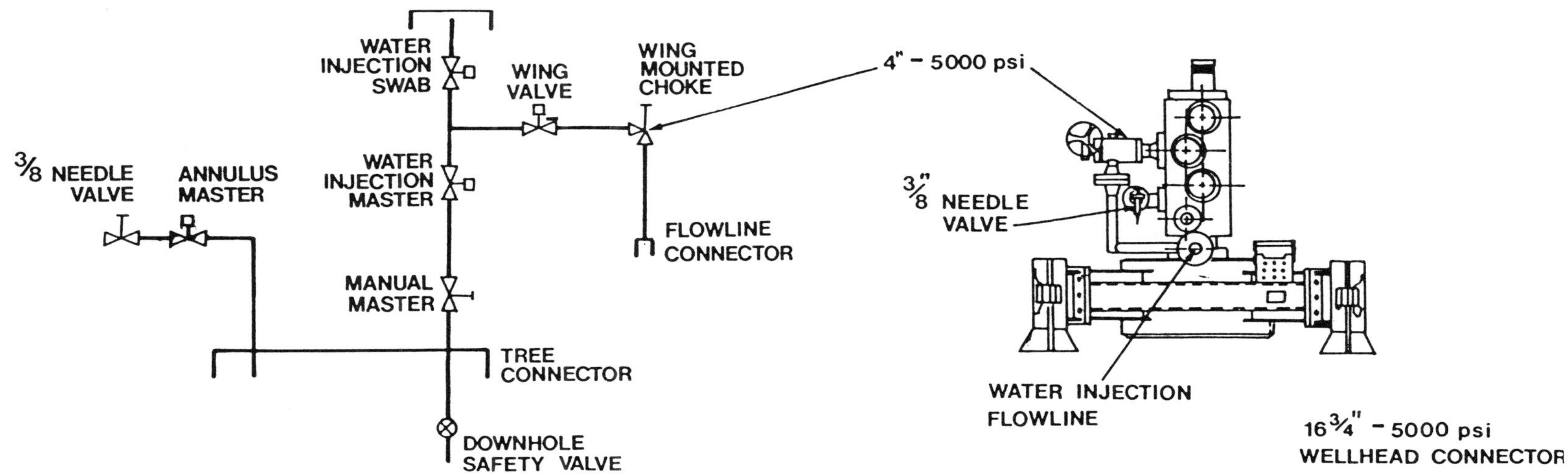

Fig. 5–3 Manual Choke on a Water Injection Tree

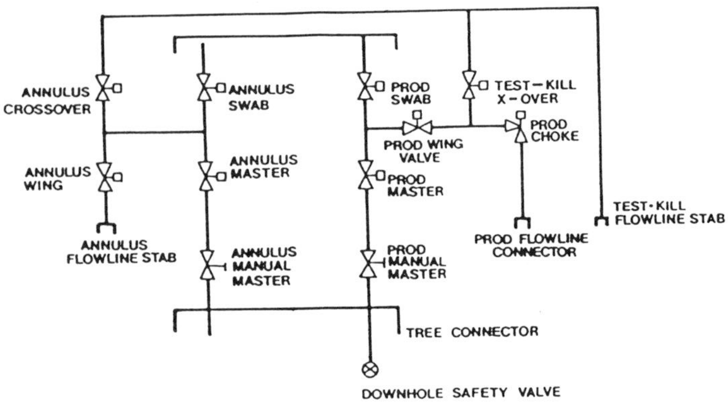

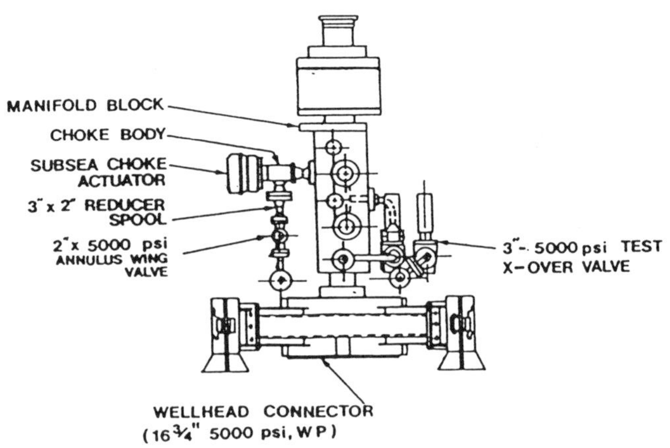

Fig. 5–4 Production Choke with Actuator Mounted on a Production Tree

Flow-Measurement Interface

To achieve a reliable method of low measurement, it is necessary to make use of differential pressure transmitter tappings across the choke body. The signal is taken from the DP cell and interfaced with the control-system lines by means of a pin or inductive coupler.

TESTING AND CALIBRATION

Owing to the critical role of the subsea choke/actuator, an extensive testing program for both chokes and actuators should be developed as early as is practically feasible in a project. The objective of this testing program is to simulate the conditions that will exist during production, gas lifting, and water injection. Without sufficient land-test data and experiences in a realistic simulation, any data provided by manufacturers will only be of a general nature, resulting in costly delays in the early commissioning period.

During testing, attention should be focused on the feedback mechanism incorporated into any actuators being evaluated. The use of potentiometers may prove unreliable as they are susceptible to signal errors, making them unattractive as a primary feedback source for position indication. A linear variable differential transformer (LVDT) has been shown to be a more reliable approach, being significantly more accurate. However, it requires more stringent testing to verify its suitability. A rotary transformer also may be used in place of the potentiometer, but again, specific data for this device must be gathered by way of testing.

A typical flow loop for correct testing and calibration should comprise the following equipment:

- Stainless-steel piping
- High-volume pump
- Turbine flow meters
- Control valves
- Computerized data-acquisition system with a display of coefficient of velocity (CV) and noise in real time (FL).
- Noise measurement, recording, and analyzing instruments, anechoic chamber.

The test circuit (see Fig. 5–5) should provide results to allow the determination of CV values at various choke settings for each of the mediums that the choke will be required to control.

Actuator Testing

The actuator should be tested to assess its torque output, and the results compared with the manufacturer's claimed values. A number of tests should be made so that the results can be quantified in terms of repeatability.

Additionally, testing should establish actuation times for the choke within various operational parameters. Extensive cycle testing also is required to show wear rates and indicate to the control-system supplier the length of time the system will be blacked out when the chokes are operated. Such testing should be carried out both at atmospheric and operational hyperbaric pressures. Additional information will be gleaned of the LVDT performance remaining within the permitted limits. This is particularly important if the results, intended to be displayed on the master station visual display unit, are to be reliable. If the LVDT is not perfectly matched to the actuator, erroneous position indications may occur. That is to say, the electronics may record, via the LVDT, that the choke is in one position whereas the actual steps moved by the choke are one short. Should this situation occur, it may be necessary to modify the master station algorithm to count the steps and compare them with the LVDT reading. Then such a comparison could be displayed to avoid possible operator confusion.

Choke Sizing

Choke sizing formulas are published by the Instrument Society of America (Publication ISA-575 01 Control-Valve Sizing Equa-

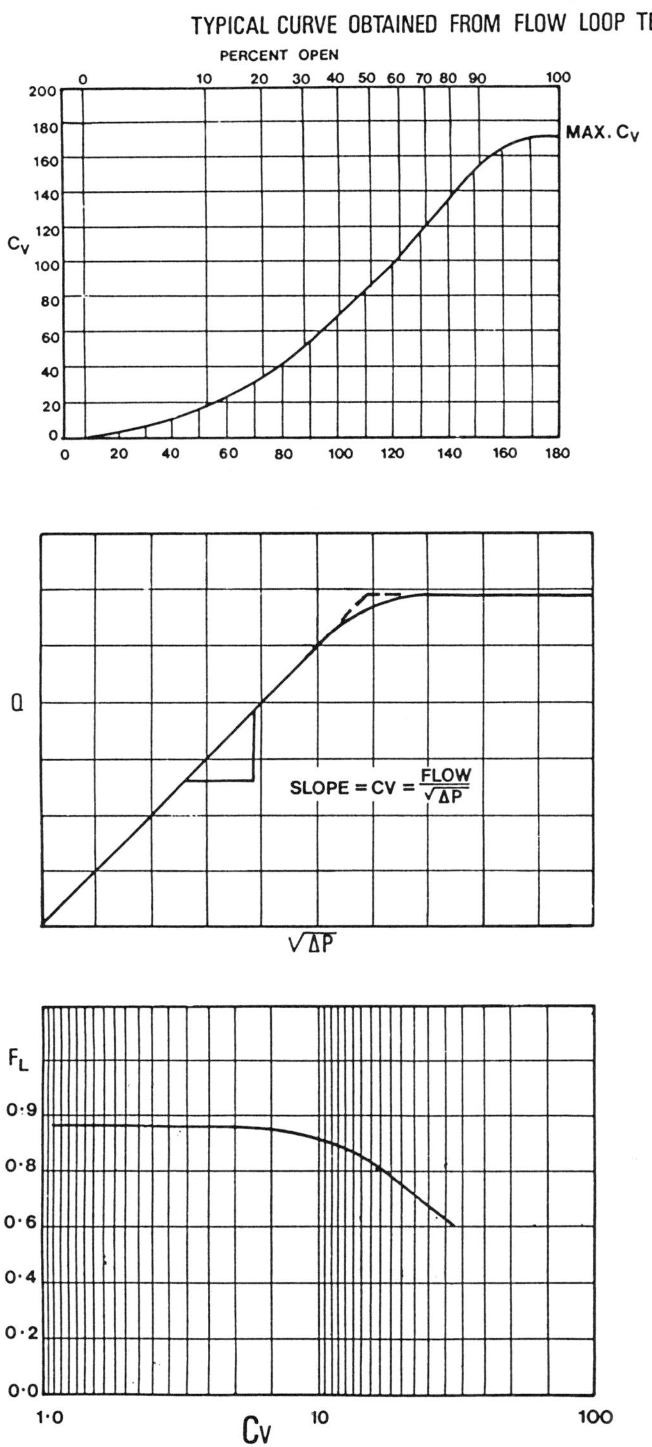

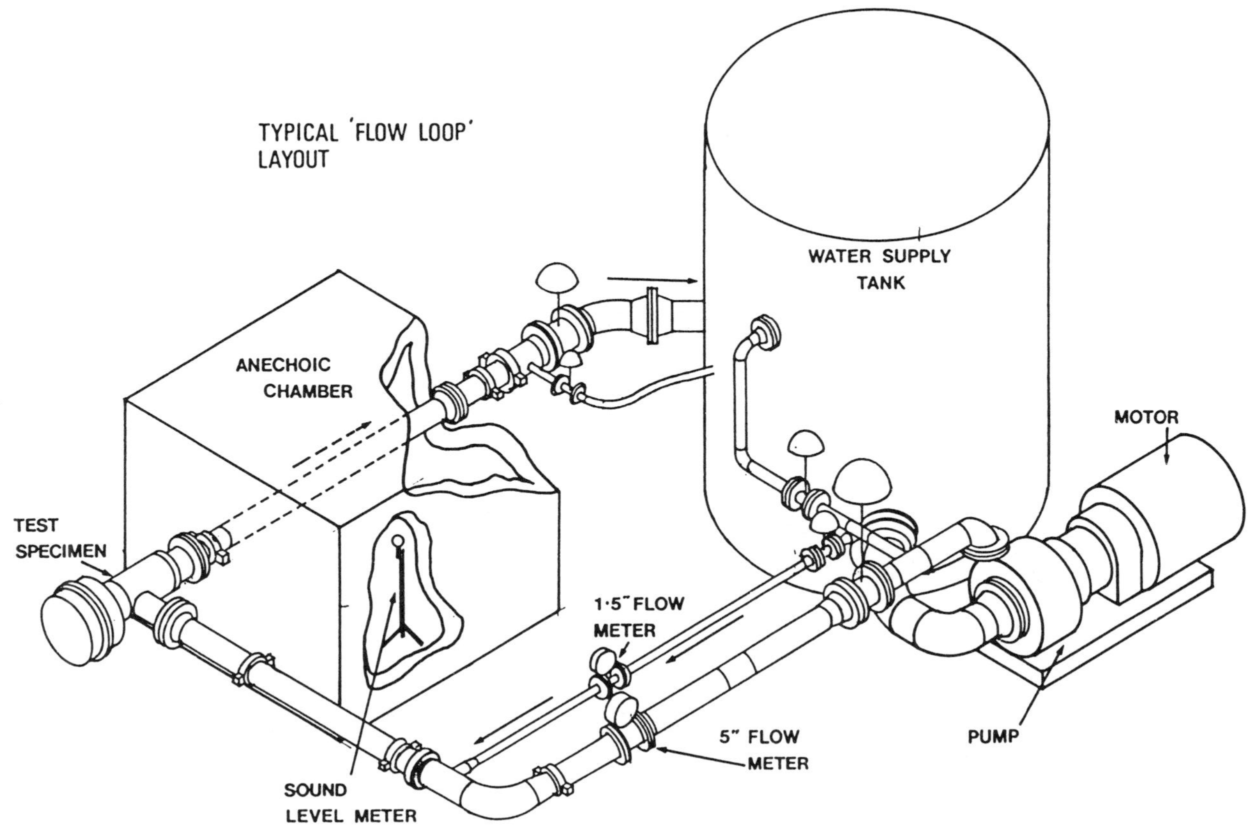

Fig. 5–5 Typical Flow Loop Layout

tions). For liquid flow only, constants given in the equations may be combined to provide a simplified form, namely:

$$CV = 0.02167.\ QL \frac{SG}{\Delta P}$$

where CV = coefficient of velocity
QL = liquid flow rate (Bls/D);
ΔP = pressure drop
SG = specific gravity of the fluid

For gas flow, the basic equations are used with the liquid-flow components equaling zero.

The accuracy of the equation is proposed by the ISA as ±5%, assuming single-phase flow with no solid content and ideally a Newtonian media. Any flow calibration tests should account for piping geometry to give a valid proof of compliance within the ±5% accuracy.

Flow-Rate Equations

To allow a satisfactory interpretation of signals from the choke instrumentation by the master control station, certain operator-inserted values must be used. These are to correct for units, geometry of the system, Reynolds number, and well-test calibration results. This correction factor will be in addition to the operator-insertable values of CV, obtained during choke testing.

Generally, the flow-rate equations will be as follows:

Water $qw = Kw \cdot CV \cdot \sqrt{\Delta P}$
Oil $qo = Ko \cdot CV \cdot \sqrt{\Delta P}$
Gas $qg = Kg \cdot CV \cdot \sqrt{\Delta P}$

where q = flow-rate (m^3 h)
CV = choke CV (operator-alterable table of values, depending on choke position)
P = differential pressure (bar)
Kw = operator-alterable value, one per choke for water
Ko = operator-alterable value, one per choke for oil
Kg = operator-alterable value, one per choke for gas

The above modified expressions must form part of the software, requiring a constant liaison be maintained between the choke and control-system suppliers. The coefficients Kw, Ko, and Kg normally would be derived from independent metering on the production facility—modified by the operator to account for changes due to wear instrument drift and gas/oil ratios.

Chapter 6

Subsea Controls

DESIGN PHILOSOPHY

Introduction

Subsea control systems are considered a most critical part of any subsea installation. At the same time, their cost often represents only a minimal part of the overall cost of a subsea completion (generally between 3% and 10%). However, the number of components which constitute a control system for a major installation greatly exceeds the total of all other subsea components.

The location of control devices on a subsea template is extremely important. Careful consideration can result in a reduction in the length of piping and cabling and the number of connections thereby facilitating installation and retrieval. Nevertheless, the control system is often specified as an add-on item, after the rest of the equipment has already been developed. Recent installations have shown that subsea control systems can be highly sophisticated, have high reliability, and afford control of subsea equipment many miles away with ease and confidence.

A subsea control system, whether simple hydraulic or more complicated electro-hydraulic, is defined as one that has some active component subsea. There are several applications for the control of systems underwater such as wellhead BOPs, riser base manifolds, direct control of subsea trees with hard-piped connections, etc. In most of these cases control is obtained at a structure by the use of a hydraulic power unit.

A subsea control system facilitates the operation of valves and chokes on subsea completions, templates, manifolds, and pipelines and also permits the retrieval of data. It can include safety features that automatically prevent dangerous occurrences and pollution of the environment. It also can assist the reservoir engineer to monitor the status of production by indicating temperatures, pressures, sand detection, etc. Production then can be optimized by the control of water injection through remotely operated choke valves. The foregoing itemizes only a few possible applications.

Presently, five types of control systems are available for the operation of subsea production systems. (Currently, additional types are being developed, with optical fibers for example, but these are not yet operational.) Three of these types, direct hydraulic, piloted hydraulic, and sequential hydraulic, rely solely on the movement of hydraulic fluid to operate subsea actuators. The remaining two types, electrohydraulic and multiplexed electrohydraulic, make use of combined electrical and hydraulic signals to control actuators.

Each new type of system that has been developed provides a control capability not present in existing systems. However, not all control systems are suitable for field developments. Selection of a system requires a clear specification of field operational requirements as a basic starting point. Such requirements then must be compared with capability, reliability, and cost before a preferred system can be recommended. Figure 6–1 shows the five alternative systems.

Types of Systems Available

Direct Hydraulic

Direct hydraulic control systems are currently the simplest and consequently the most reliable available. Each subsea function requires a hydraulic flowpath from the surface. Then, actuation of a surface-located mimic display panel results in pressurized fluid being routed through the dedicated flowpath to the selected subsea hydraulic actuator. This system, with one line per device, is suited best for applications where control distances are relatively short and a limited number of subsea functions are required.

As distances from the control source get longer, response times drastically increase. For example, actuation of a subsea choke from fully closed to fully open through approximately 4,000 m (13,000 ft) of umbilical could take as long as three hours.

As the number of functions requiring control increases, so

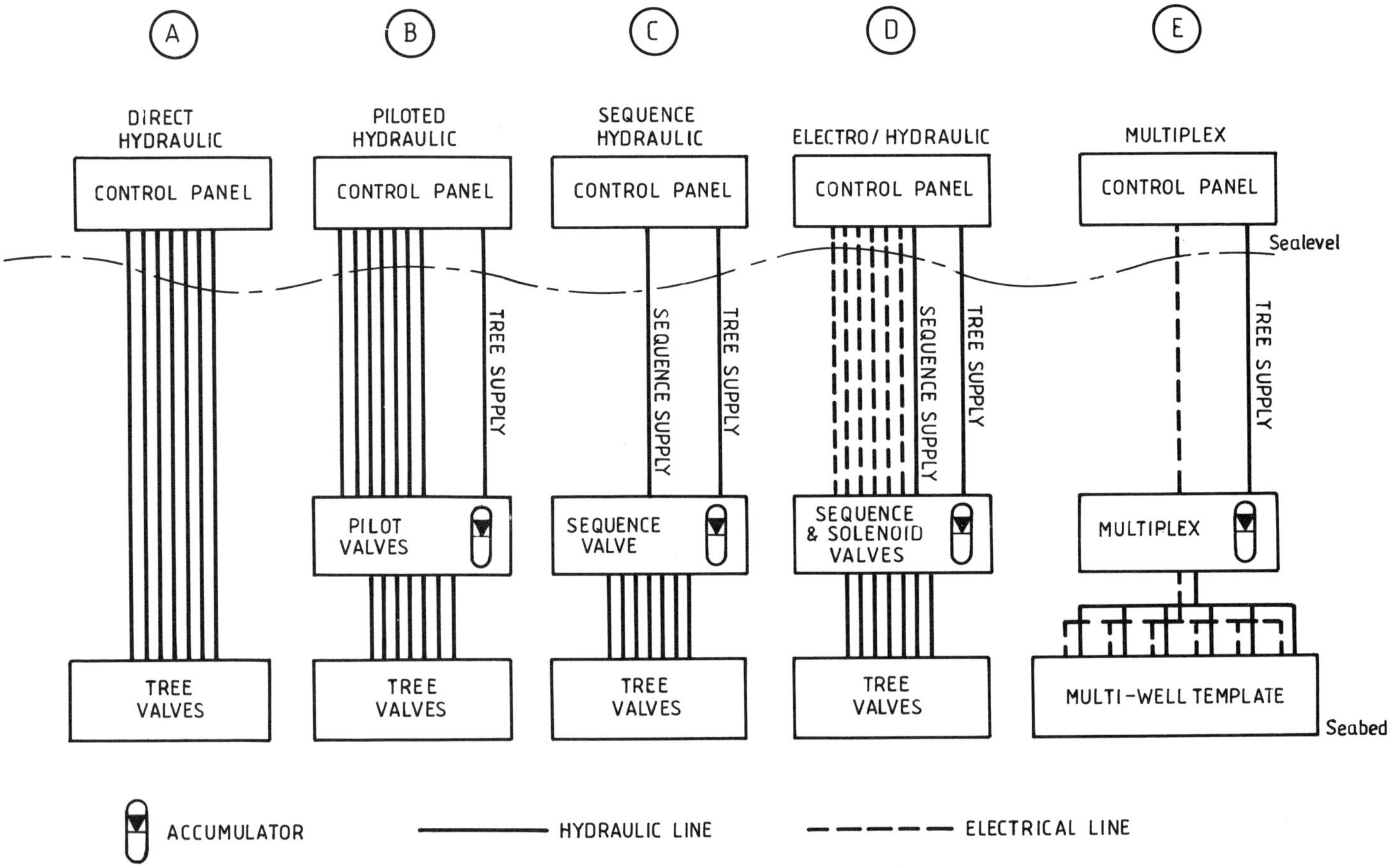

Fig. 6–1 Choice of Control Systems Available

does the complexity and cost of the umbilical. Consequently, umbilical size and cost quickly rise to dominate the cost of the entire system, in some cases reaching 50–60% of the total system cost. For a 10–20-well reservoir, a direct hydraulic system would require very large umbilicals to control the field. These could be a minimum of 5 in. in diameter and require a large area in the vicinity of the moonpool for the reel units to pay out and retrieve them.

Apart from the space and weight problems of the surface equipment, a direct hydraulic system as described would provide only a marginal primary system with no growth capacity and without back-up capability. Because of the long umbilical lengths, system response time would be extremely slow. Although this is not as serious for a production system as it would be for a BOP system, it should be avoided.

An extensive direct hydraulic system is used on Buchan Field in the North Sea, as shown in Figure 6–1A.

Finally, subsea chokes require some monitoring capability to track choke position. To accomplish this and any other desired monitoring function, some form of electrical feedback is necessary. A direct hydraulic system cannot provide this facility.

Piloted Hydraulic

A piloted hydraulic system improves response time over a direct system by storing the hydraulic pressure energy at the work site with pilot valves to effect actuation. Umbilical size is roughly equivalent (although slightly smaller) to that of a direct system because each valve operated subsea requires a pilot (or signal) line in the umbilical. In addition to these signal lines, the umbilical also carries a main pressure line.

Actuators are activated by sending a hydraulic signal to the pilot valve. The pilot valve opens and allows fluid to flow from the supply line and storage accumulators to the selected actuator.

The piloted system offers a good reponse time. With other aspects, such as monitoring capability, umbilical size, and suitability for the total field development, it has similar limitations to the direct hydraulic system. Additionally, when compared with the direct hydraulic system, it is more complex and less reliable. To alleviate problems created by large umbilicals, long control distances, and large numbers of subsea functions, the sequential hydraulic system was introduced.

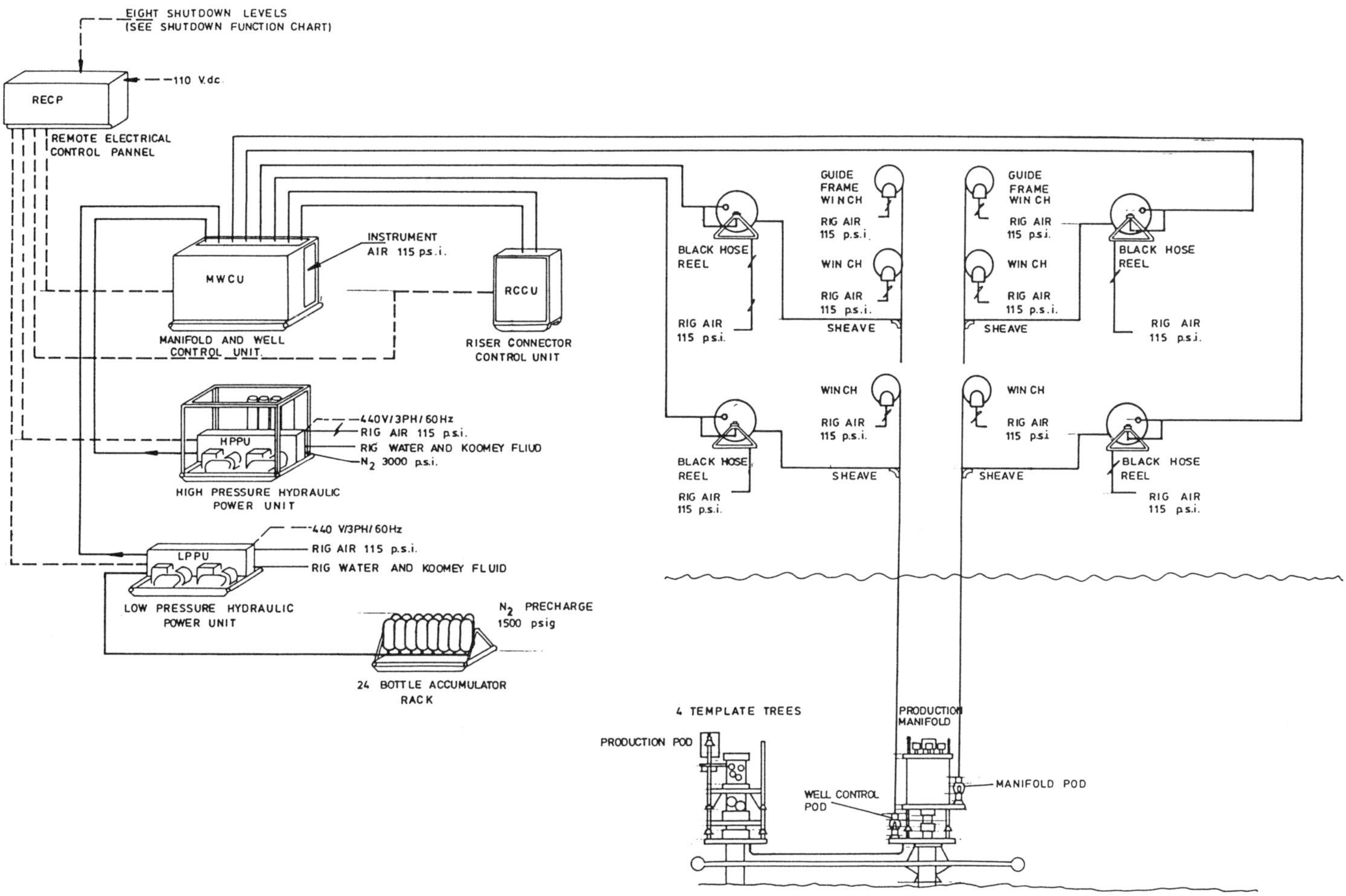

Fig. 6–1A Buchan Field Direct Hydraulic Control System

Sequential Hydraulic

As with the piloted system, the sequential system has accumulators and valving subsea, but the umbilical size is reduced significantly. The umbilical bundle is made up of a single supply line, a single line and, in some cases, a reference line. However, the reference line is sometimes eliminated by deriving the reference pressure subsea from the supply line pressure or springs.

Subsea control is obtained by sending a pressure signal to a sequence valve that is pre-set to operate at a specific signal pressure (for instance 500 psi). At this pressure, hydraulic fluid is routed to energize selected actuators. Then the signal pressure is increased in a series of discrete steps resulting in additional pre-set functions being accomplished at each (500 psi) step.

The major drawback to the sequence system is that once the design is implemented it is fixed, and no operational program changes can be achieved, other than by significant changes in the subsea hardware. That is, once the sequence of valve operations is established, it is not possible to make even minor changes. This is a potentially serious problem in large or commingled subsea production systems.

The system also is limited in the number of combinations of valves that can be operated, because the number of incremental pressure steps is limited by both the maximum safe hydraulic operating pressure and the size of pressure increase needed to differentiate between one step and the next. It is further limited by its inability to monitor subsea pressures and choke positions.

As subsea oil production systems became larger and more sophisticated and developments took place in greater water depths, a need was identified for improvements in control system technology. This resulted in the emergence of electro-hydraulic and multiplexed electro-hydraulic control systems.

Electro-Hydraulic

Basically, electro-hydraulic controls are piloted hydraulic controls with the pilot function replaced by an electrical signal. Each subsea actuator requires a signal conductor from the surface, in the same way that a hydraulic pilot requires a signal line. This affects umbilical complexity and subsea interfaces, so that electro-hydraulic systems are suited best for use with only limited numbers of subsea actuated devices. A typical electro-hydraulic control system is shown in Figure 6–2.

Multiplexed Electro-Hydraulic

An improved version of the electro-hydraulic system is called the multiplexed electro-hydraulic system. This system is capable of controlling a large number of actuators, very quickly, using only a relatively small supply umbilical. To accomplish this, a single electrical power line is used, together with a signal line coupled to a multiplexer to operate a multiplicity of subsea solenoid pilot valves. Hydraulic fluid at pressure is made available via a bulk supply line with the umbilical.

When a digital signal is sent to the multiplexer, it energizes the selected solenoid valve, thereby directing hydraulic fluid from the supply umbilical to the associated actuator. The presence of a multiplexer introduces the capability of monitoring pressures, temperatures, and valve positions by means of electrical signals, without further complicating the electrical connections through the umbilical. Operationally, the multiplexer system reduces deck space requirements over all other systems, and it provides a monitoring capability which is not available in any other of the hydraulically based systems. The speed of operation is also superior to that of hydraulic systems. These reasons form the basis for the recommendation of a multiplexed electro-hydraulic multiplexed system. This recommendation also is based on experience controlling a 10–20-well reservoir/field. Note that inductive couplers are used for electrical connectors subsea. (Refer to the Electro-Hydraulic Systems: Major Components Description section in this chapter.)

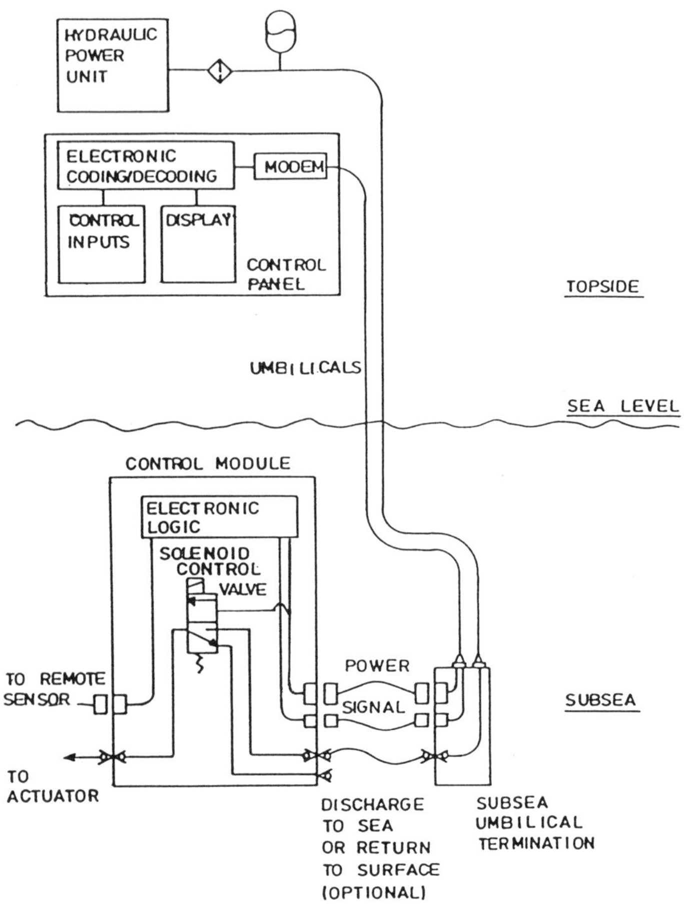

Fig. 6–2 Simplified Electro-Hydraulic Control System

Past and Present Experiences

Past Experience

While subsea completions can be traced back to 1943 with the Lake Erie completion, it was not until the 1970s that they were considered as a viable alternative to fixed platforms. The rapid increase in the cost of oil and the earlier (1963–1964) oil and gas discoveries in the North Sea made oil companies think hard about subsea developments. By this time, Exxon had already started a development program with their Submerged Production System (SPS), which was initiated in 1968 and eventually installed off the Louisiana coast in 52 m (170 ft) of water.

From a control system point of view, SPS was probably the most advanced system in the world. It used an electro-hydraulic control system and included an automatic fail-safe feature which allowed the system to be shut in if parameters departed from their predetermined values. The system was "wet," included a remote maintenance vehicle, and used some novel features which are now considered "state of the art." Although a lot of work was carried out on an oil-filled conductive coupler at the time, all electrical connections to the control modules were ultimately made through inductive couplers.

Some of the main features of the Exxon control system are:

- Extensive use of inductive couplers for both power and signal transmission; no failure occurred throughout the whole project.
- Use of metal-to-metal seals for all hydraulic connections.
- Use of a retrievable trunking for the electrical and hydraulic interconnection.

During four years of operation, Exxon gained sufficient confidence to consider this approach for the joint Shell/Exxon UMC.

Enchova Field. The Enchova Field in Brazil was primarily designed for early production and used a multiplexed control system for multi-well completions. The main electrical connections were made with conductive couplers, and some problems were experienced. Nevertheless, this completion showed that an economical system could be used to rapidly develop a subsea field with floating production facilities.

Tazerka. Following this philosophy, Shell successfully developed the Tazerka field discovered in 1979, with first production obtained in December 1982. An all-hydraulic control system was used for the multi-well completion.

Cormorant. This 1980 single-well completion, in the North Sea Cormorant field, was used by Shell. Their aim was, essentially, to prove the reliability of subsea equipment which included the control system, and thereby decide whether the UMC project was viable or not. The multiplexed electro-hydraulic control system used both power and signal inductive couplers and self-sealing hydraulic connectors with metal-to-metal seals.

The system proved extremely successful and performed satisfactorily, thereby giving Shell and Exxon confidence for the multi-well UMC system which uses the same principles.

Some of the experiences gained in the projects mentioned above showed that a planned control system, for a multi-well completion, would average a three-year period between the preliminary feasibility study and final commissioning. The cost of a 10–15 multi-well control system, including the control lines, can vary between 10 and $15 million (1985 prices).

Present Experience

To give an overall view of any transition into the present status, five significant examples can be reviewed. Each uses a different approach to the control of multi-well subsea completions.

BP Magnus. Magnus is considered a major North Sea field development using the concept of single satellite completions located radially around a central platform. The electro-hydraulic control system is designed to control up to 11 satellite wells, with wet Christmas trees.

Power is transmitted at 400 Hz, and the signals use a Simplex mode of transmission in the 1,024 Hz bands, with a Frequency Shift Key (FSK). Electrical connections are made subsea by means of inductive couplers. Primary control is electro-hydraulic, but a back-up sequential hydraulic system is available in case of primary system failure. The field became operational in the latter part of 1983.

Shell-Esso Underwater Manifold Center (UMC). Shell-Esso UMC is located between the North and South Cormorant platforms, at a depth of approximately 180 m (580 ft). The UMC has facilities for the connection of up to nine satellite tree wells. The control system, originally designed for depths down to 200 m (650 ft), is electro-hydraulic using high reliability components with a duplication of all the power, signal, and hydraulic supplies.

A total of 292 inductive couplers of various dimensions are employed. All the hydraulic connections are self-sealing, using metal-to-metal seals. The hydraulic fluid is a mineral oil, and subsea pressure accumulators are used on the supply and return lines to minimize response times. Subsea sensors monitor pressure and also are used to detect the presence of hydrocarbons. Several types of fail-safe shutdown devices are available with different features, from the isolation of single lines to a complete manifold shutdown.

Power is transmitted at high voltage through a subsea cable which also contains signal conductors. Then it is transformed and distributed, at various voltage levels, through inductive couplers. Signal transmission is by means of FSK modulation, using a single pair of conductors.

The control system was delivered in July 1980 and then was integrated and land tested for over a year. Installation of the UMC structure was in the first half of 1982, and oil production commenced in the spring of 1983, the control system having been fully operational from the beginning.

Elf-Aquitaine NE Frigg. Elf-Aquitaine NE Frigg development is a six-well template with the control station located 150 m (490 ft) away on an articulated column. This is used to locate a hydraulic power unit which delivers power directly to the subsea valves, avoiding the need for an electro-hydraulic subsea system.

In a certain sense, it shows little confidence in the capability of subsea electro-hydraulic systems; this scepticism may have stemmed from lessons learned by Elf during their Grondin experience.

Sun Oil Balmoral. The Sun Oil Balmoral development makes use of a purpose-designed floating production vessel moored over a 14-slot drilling template. Initial development involved 10 template producing wells, 3 satellites and 6 satellite injection wells. The production is commingled subsea, using a template header system and 3 task-separated manifolds. This is one of the first of a new breed of marginal-field developments that paves the way for the systems of the future. The water depth is 137 m (450 feet) and thus allows the extensive use of divers where necessary.

The production control system is of the hydraulic multiplexed type with 235 signal and 97 power-inductive couplers. It is also a commanded-state system, with no positional feedback except on the production chokes, where the positional signals are compared with differential pressure signals to monitor production and injection flow rates. The commingled production is taken via flexible risers to the topside floating production facilities. The subsea control system is microprocessor-controlled and integrated with the process-management system via an interprocessor data link.

Single Valve Control

One interesting application of subsea control systems has been the operation of pipeline valves for the Northern Leg Gas Pipeline in the North Sea. In this case, the system controls and monitors the status of the isolation and spur valves. Features of this project are the use of a piloted hydraulic control system combined with a data retrieval system which uses inductive couplers. The control modules require diver-assisted installation.

Notions of Reliability

Probability of Failure

The various developments described above indicate ''state of the art'' subsea control systems. ''State of the art'' in this context means equipment which can be used reliably during the life of the system—at least 10 years and in some cases as long as 30 years. However, as well as identifying those items which can be considered fully acceptable, it establishes those areas requiring further engineering.

When a control system is specified, it is common practice to quantify a reliability value. This is usually expressed as a percentage probability of the system performing without failure over a period of five years. It can be accepted that, for the reliability calculation, a bath tub-shaped curve can be used. Figure 6–3 shows such a curve.

The first period represents ''infant mortality'' of the system. Failures occur during this period due to faulty components, bad assembly of the subsea systems, quality control problems, etc. (It is normal to ''burn in'' the components and the system for a certain period of time prior to installation.) The difficulty is to assess the duration of this period. With inductive couplers for instance, it has been found that 48 hours are sufficient to determine infantile faults in a production run. Additional safety margins are obtained during testing by doubling and working pressures and by the testing of individual components, etc.

While most electronic components have a constant failure rate after the first period, this cannot be true for mechanical components used in a dynamic mode, such as control valves. The failure in such cases is dependent upon the number of operations, involving a wave factor, and other aspects such as corrosion, overstressing, etc. For such critical components, it is normal to qualify them by subjection to comprehensive testing.

It is important during such testing to use conditions which simulate those anticipated in actual service. For instance, it has been found that the life of a control valve is extremely dependent upon the type of fluid used. In some cases, valve life has been 10 times longer with mineral oils than with water-based fluids.

Reliability Analysis

For the reliability analysis of the control system to be effective, it also should cover the control lines. Any reliability analysis starts with the collection of statistical data on line types, failures,

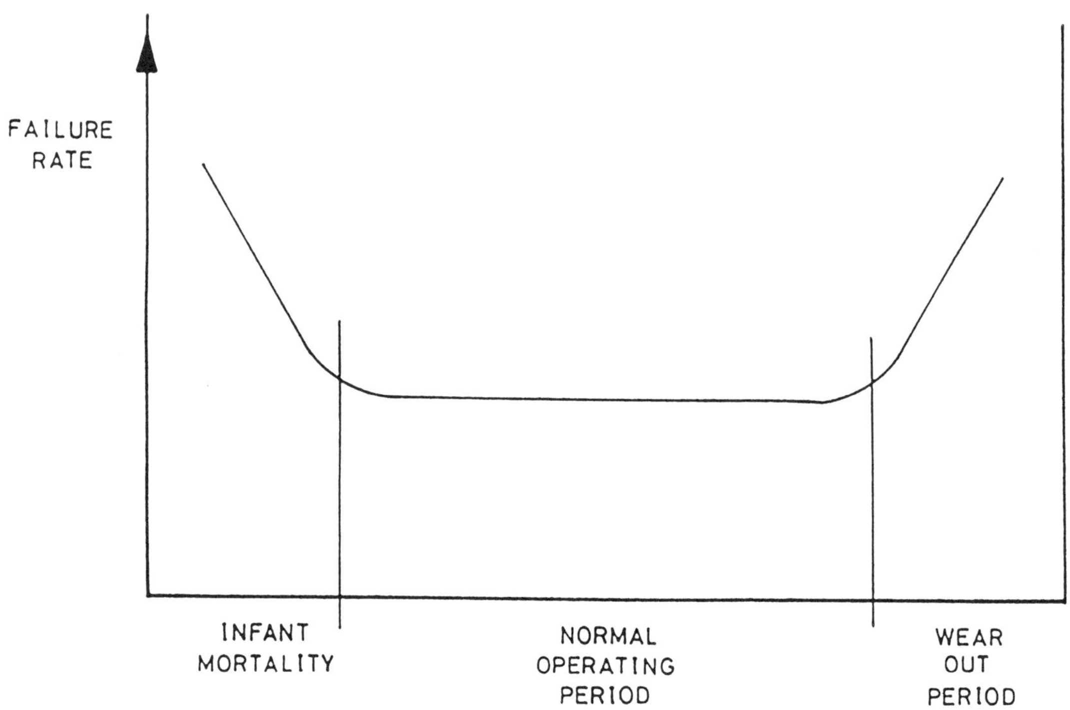

Fig. 6–3 Failure Rate Curve

etc. However, this data is particularly difficult to obtain on seabed cables. This is because of the variety of types that have been used and a general lack of data from the industry. Most reported cable failures can be attributed to extraneous causes, such as anchor lines, trawl boards, etc., and have no significance to reliability analysis.

Reliability evaluations. Reliability evaluations should, therefore, only be used in a relative way—to assess one option with respect to others and to identify weak points in design.

Critical Subsea Components

Electrical Connectors

Electrical connectors are considered by many people to be the most critical item in any subsea control system. The mistrust of wet make and break connections has, in the past, delayed the acceptance of electro-hydraulic systems. There are two distinct approaches to this problem.

Conductive Couplers

The most familiar concept is the plug and socket. Many attempts have been made to design and manufacture a satisfactory conductive connector (metal-to-metal contact). Although the majority of designs appear to be reasonable, they have a poor record of long term reliability. There are many designs adequate for short mission applications and for relatively low voltage levels. However, a design that can be confidently installed and used for long periods with complete reliability is still lacking.

Inductive Couplers

Inductive couplers use the transformer effect in place of a plug and socket and have replaced conductive couplers in most of the advanced subsea control applications. Inductive couplers are ideal for making and breaking underwater, since the transmission of energy is based upon an electro-magnetic field. A magnetic field is not affected by water and at no time can conducting metal become wet. The electrical principles involved are simple and well understood, but the behavior of an inductive coupler underwater can be defined only by experience in the long term.

The design of inductive coupler application must be considered very carefully. It also is essential that they be manufactured under conditions of high quality control. The method of mounting and the tolerances in gap dimensions between the two half-couplers are all design parameters which have been established by experience and are known to be quite critical.

The history of inductive couplers is one of comparative success. TRW Ferranti, for example, has produced more than 500 couplers for subsea installation. Only one failure has been recorded (in 1978) during more than 13 million unit hours of service. This failure was found to be due to a material incompatibility which has since been corrected.

Shell included a diver mateable pin to pin (conductive) connector option at the main power inlet to UMC stations in case of inductive coupler failure. These large couplers (21¾-in.) were "doubled" so that if one half failed, the coupler could be rotated without complete disconnection and the second half used.

The range of coupler sizes is given below as a comparison of the power transmitted:

Size (inches)	Power Transmitted (Volt Amperes)
4	200
5	500
7½	1,000
10¾	5,000
21¾	9,000

A feature of the inductive coupler is its ability to operate as an isolating transformer, which can be utilized to eliminate the need for fault isolation devices in a circuit. The transformer effect can be used to good purpose as a current limiting device.

Several types of signal couplers exist, and some can be used to transmit both power and signals at different frequency levels.

It must be mentioned that there are drawbacks in the application of inductive couplers. Most applications to date (on power circuits) have required power to be generated for transmission at higher frequencies, e.g., 200 and 400 Hz. Moreover, there is a power loss at each coupler due to normal iron and gap losses associated with the transformer effect.

This generates a heat loss at the coupler which must be dissipated. The gap between the mating faces of the coupler is isolated from the water by a skin or membrane, but this must be as thin as possible to minimize the gap in the iron circuit and consequently is quite fragile.

Any variation in the gap when mated, caused possibly by a sand particle, can have a detrimental effect upon coupler efficiency. Due to the induction of the coupling transformer, the circuit power factor can suffer severely, although this can be calculated and compensated by suitable connecting cable design. For example, use of "insert" type couplers, a mating male and female, can reduce the problems of entrained dirt and mechanical gap, but these require a greater degree of mechanical alignment than for face-to-face connections.

For applications requiring higher power levels, the inductive coupler still has considerable limitations, and it cannot work with direct current.

Although the use of inductive couplers is now the standard for wet make and break of control system connections in the

North Sea, at least, they must be applied with very careful engineering. There is still a considerable need for a reliable conductive coupler to be developed and, unless this occurs, installations operating at higher voltage and power levels will be restricted.

Sensors

One of the main advantages of an electro-hydraulic system is that it allows subsea data monitoring, though it is important to establish what data requires measuring. Any proliferation of sensors will affect the reliability of a subsea control system and so, as a general rule, they should be kept to a minimum.

Contrary to what might be expected, there are at present not many types of sensors with proven subsea records that could be used. Currently, various manufacturers have developed the following types for subsea completions:

- Pressure sensor
- Hydrocarbon detector
- Temperature sensor
- Wellhead inclination measurement system
- TFL passage detector
- Valve position indicator

Note that the word *sensor* and not transducer is used. In most cases the transducer represents only a component part of the sensor—together with housing, cables, connectors, and electronics.

If the sensors are located externally to the main control modules, they need be connected to them. A system which uses one inductive coupler only, for both energizing and transmitting signals, has been developed and is being integrated into some present subsea projects.

The sensors which still need to be developed for subsea use include:

- Water cut
- Sand detectors
- Flowmeters

As we have seen, pressure, temperature, and valve position sensors already exist with the required reliability. However, it may be necessary to package them for the intended use.

More important is the sensor installation and retrieval system and the various options (discussed earlier in this section). Eliminating the cables and making the sensors easier to install by ROVs must be a requirement.

The use of short-range acoustic control communication across a distance from a central control pad to an adjacent subsea instrument cluster, or Christmas tree control system (on a template for example), is feasible as distance is minimal and hence signal integrity enhanced.

Hoses and Hose Bundles

This area still remains one of the most critical in any control system. Manufacturing and installation costs, risk of external damage, and failures in service are all problems which need to be solved. Also, the number and size of hoses must be minimized to reduce costs and increase reliability. Duplication is most important, to avoid loss of production upon loss of one single hose.

The cost element of umbilical construction is often stressed but rarely considered. The cost of installing the umbilical far outweighs the actual umbilical cost. Similarly, the umbilical purchase cost constitutes a limited percentage of the overall control system price. Agreed, these costs must be minimized, but the prospect of compromising the overall operability/reliability/back-up availability by reducing the content of the umbilical could be considered false economy—risking the ship for a few pennies.

When specifying a main control bundle the following criteria apply:

- Reduce number of hoses
- Include spare(s)
- Reduce diameter of bundle
- Reduce number of joints
- Protect bundle (e.g., armoring)
- Specify at least 4:1 safety factor
- Qualify hoses
- Test hoses and bundles comprehensively

One technical problem, which exists mainly for hydraulic systems, is the response time of hoses, i.e., the time which is taken for a certain pressure to appear at the far end of a hose when the near end is pressurized. The response time follows a curve as shown in Figure 6–4.

This delay, due to volumetric expansion of the hoses, can be reduced by specifying the correct design prior to hose manufacture. The effect, however, can be rendered unimportant by using electro-hydraulic systems with subsea accumulators.

As an alternative to hoses, metal-cored umbilicals could be used. However, the laying of such umbilicals requires significant attention to the back tension, and this may incur a cost premium during deployment.

Electronics

A phrase quite common a few years ago and which is still heard even today is, "If it has electrics, I don't want it underwater." This is understandable. Failures have occurred in the past due to unreliable electrical connectors, badly designed electronics, poor housings, and a general lack of knowledge regarding electrical applications underwater. Electronic devices can be designed to be extremely reliable with several levels of redun-

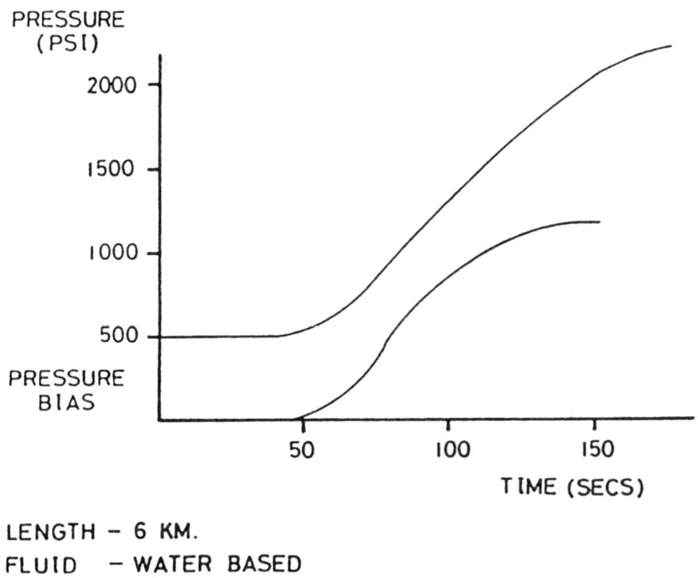

Fig. 6–4 Two Typical Hose Response Times

dancy easily included in a system. It also is easy to provide fault isolation, switching, etc., and all with a single pair of wires.

Nowadays it is standard practice for the industry to use electronic subassemblies.

It is, however, necessary to make "reliability" a serious consideration.

With a single electronics subassembly for the control of a well, a probability of success of 90% or better in a five-year period could be expected. For an electronics package with redundancy, this value should easily be exceeded.

The Shell UMC and BP Magnus, Texaco Highlander, Sun Oil Balmoral, Statoil Tommeliten, Norsk Hydro Togi, and other projects are all using electro-hydraulic systems.

Use of the American military MIL specifications to qualify electronics, together with better packaging, is making electronics more acceptable. Additionally, the simplification of the role of the pod electronics, by absorbing the complexity into the software functions, has reduced the amount of electronics deployed and hence increased reliability. This effort is currently being pioneered by BrasNor.

While the rate of advancement in the electronics industry greatly exceeds the pace at which subsea control systems develop, the need to have a proven system, with a long field record and successful electronic components, has to date restricted the use of the latest electronic technology. Even accounting for such delays, control systems are nowadays making use of the newer components and technologies—for example, complementary Metal Oxide Semiconductor (CMOS) technology.

The use of Large Scale Integration (LSI) and Very Large Scale Integration (VLSI) not only minimizes the electronic subassemblies but increases reliability.

Hybridization of circuit boards will give an economic solution for small production batches; the use of microprocessors subsea will be delayed until very high reliability chips are available.

A suitable microprocessor will allow fault isolation, control or power loading, effective communication, selective interrogation of sensors, and many other features which will simplify control systems.

Hydraulics

If it can be achieved simply with an all hydraulic system, then this solution would appear to be preferable—single satellite wells for water injection being typical application of direct hydraulic systems. But as soon as one of the following factors applies then serious consideration should be given to electro-hydraulic systems:

- Large number of functions, i.e., large hose bundles
- Data retrieval
- Long transmission distance
- Emergency shutdown requirements

Electrohydraulic systems also have hydraulic components.

At present there are three such main critical components in subsea control systems:

1. Control valves
2. Hydraulic connectors
3. Subsea accumulators

Operational installations prove that these components are both available on the market and work satisfactorily.

Purely hydraulic systems will continue to be developed for simple applications, but the long-term aim must be to reduce the size and cost of umbilicals. For example, there are mineral oils being tested with viscosities of less than 10 centistokes. Their use will maximize the advantages of using a mineral oil, without the problem of high viscosity. Also, work is being carried out on water-based fluids, and these are becoming more and more acceptable.

For details of hydraulic fluid selection, refer to the section on Control Fluids.

Hydraulic components also will need improved performance and reliability. For example, while operating pressure levels of 1,500 and 5,000 psi are specified for most hydraulic systems, obvious advantages would result from higher system pressures. At levels of 5,000–7,500 psi, present control systems would require only minor modifications.

An increase in pressure would allow a reduction in actuator size, faster response times, and smaller subsea pressure accumulators and smaller bore control lines.

Control Valves

Control valves should be as simple as possible, and a two-position, three-way valve is preferable. The valve selected must be well proven, but its life will depend on the selected hydraulic fluid.

With a cleaner fluid, the longer the life; for instance, National Aerospace Standard 1638, Class 6 is being specified for most control systems in the North Sea. Also, it is essential that these components be assembled with the greatest care. Reliable control valves already exist, but the aim of manufacturers must be to make them more tolerant of dirty operating fluids. In the case of solenoid valves, another aim is a reduction in power consumed.

Shuttle valves are a very critical component and one which will be used increasingly subsea. Therefore, they must be developed to meet reliability requirements. These valves have a historically poor performance in subsea applications. Nevertheless, they are probably the best available current solution where secondary or backup systems are used.

Hydraulic Connectors

These should be self-sealing to avoid the ingress of sea water and dirt, as well as allow the hydraulic circuit to remain pressurized when unmated. Metal-to-metal seals are preferable and particularly important when used with aggressive fluids, e.g., injection chemicals or crude oil. Correct design, manufacturing tolerances, and proper installation of hydraulic connectors will determine their success.

Subsea Accumulators

Bladder type accumulators are now the standard in subsea use. Their main drawback is gas permeation which, with time, reduces their precharge—a factor to be accounted for. Low permeability bladders must be specified for use subsea, and even then it may be necessary to replace accumulators after 5–10 years of service. In some applications, the umbilical line retains sufficient accumulation to make the separate deployment of accumulator packages unnecessary, if the depressurizing of the umbilical can be controlled.

Electrical Power

Electro-hydraulic systems are finding increasing application but, even so, various developments will have to occur. It will be necessary to install and recover equipment by ROV with consequent attention to modular design.

A wider range of sensors will be needed. The power requirement for control functions will have to be reduced in order to minimize the size of connectors. The use of rechargeable subsea batteries will be more common and several years of work have already gone into battery testing. If batteries are to be used, however, without the complication of electronic invertors, then a conductive coupler will have to be developed.

Subsea electrical power generation subsea will have to be considered in order to reduce the need for long cable connections.

Although the application of a thermonuclear power unit may be unacceptable for environmental reasons, the use of thermal, mechanical, or chemical energy from wells is feasible. A considerable amount of development and trial work has already been carried out.

Communications

The communication requirements of a normal subsea control system are minimal. A single pair of wires can be used to transmit signals for the most complex manifolds over distances of 20–30 m (65–98 ft) very reliably. However, new developments may arise requiring the use of permanently installed equipment, e.g., subsea cameras or greater distances may be envisaged.

The current restriction on the application of optical subsea links involves the difficulty of making connections of fiber optic lines with the necessary degree of mechanical accuracy to avoid an unacceptable degradation of the signal. This requires alignment accuracies an order of magnitude better than those currently achieved by remote mechanisms.

Acoustic systems have so far been disappointing because of problems associated with multipaths, subsea currents, water thermal gradients, etc. However, simple installations, such as a single valve control, could well use an acoustic system. And, acoustics also could be used as a backup system or for the retrieval of data (in small quantity) in hydraulic systems, thus eliminating electrical cables.

Materials

The cost of materials in subsea control modules is minimal and irrelevant when compared with the consequences of corrosion failure. Low corrosion stainless steel is increasingly being considered for subsea equipment. Monel is used quite commonly, especially for hydraulic connections and piping. Ferrliumis is a material which has exceptional corrosion resistance properties.

Surface corrosion is only one of the aspects of the problem. Pitting and crevice corrosion can affect some of the most commonly used metals.

Figure 6–5 shows that some materials have good general corrosion resistance but can exhibit poor resistance to pitting or crevice corrosion.

When selecting materials for equipment enclosures to be used subsea, the electrolytic corrosion effect of dissimilar metals, either in contact or very close proximity, must be seriously considered.

The use of non-metallic enclosures, such as glass fiber reinforced plastic plates suitably pressure-balanced (oil-filled), have a great deal to commend them. Many years of experience have already demonstrated the suitability of this approach.

Metal	General Corrosion	Pitting	Crevice
Stainless Steel	VHR	VLR	VLR
Nickel-Chromium Alloy	VHR	HR	LR
Nickel-Copper Ally 400	HR	HR	HR
Copper-Nickel-Iron Alloy	HR	VLR	VHR
Carbon Steel	VLR	HR	HR

Key: VHR very high resistance to corrosion
HR high resistance to corrosion
LR low resistance to corrosion
VLR very low resistance to corrosion

Fig. 6–5 Indication of Corrosion Metals in Seawater

CONTROL FLUIDS

Aqueous Control Fluids

The selection of a control fluid is dictated by the configuration of the control system, the fluid selected being either mineral-based or water-based. In this section, water-based fluids only are considered, as they provide two advantages over mineral fluids. The major advantage is their lower viscosity, which improves response times. And, their water base allows them to be dumped at sea, thus eliminating the need for a return line.

At the 1989 Offshore Technology Conference in Houston, Norsk Hydro announced that following successful testing of a mineral-oil control fluid for dispersal and environmental damage, they would be dumping it into the sea—with the blessing of the Norwegian Petroleum Department. Their reasons for selecting a mineral oil are:

- Long-term stability
- Ability to absorb up to 5% water without degradation
- Lubrication characteristics
- Inhibition of biological growth

This approach naturally negates the main advantage of return fluid systems. Once filled, the system requires minimal topping off, whereas water-based fluids are continually being replaced.

Although many types and mixes of water-based or aqueous control fluids are available, the most generally acceptable for production control systems are the glycol additive fluids. These may contain up to 50% glycol with 5% proprietary additives, the balance being deionized water. The additive package normally contains chemicals to inhibit corrosion of ferrous and yellow-metal alloys, to give lubrication properties, reduce foaming tendencies, and to inhibit microbiological growth. The additives should be balanced to ensure that the resulting mix is tolerant toward ingress of seawater.

Generally, the control fluids should exhibit the physical properties as given in Figure 6–6. Stability graphs also are given relating to the same fluid (see Fig. 6–7 and 6–8).

TYPICAL PHYSICAL PROPERTIES		
Viscosity (cS)	@ −20° C @ 0° C @ 20° C @ 40° C	10.5 7.2 5.2 4.2
Pour point	(° C)	−25
Specific gravity	(@ 20° C)	1.055
pH value		9.3
Flash point		None
Fire point		None
Cleanliness level		NAS 1638
Bulk modulus (N/m²)		2.3×10^9

Fig. 6–6 Water Based Control Fluid With 40% Glycol

Aqueous fluids can be supplied with very low viscosities (often approaching 1cS at 4°C) and have a compressibility of below one-half of that of mineral oils. When correctly formulated, they are much more tolerant of ingression of seawater, a condition which can make mineral oil-based products break down severely, destroying many of their system-protective properties. Aqueous fluids have been used for many years as the

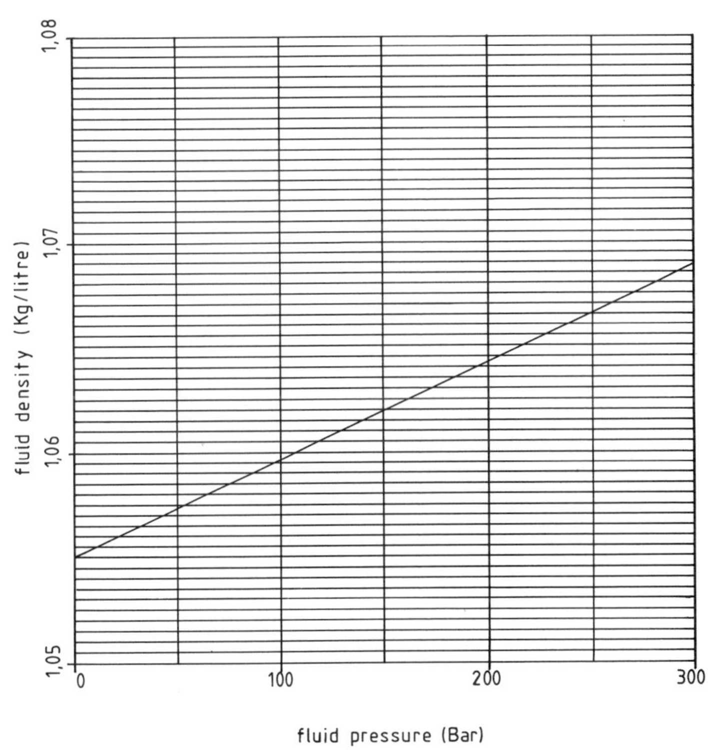

Fig. 6–7 Density-Pressure Relationship for 40% Glycol Control Fluid

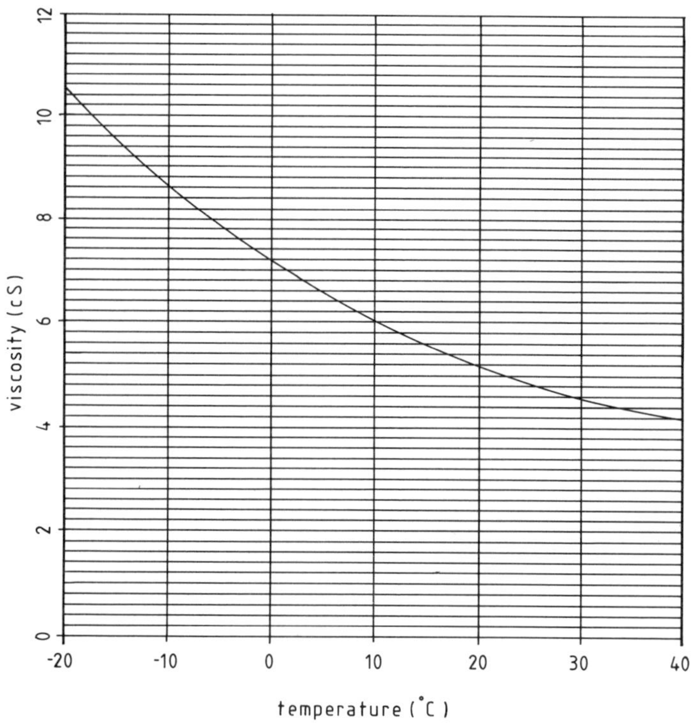

Fig. 6–8 Viscosity-Temperature Relationship for 40% Glycol Control Fluid

operating medium for BOP systems, and several subsea control systems now are used with aqueous BOP fluids. The main drawbacks in the use of these dilutable fluids result not from any lack of performance in the concentrate itself, but in the difficulties in maintaining adequate control procedures with the final fluid mixture.

Fluid Selection

Total-loss hydraulic subsea systems involve a continuous replenishment of the control fluid. Factors such as the degradation of chemical additives in service are therefore largely irrelevant. For example, ESD valve systems could remain unfunctioned for several years, with the hydraulic fluid remaining dormant in a spur line or actuator. Many subsea systems are, however, either of the closed-loop or dead-ended design format. When these designs are coupled with the difficulty and expense of carrying out even routine maintenance work, obviously it is necessary for the control fluid to have as long a lifespan as possible. This lifespan depends on two main factors:

1. Make-up of the fluid itself
2. Environment in which it operates

The first factor, the make-up of the fluid, is relatively simple to evaluate. The primary function of any hydraulic medium (after the requirement of power transmission) is to protect the hydraulic system. Main causes of hydraulic system malfunction are wear and corrosion, and each of these undesirable effects can be the result of several conditions.

The operating fluid should contain additives to combat the corrosion of all metals used within the system. There are some metals whose use is not recommended with any water-based fluid, notably zinc and cadmium, a point to be considered when designing new systems. The fluid also should contain additives to reduce system wear. Generally, the additives used react with the metal surfaces at high contact temperatures, reducing intersurface welding and scuffing. The performance of the fluid in either of the aforementioned areas (corrosion and wear) can be readily determined by using standard test procedures, and a ready comparison can be made between the performance of two or more fluids. Difficulties arise when attempting to evaluate the performance of fluids under the effect of various environmental conditions which may not, on first examination, be anticipated.

Hydraulic-fluid performance can be adversely affected by several factors in service. Ingression of seawater is a condition that is almost inevitable to some degree when considering the need to include some form of subsea connector mechanism. Many fluids are severely degraded in the presence of even low levels of seawater, as this can reduce the stability of the fluid to a level at which some additives may fall out of solution. The additive content should be high enough to ensure that long periods of use do not allow them to be consumed to such extent that they are no longer present in sufficient quantities to perform their function satisfactorily.

Also, when a dilutable concentrate is used, it is essential that the equipment used for preparation of the final product is of adequate accuracy and that it allows thorough homogenization. For this reason, and for others dealt with below, it is far more desirable to use a ready-mixed fluid.

By far the greatest cause of fluid degradation is the action of micro-organisms such as bacteria and fungus. These organisms can feed on and consume virtually any organic material and are hence capable of damaging the protective properties of the operating medium. In addition, their by-products can be highly acidic, causing corrosion and in some cases, serious structural damage.

Dead bacteria are often abrasive also and so can cause accelerated wear of moving components. All these organisms can block filters also and jam closely fitted parts such as solenoid valves. Obviously, the presence of these organisms is highly undesirable, and any fluid used should contain an effective biocidal and fungicidal agent. This agent should be of such a type that it only operates in the presence of micro-organisms and is not used up before any contamination takes place.

Micro-organisms can enter a system in several different ways, the most common being seawater contamination. All seawater contains micro-organisms but, provided that the level of ingression is not excessive, the fluid additives should be adequate to prevent rapid growths.

Any antibacterial additives will only be capable of dealing with a certain level of contamination, and continuous attack eventually results in a rapid growth which soon degrades the fluid.

Such a continuous ingression of micro-organisms is frequently the result of the water used for diluting concentrates being contaminated. Over long periods of time, such low levels of bacterial contamination can be detrimental to the system. Many of the potable water systems encountered on offshore installations deliver impure water, owing to infrequent or incorrect maintenance. The other main source of bacterial contamination is the system itself. Many components are manufactured before installation and become contaminated during storage and even at the manufacturing stage if the machine's coolant contains bacteria. The completed system may therefore contain inaccessible pockets of heavy growth. To minimize this risk, a system cleaner can be used in solution with the initial change of fluid. This is a biological shock treatment which cleans and flushes the system before normal operation commences. Also recommended is the use of a preservation fluid to protect equipment in storage from bacterial and corrosive attack.

Viscosity and Temperature Requirements

The most important factor in determining the response speed of a long-range remote hydraulic system is the viscosity of the operating fluid. For most cases, the viscosity should be as low as possible, although occasionally some damping effect is necessary. Virtually all subsea control fluids consist of an additive pack in aqueous solution, along with ethylene glycol to prevent freezing of fluid in parts of the system exposed to sub-zero temperatures. Subsea system areas below a depth of about 100 m (328 ft) may be considered to be at a stable temperature of 4°C (39°F).

Where ambient temperatures never fall below zero, no glycol is required, and the operating fluid will have a viscosity of approximately 1 cS. For lower ambient temperatures, however, increasing percentages of glycol are included in the fluid. Therefore, as glycol has a higher viscosity than water, the overall fluid viscosity increases up to a maximum of about 10 cS at zero for a fluid freezing point of minus 60°C. Therefore, it is important to accurately anticipate the operating conditions to optimize response time while avoiding freezing.

Another important although less obvious factor should be considered carefully when designing systems in which various components are being supplied by subcontractors. It must be made clear to the supplier that their own particular piece of equipment is only part of an overall system, as it is not uncommon to come across two or more types of valves with different fluid requirements, while they are in fact to be operated on the same hydraulic circuit. Some co-ordination is obviously required to arrive at a simple fluid which will satisfactorily operate in all parts of the system.

Material Compatibility

Many equipment suppliers have their own preference for components, such as seals and wear strips, making it essential to consult with the field manufacturer to ensure that there will be no adverse effects. Most synthetic elastomers, for example, are satisfactory with most aqueous fluids, although there are a few exceptions.

Compatibility testing of elastomers and plastics is carried out by totally immersing a sample of the relevant material in the fluid. A clean, dry sample of the material to be tested is weighed accurately and its thickness recorded. Then the sample is placed in a container which is filled with the test fluid. The container is sealed and placed in a controlled environment at the desired temperature.

The samples are removed at weekly intervals, dried, weighed, and dimensions rechecked. Then the samples are replaced and the test continued. Testing terminates once the readings stabilize. The data are presented as percentage changes in sample weight and thickness over an eight-week period at each test temperature.

Interpretation of results obviously depends largely on conditions of use but, in general terms, changes in properties of below 5% prove to be acceptable, provided that stability has been reached before the end of the eight-week test period.

One important point, however, is that some porous materials, particularly paper gaskets and filter elements, will absorb moisture readily and can become structurally unsound. Where these materials are encountered, suitable substitutes should be employed.

All control fluids should afford good anticorrosive protection to ferrous and yellow-metal alloys but, as previously mentioned, zinc and cadmium plating should be avoided. Preferably, aluminum should be anodized.

Most paints used in the offshore industry are resistant to aqueous fluids (recommended paints are epoxy, phenolic, nylon, and poly-urethane), but the painting of internal hydraulic-system surfaces (e.g., reservoirs) is not recommended, as, although no chemical attack will occur, the coating can flake off if cracked, contaminating the control system.

Although at first glance there may appear to be a great deal to consider before aqueous fluids for subsea hydraulic control systems are adopted, the long-term advantages in efficiency of control are obvious.

ELECTRO-HYDRAULIC SYSTEMS: MAJOR COMPONENTS DESCRIPTION

A typical multiplexed electro-hydraulic (EH-MUX) system should be designed to meet the functional requirements of the production system. It should provide the needed field-development flexibility and be as reliable as possible. Figures 6–9 and 6–10 are schematic representations of typical systems and show a practical representation of the key components.

The system would be designed to operate in a safe manner without significant risk of pollution and with a minimum of lost production time. Systems must be able to function under "re-configuration," i.e., if one element (well) is off-line because of workover, the remainder must remain fully functional without requiring any major re-configuration of surface or subsea control equipment. This is accomplished by designing the controls to operate only as long as the system is safe and not leaking, and to shut in production and automatically stop injection when any part of the system is in danger. Subsea equipment must be of the highest possible reliability, which is achieved by using proven high-reliability components in the simplest possible circuitry.

Equipment mounted at the surface would include: control-station hardware, electrical and hydraulic power-supply units and, for retrievable installations, umbilical reel units, etc.

Surface-to-subsea supply riser umbilicals would be provided for electrical-power and hydraulic-fluid supply. In addition, a telemetry link must be provided to communicate with the subsea control modules. All these links may be bundled together to form one riser umbilical, either armored or unarmored, depending on the deployment method.

Subsea components which typically require individual control modules are:

- Template trees
- Satellite trees
- Template manifolds (of different types)
- Special valves (export pipeline, etc.)
- Subsea pigging system

Each component of the system may have a dedicated control module for actuation of tree valves and for monitoring selected flow pressures and choke positions. In certain instances, a module may share its controlling functions with other components for the sake of economy.

Therefore, to minimize the number of control modules, functions may be combined according to the detail design of the subsea production system. In addition to the surface-to-subsea supply, umbilical connections are required from the template junction box to the individual control modules and out from the template to the various satellite trees or manifolds.

Topside Master Station/Computer

Control Room

The control-room equipment should be designed to make the system as simple and convenient as possible to operate, while maintaining a high degree of availability. All items should have redundancy to avoid interruption of services. Also, provision should be made for keeping permanent records of system activities to permit analysis of performance and to provide information for troubleshooting. It is important to ensure that all equipment housed in the control room is designed to respect the various environmental limitations set by certifying and statutory bodies.

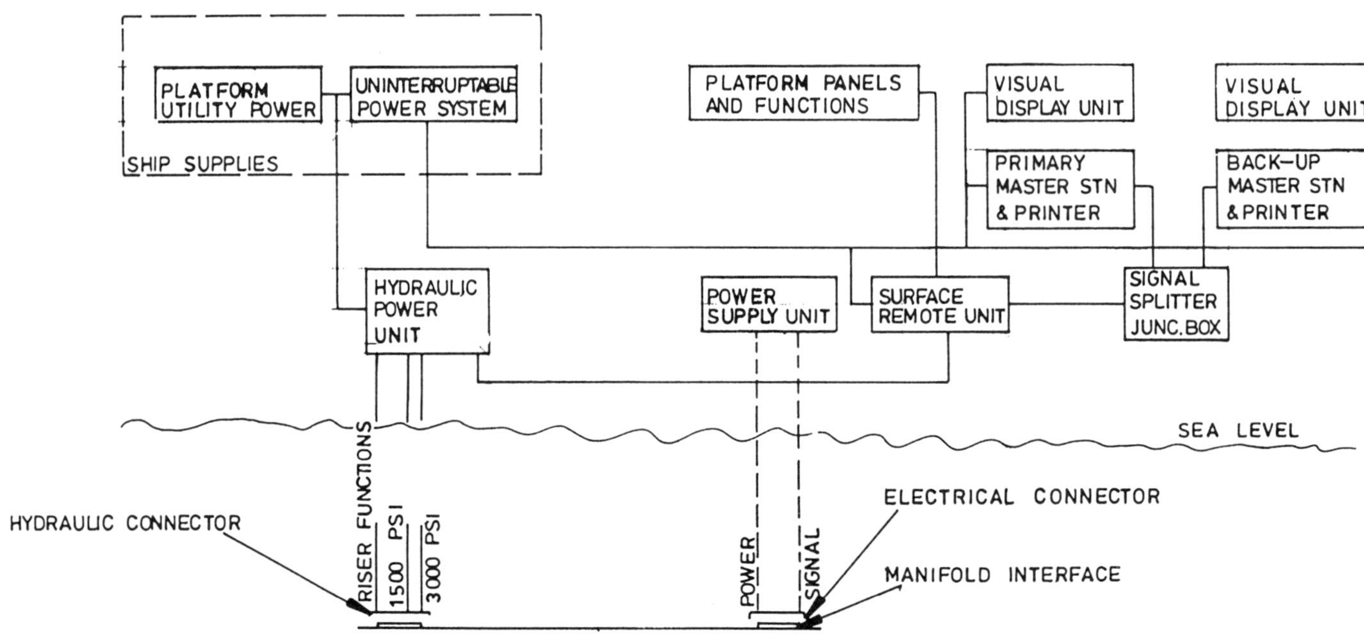

Fig. 6–9 Electro-Hydraulic System Components

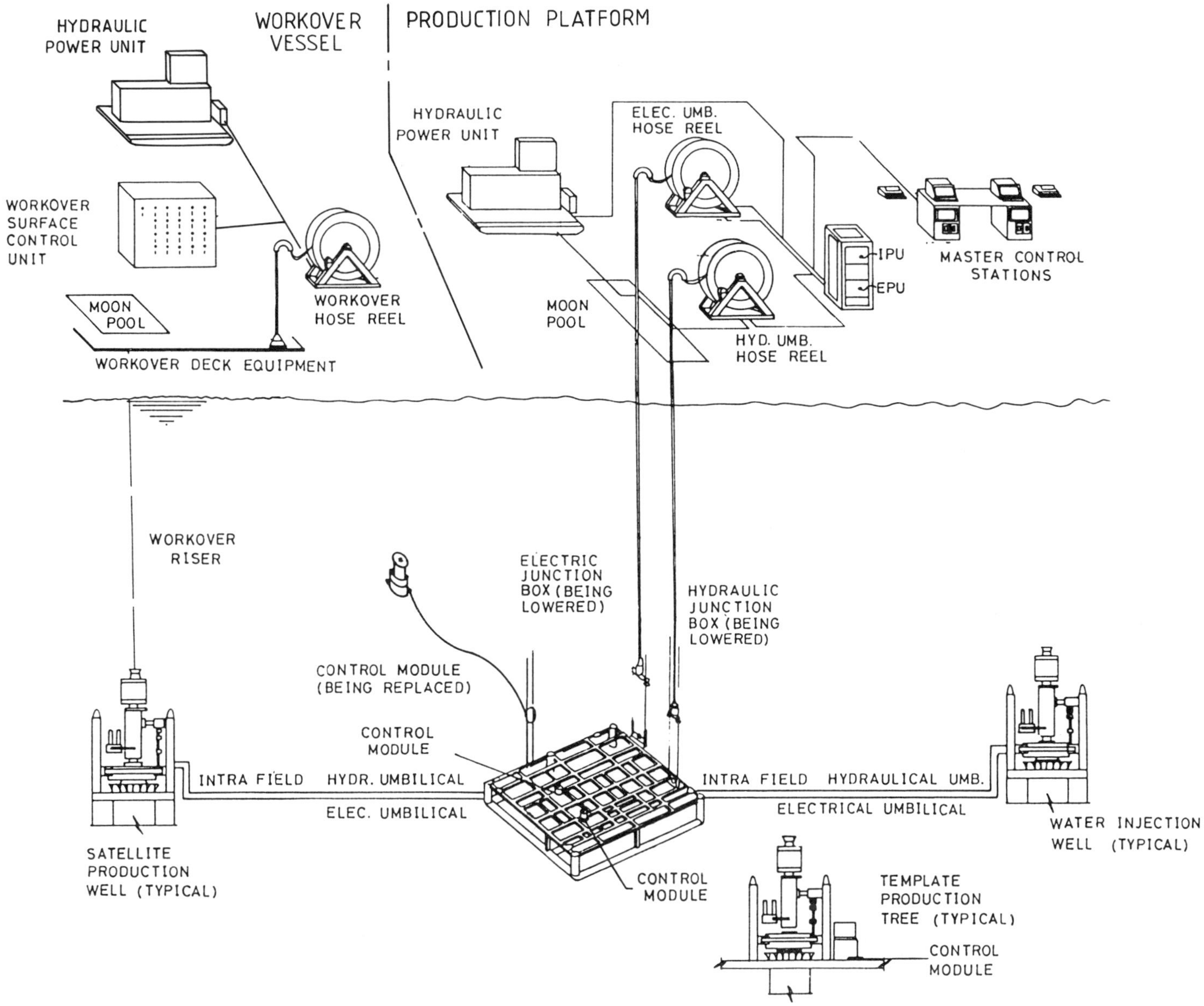

Fig. 6–10 Electro-Hydraulic System Components

This particularly applies to temperature and vibration levels for the input/output racks and the microprocessor.

Master Station

The master station is used to operate the subsea system and includes a visual display unit consisting of a cathode ray tube (CRT) display and a keyboard. This is the primary interface with the operator and, as such, the operational software is designed for operator convenience. The visual display units are identical, and each can be connected to either computer. Figure 6–11 shows a subsea control master station and its relation with the platform's distributed processing system using a data highway principle.

Screen formats are provided to allow the operator to view all status points, analogue values, and control output states. The preferred method of presentation for data read-back from the templates is a graphic display showing simplified piping diagrams of valves and transducers. All alarm conditions or deactivated functions should be clearly indicated wherever such points are displayed. Figures 6–12 and 6–13 show typical CRT displays for the control of the field and a tree.

For operator convenience, formats may break the system into smaller units for display. The most logical division is a format

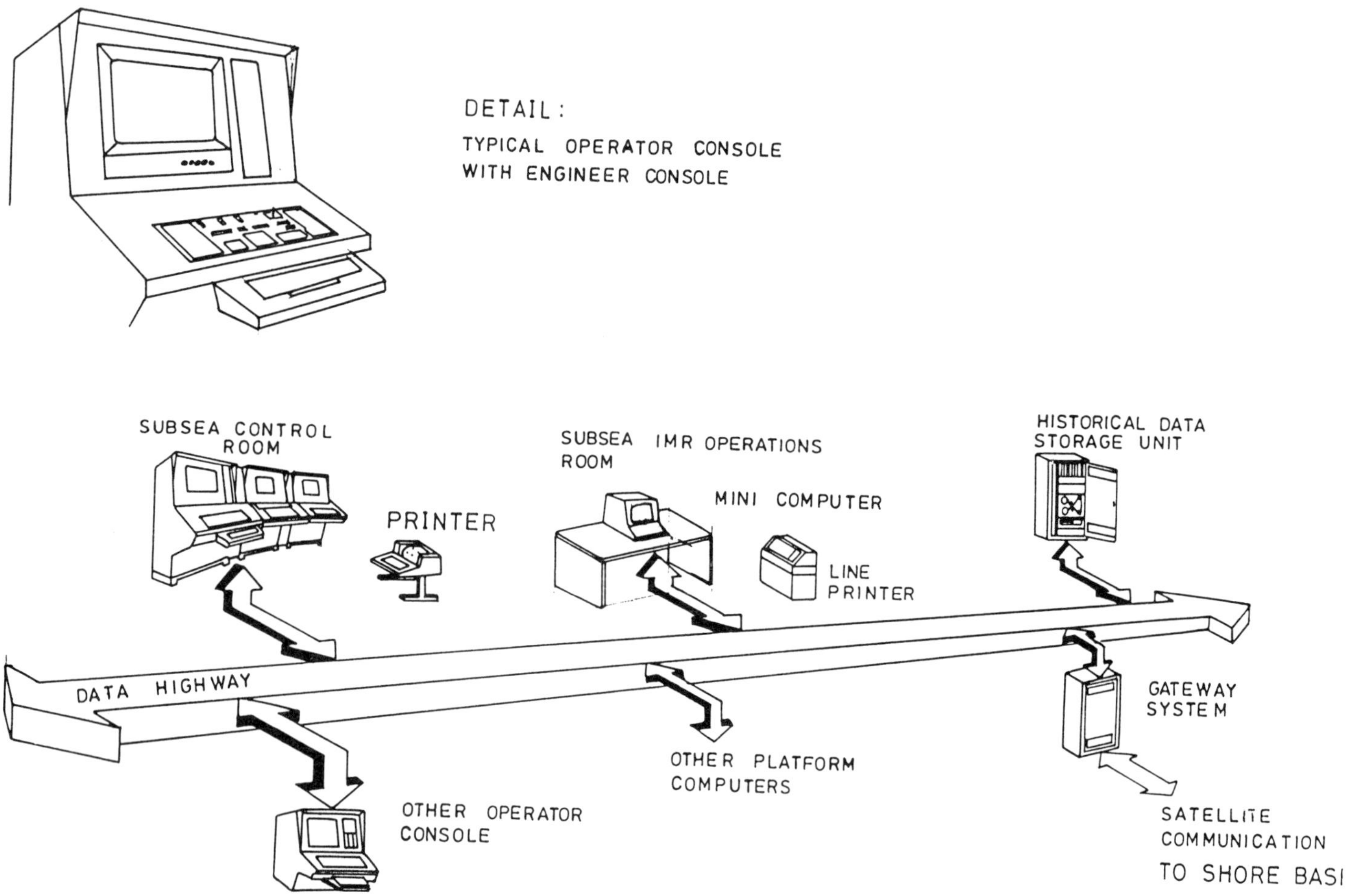

Fig. 6–11 Subsea Control Room and Distributed Processing System

for one well, another for platform functions, and yet another for the common template functions. This division is recommended. Other formats may include communications, error statistics, and a summary of functions in alarm states.

The master station includes a printer readout for keeping permanent hard-copy records of system operation. In practice two printers would be used, identical with each other, capable of being connected to either computer (one at a time). The printers provide a copy of operator commands, event alarms, and other information—in particular, analogues from the modules to provide housekeeping data for the module-maintenance program.

Subsea Computer

Full redundancy should be provided to allow for the automatic (bumpless) change-over between the on-line computer and a standby—the standby computer being live at all times and linked with the on-line computer through a modem change-over unit. The modem takes a series of analogue values generated by the computer and converts it into a digital signal suitable for long-distance transmission through the signal lines. The modem also has circuitry to receive digital signals from the subsea modules and convert them back into an analogue sequence for use by the computer.

All communications between the computer and the remote units take place over a single pair of wires: the data bus. Data traffic on this bus is controlled by the on-line computer, since no remote unit will transmit unless told to by the computer. However, the subsea pod may generate an interrupting up-link signal to the topside controller, if an out-of-range valve is read.

Usually, if an out-of-range valve occurs, the subsea pod will re-poll the sensor/instrument to verify that this is not an error or transient and reports accordingly. This facility is optional.

The computer periodically polls the subsea modules sequentially to gather data on an automatic basis.

However, if the operator enters a command, the polling process is interrupted to send the command to the appropriate subsea module as soon as possible. The computer is responsible for timing control commands so that proper sequencing is observed

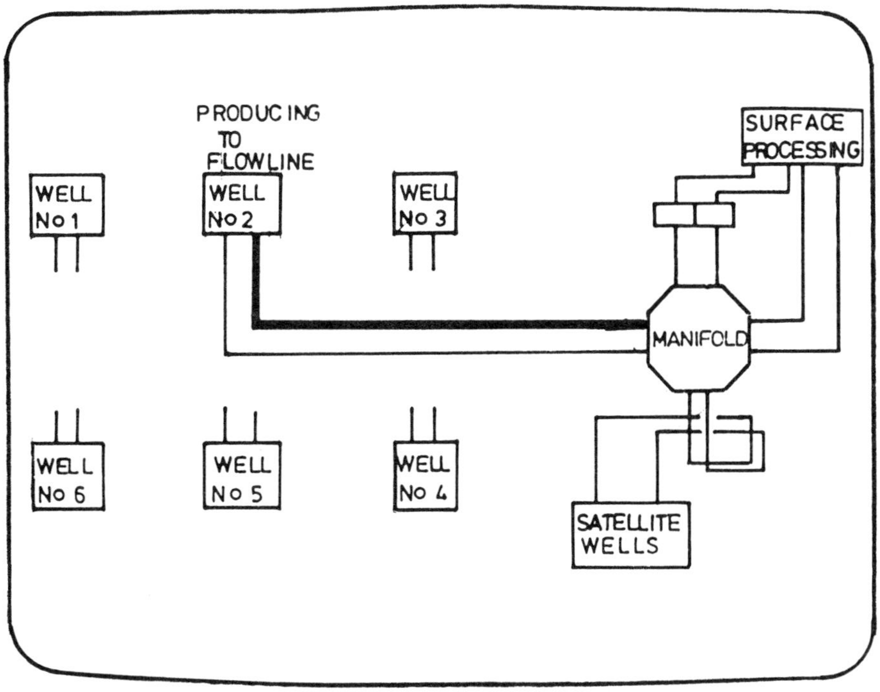

Fig. 6–12 Field Control—Typical VDU Display

and subsea power consumption limits are not exceeded. Figure 6–14 shows a typical message format between the master station and a control module.

The communications-link signals contain an error check mechanism deliberately generated to verify the integrity of the report.

Power Supply

Owing to the nature of electronic components, normal platform power, with its spiking tendency, is unacceptable as a power-supply source. For this reason, and to allow for an orderly

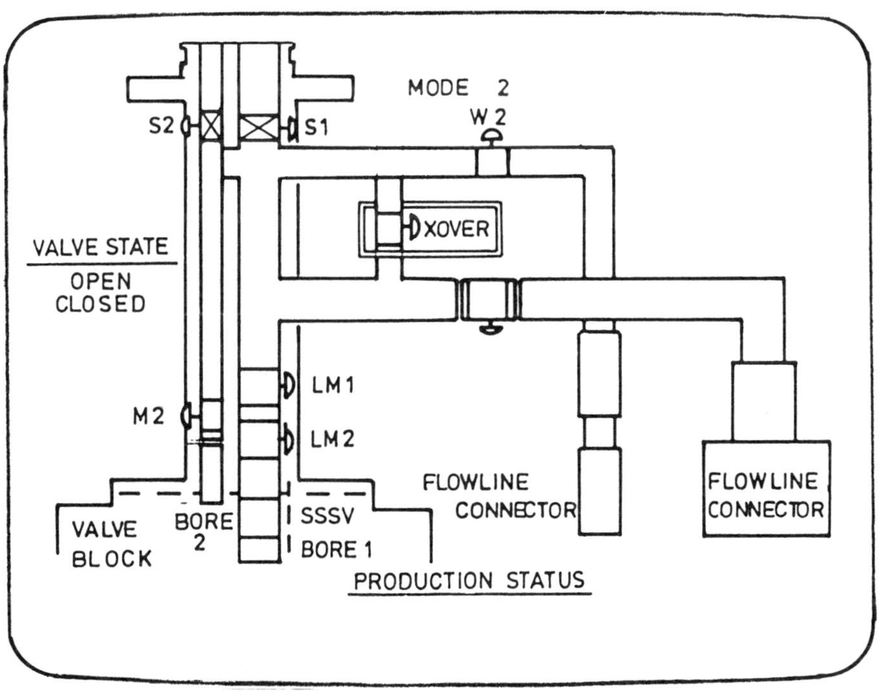

Fig. 6–13 Tree Control—Typicial VDU Display

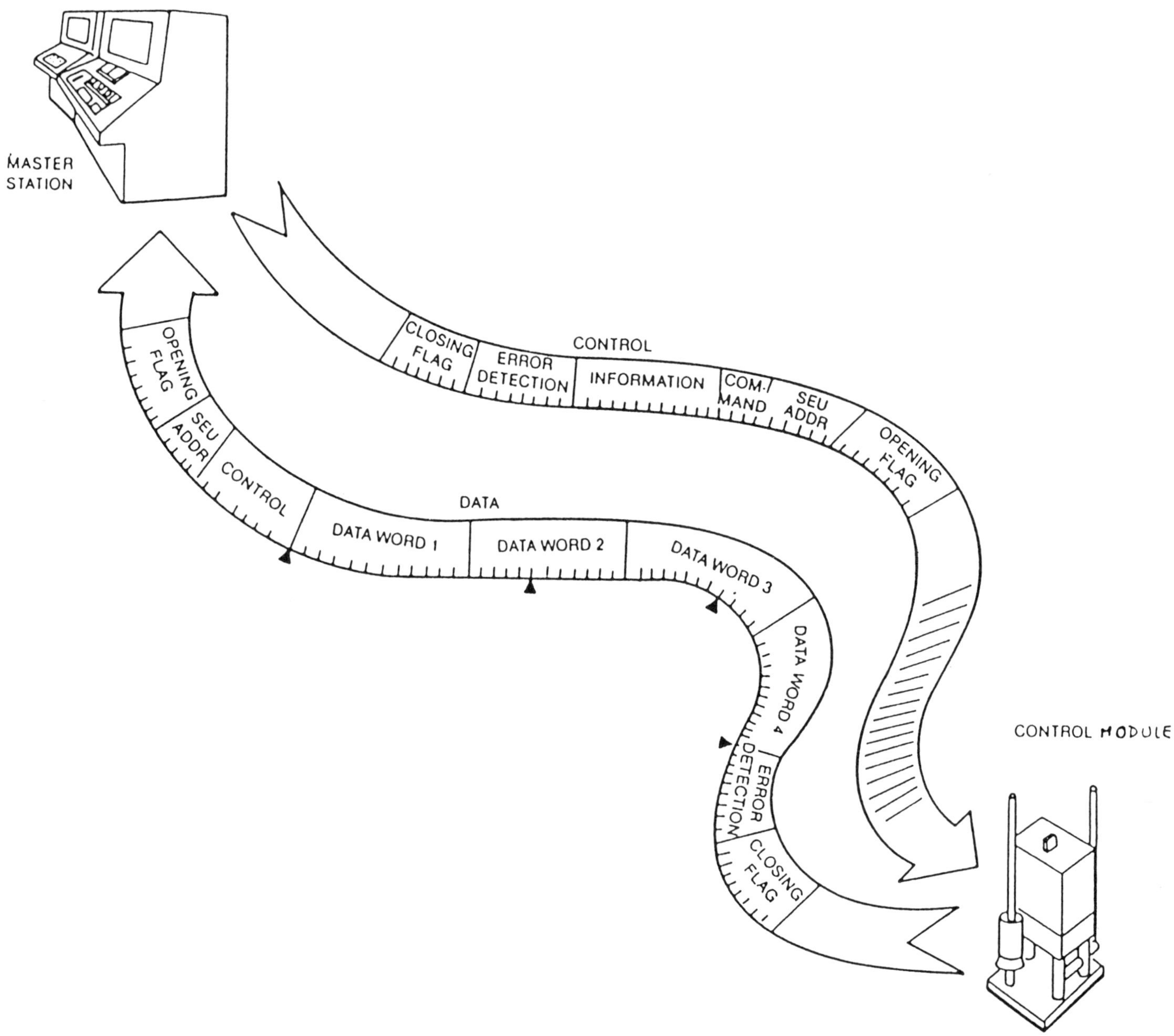

Fig. 6–14 Message Formats

system shutdown, the power should not be instantaneously interruptible. To ensure continuous, smooth power, the electricity is taken from a regulated supply with a battery backup to provide power during platform power failures.

Subsea Junction Box

A junction box may be provided on the template for the distribution of electric power and signals to the subsea control modules. This junction box is end-fitting connected or directly spliced onto the end of the umbilical.

Within the junction box itself, a bus bar distributes input power and signals to output cables. The bus bar and all connections are well off bottom inside the oil-filled junction box.

The junction box contains two compartments which isolate the umbilical/bus bar connections from the cable harness connections. These compartments are pressure interconnected but sealed off electrically by pre-wired bulkhead-type connectors. The junction box should be oil-filled and pressure-balanced through the diaphragm to the external ambient.

Figure 6–15 shows a possible design for an electric junction box, and Figure 6–16 represents a hydraulic schematic via the

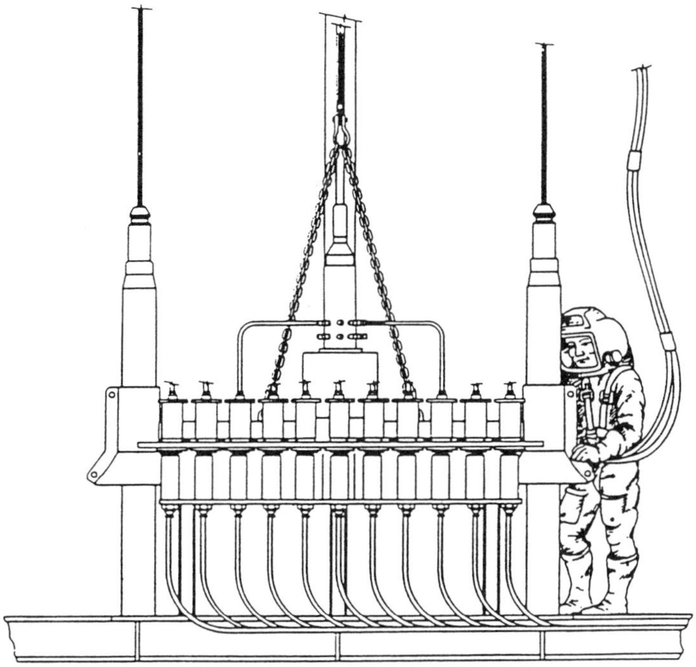

Fig. 6–15 Electric/Umbilical Junction Box

junction box. More usually, the hydraulic umbilical is connected directly to a hardpiped template/tree distribution and is not connected via the junction box.

This type of design minimizes the risk of water leaking into the box to short out the system. It also will help prevent faults in the cable harness or inductive couplers from shorting out the entire bus bar system or damaging the bus bar/umbilical connections.

Subsea Control Modules

Individual tree and manifold control is provided via subsea control modules. These modules contain the valving and associated electronics/electrics required for routing the hydraulic fluid to the various valve and choke actuators. Additionally, all monitoring of subsea system status is accomplished in this module. A universal module (i.e., all modules having identical configuration) may be considered to allow module interchangeability and to reduce system cost.

In normal operation, the subsea modules should be quiescent except when commanded from the surface; only the receiver circuitry and minimum power supply will operate continuously. When a carrier tone is placed on the communications line by the surface control, the logic in the subsea control module automatically powers up. If the downlink signal is addressed to another module or if the carrier tone is removed, the module tunes out and turns off.

In some systems, a second signal is required to verify that the correct command has been sent before the function executes—the ''check before execute'' function.

If a correctly addressed signal is received, the logic analyzes the signal and executes it. The surface unit will be programmed to automatically read appropriate pressures on the tree or manifold. This initiates the topside controller to make a decision or raise a flag if the command has not functioned or the operator has executed the wrong command. Unless specifically programmed, this is not usual. The topside controller display only indicates the last commanded function, and unless there is a valve position readback (not common), the operator is assumed to be correct.

Some subsea pods monitor certain pressures (flowline or hydraulic supply), and if that pressure goes out of range, a flag is raised and/or the pod shuts the tree automatically.

In abnormal operation, when electric communication is lost, all pilot valves fail ''as is.'' Then the system can be kept in production by use of the hydraulic supply lines. Some systems utilize a single ''fail-safe'' valve which is electrically latched and hydraulically isolates all other lines to the various valves. Thus, if there is a power failure, the pod fails and the tree closes. Some functions may be designated outside the fail-safe valve control and these fail ''as is.''

All production valve actuators are fail-safe when hydraulic pressure is removed. Thus, if electric communications are lost

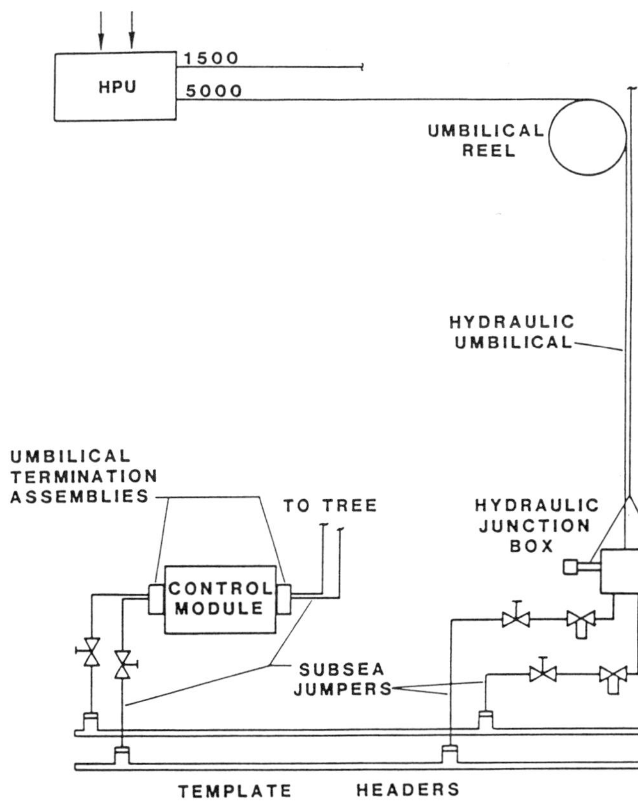

Fig. 6–16 Hydraulic Junction Box—Schematic

to one or more modules, production can be kept as it was by continuing to supply hydraulic pressure to the subsea template. The system can be shut in by removing the control supply pressure to shut all tree and manifold valves, and secured by removing the high-pressure supply to close the downhole safety valves.

For system backup, where a universal type of pod is adopted, only one fully equipped module is kept on board the production facility.

No redundancy is normally provided in the individual modules; BP however, has pods with EH MUX primary control and pilot or direct hydraulic backup.

During final design, it will be necessary to assign hydraulic and electric connectors to allow the universal module to be used as a spare to any of the specific modules.

Each module includes an electronic communications and control unit, pilot valves, and transducers. In systems such as Highlander, the sensor and conditioning electronics are external to the control pod, and the pod simply digitizes the analogue signal and transmits it.

The electronic unit is capable of receiving signals, analyzing them for proper address, selecting the correct function and activating a pilot valve or transducer, and transmitting data to the surface. The electronic unit need not be completely redundant but should at least contain enough internal redundancy to provide a highly reliable system. Only one set of pilot valves is included; shuttle valves are usually used to select one of two hydraulic supply networks and have nothing to do with the sensing/transducer system.

Figure 6–17 and 6–18 show typical schematics incorporating downhole sensors and all the functions necessary to control a tree. Receivers, transmitters, and power supplies in each module

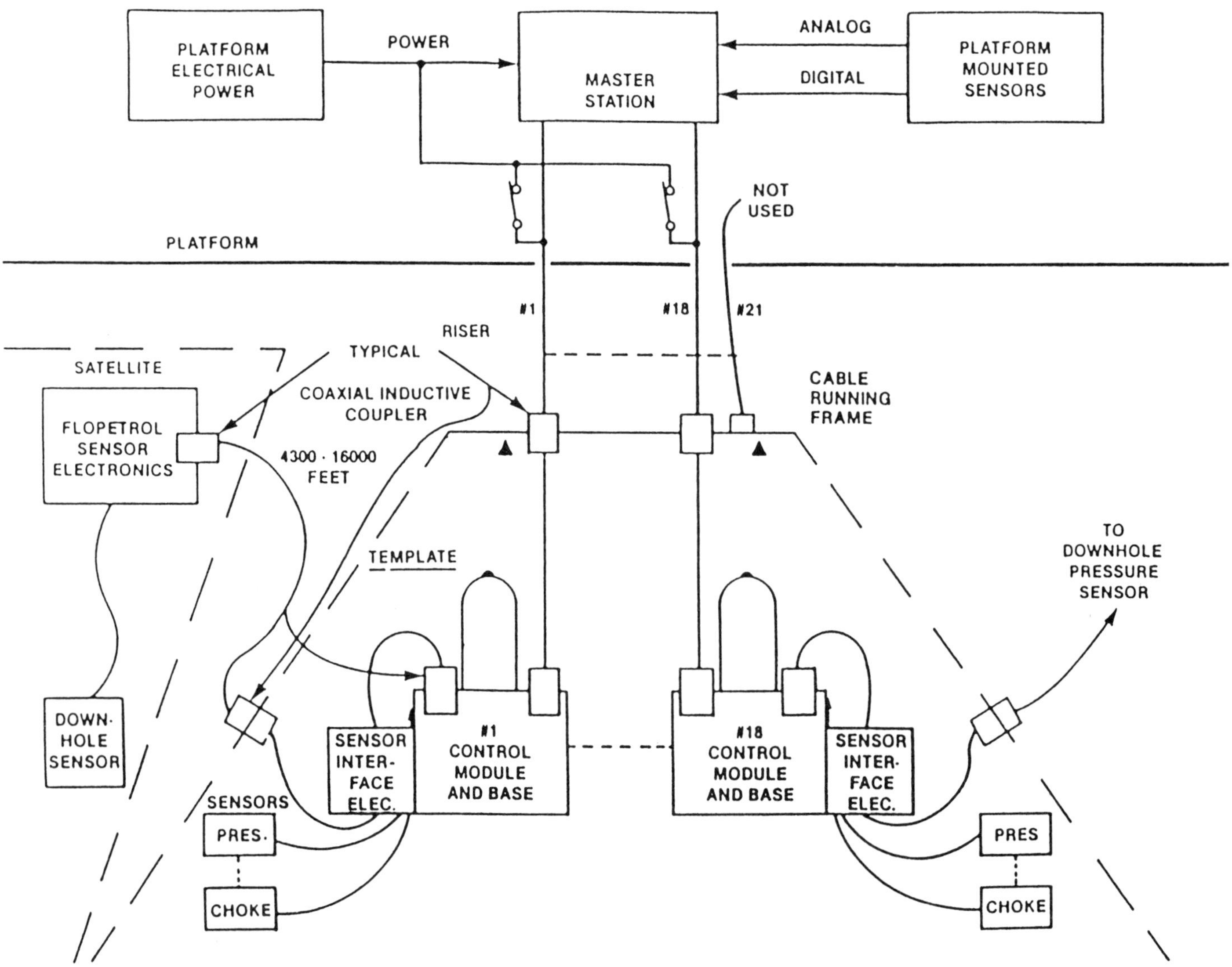

Fig. 6–17 Sensors—System Diagram

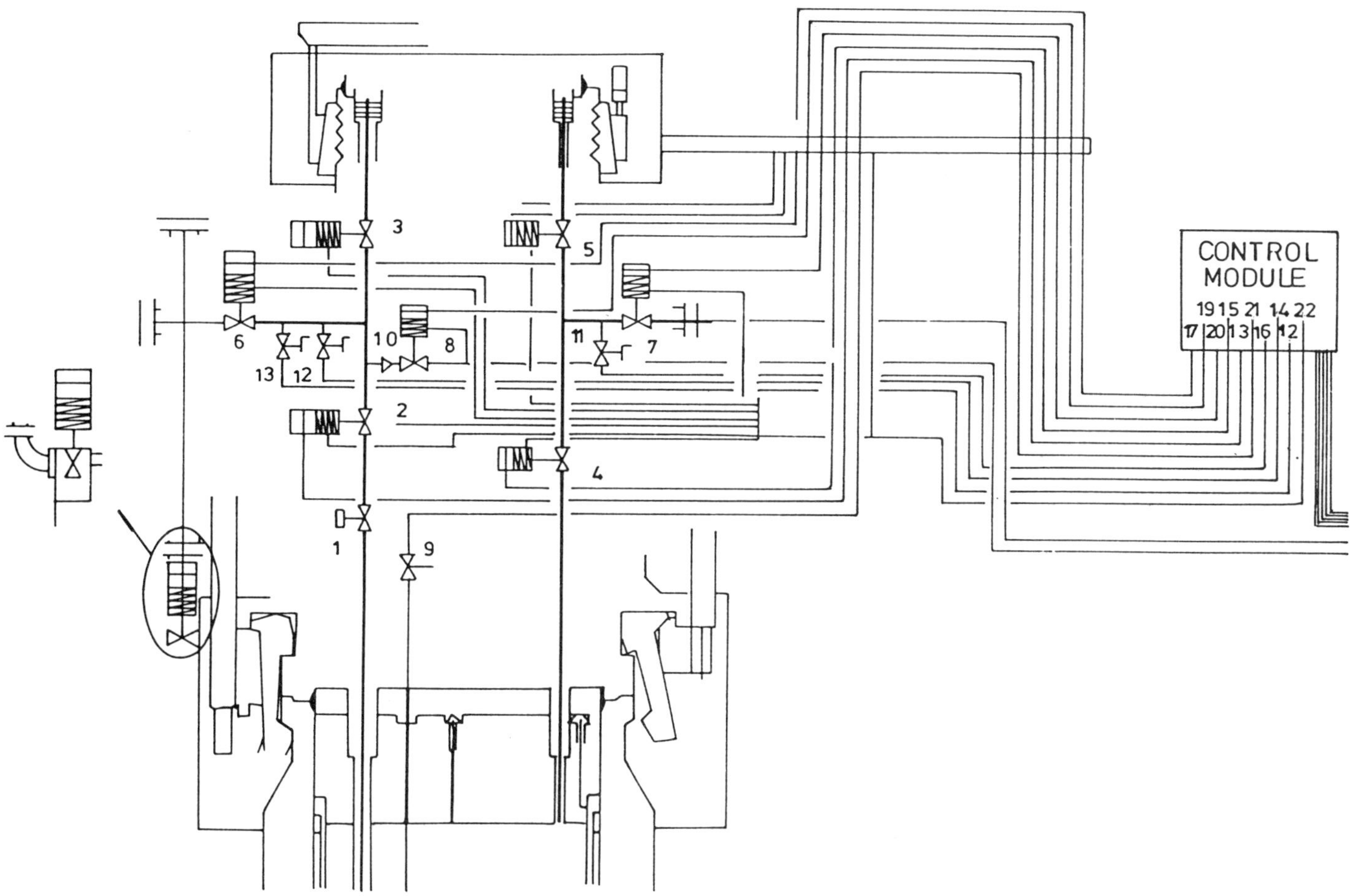

VALVE DESIGNATION

1. 5 1/8" LOWER MASTER VALVE
2. 5 1/8" UPPER MASTER VALVE
3. 5 1/8" SWAB VALVE
4. 2 1/16" ANNULUS MASTER VALVE
5. 2 1/16" ANNULUS SWAB VALVE
6. 5 1/8" WING VALVE
7. 2 1/16" ANNULUS WING VALVE
8. METHANOL INJECTION VALVE
9. SCSSV CONTROL LINE ISOLATION VALVE
10. METHANOL INJECTION CHECK VALVE
11. ANNULUS PRESSURE SENSOR ISOLATION VALVE
12. PRODUCTION PRESSURE SENSOR ISOLATION VALVE
13. PRODUCTION TEMPERATURE SENSOR ISOLATION VALVE

Fig. 6–18 Typical Well Flow Control System Schematic

must be able to operate with high noise levels and over a wide range of signal levels and power voltages (variations of +20% may be expected). This is necessary to make the system tolerant to the use of inductive couplers.

The electronics chamber may be dry nitrogen filled and the remainder of the pod filled with transformer oil, or a single atmosphere could be used.

Template Cables and Intrafield Umbilicals

Electrical Cables

Electrical distribution on a template would be provided by jumper cables extending from the electrical junction box to each subsea module. However, on the UMC, both hardpiped hydraulic and permanently installed electrical cabling was used.

Hydraulic Hoses

Hydraulic fluid would be supplied to each module via high-pressure and low-pressure supply headers in the template piping. Jumper hoses connect the supply headers to the riser interface and the various modules. Careful analysis of the hydraulic circuits may allow the exclusion of a subsea accumulator pack. This can be accomplished by using the inherent elasticity of the header and hose circuits to provide the necessary accumulation.

If system design requires only hydraulic-supply lines to satellite-well locations, then the required hydraulic lines would be laid along with, or as, the composite part of satellite-tree flowlines, all electrical functions remaining on the template with the tree control module (depending on the out-step distance).

Satellite Electrical/Hydraulic Umbilicals Routing

Where provision has been made in the design for both an electric and hydraulic power supply to the satellite-reservoir control systems, depending on the configuration and location of this equipment, the required umbilicals could be laid with the flowlines.

Connections

To minimize cabling and connections and to permit trees or manifolds to be removed separately from the control module, etc., each control module is located on the template near the well or manifold which it controls. Each module is connected directly to the template power and communication cables and to the hydraulic supply headers by short jumper hoses.

SURFACE EQUIPMENT

Hydraulic Power Unit (HPU)

At its heart, the subsea system has the Hydraulic Power Unit (HPU) which provides the high-pressure and low-pressure hydraulic power. Generally, the unit is constructed to provide an accumulated supply by using a bank of accumulator bottles and various let-down valves.

The hydraulic unit is powered by both air and an electrical supply. All arrangement provides two services: 1) as a means of backup if the main supply, either air or electrical, is cut and 2) as a means of topping up the pressure level for the subsurface safety-valve header. Emergency shutdown facilities are provided which let down the hydraulic pressure, resulting in the fail-safe valves on the subsea components shutting in. The HPU unit is normally skid-mounted and comes as a self-contained package for ease of installation and repair.

Electrical Power Unit (EPU)

Power is usually supplied to a subsea system at a high voltage to allow for losses and improve efficiency. On arrival, this power enters the Electrical Power Unit (EPU) and is transformed.

Within the subsea system, turbo machinery requires high voltage power or hydraulic power, whereas the control system needs only low power. Therefore, the EPU contains the transformer and electric motors which either supply direct electric or convert to hydraulic power.

The EPU should be a self-contained unit which is either skid-mounted or modular in form. Maintenance philosophy should be reviewed during the design of the unit, with adequate provision for inspection and repair.

Surface-to-Subsea Umbilical

A surface-to-subsea umbilical is required to supply the hydraulic and electrical power, together with the necessary signal pairs for the telemetry link between the subsea computer and the template and field-mounted equipment.

This umbilical may be either armored or unarmored, depending on the method of deployment and the need for mechanical protection.

Should the deployment be of the conventional tensioned design, the armor requirements are of a less stringent nature. However, the trend is toward the deployment of umbilicals in an untensioned mode by using a straight-down deployment. In this instance, great care must be taken to ensure that the armor is capable of withstanding the added fatigue imposed, due to umbilical motion and form. The current approach is to provide a lubricated layer between the inner and outer armor, to resist the chaffing and binding encountered in service. The umbilical hoses and conductors should be bundled, wrapped, and located within the envelope of the armor layers. The wrapping for the individual hoses and conductors should be of a mylar-type material.

Umbilical Reel Unit

The electro-hydraulic umbilical is mounted on a surface reel unit, being either pneumatically or hydraulically driven. Generally, the unit is skid mounted with the reel held in pedestal bearings and equipped with locking dogs and a friction brake. Connections between the rotating reel and the topside services are either through live shaft centers or a secondary reel jumper connected after the umbilical is deployed.

The reel is located at the side of the vessel moonpool either through a shared main moonpool or dedicate-controls moonpool. Associated with this reel unit would be the umbilical turn-down sheaves, tensioners, and constant-tension winches for the junction-box guidelines.

Chapter 7

Subsea Wellheads and Trees

DESIGN PHILOSOPHY

Introduction

The number of oil and gas wells that cannot be produced economically from a platform completion multiplies in line with the demands posed by offshore exploration. Application of subsea completions in the near future will accelerate for three basic reasons.

- Confidence in this type of installation is now established, allowing operators to view new applications in a favorable light.
- A dramatic increase in the development of truly marginal fields, particularly in the U.K. North Sea, has dictated the use of low-cost schemes. This is borne out by the published statistics for the U.K. sector which show that while 5 fields were added in 1983, 15 were developed in the 1985 and 20 fields are under development. In the majority of cases these fields are marginal.
- Deep waters in the Norwegian sector of the North Sea, off Brazil and in the Gulf of Mexico, will require novel remote schemes which use subsea completions.

In recent years, the equipment used to achieve efficient subsea production has increased. In many instances untried and vague design premises have been used with little attempt at standardization. The reason for this is clear when the numbers of subsea completions worldwide can be counted in the hundreds (see Fig. 7–1), each having its own unique requirements. Of the total number of worldwide subsea completions, 91% are in a water depth of 180 m (600 ft) or less.

The first subsea trees were configured with stacked valves, and a number of these are still in use. As the water depths and workover requirements increased, so the tree structure was subjected to increased loads. This led to the first composite valve block trees and the marine wellhead system. Subsequent to this development, trees of varying complexity have emerged, each providing a variety of features. The extreme is the immensely complex TFL satellite tree used on the Central Cormorant UMC project. This tree is electro-hydraulically controlled with TFL facility for workover and twin 3-in. tubing strings for completions into two strata.

Of a simpler nature are the trees used for the Buchan development (see Fig. 7–2). While being of a more basic design, these are of a hybrid design with components from two major manufacturers (National Supply and Cameron Iron Works). This type of tree uses vertical access for workover and is split allowing the lower master valve to be left subsea while pulling the tree during maintenance or major workover.

Attempts have been made to provide a simple base-case tree, and a very successful tree design was produced by National Supply in conjunction with Hamilton Brothers for use on the Argyll field. This arrangement uses a single string composite block, made possible by using a sliding sleeve valve/tubing hanger to give hydraulic access to the annulus.

Unfortunately, many operators prefer to use trees with vertical annular access for plug insertion; therefore, this approach has not been widely accepted. In addition, the lack of a large annulus circulation flow area detracts from its obvious advantages.

Some basic marginal field philosophies have evolved from various existing developments:

- Reliability through simplicity
- Maximization of field-proven equipment and techniques
- Diver-assisted installations
- Design compatible with field projections and characteristics

Each development will obviously amend these basic philosophies. However, dramatic departure from their intent will obviously be incompatible with the majority of marginal field economics.

Wellheads

The wellhead system selected during the design phase cannot always be a matter of free choice; it very often happens that a field may have a number of suspended exploration wells that

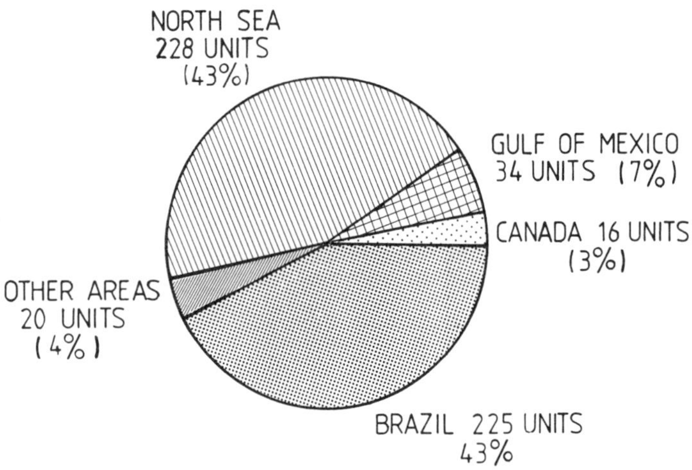

Fig. 7–1 Forecast of Subsea Completions (Source: IFP)

are suitable for conversion into producers. It is not uncommon for a field under development to have anything up to three types of wellhead systems. This demands a decision to either redrill or convert the wells into a common wellhead profile. This applies particularly to a number of the shallow, southerly North Sea fields, where the need arises to convert suspended mudline suspension systems into subsea completions.

It is obvious that wellhead selection may be influenced by more factors than just the most technically suitable system, so that trade-offs caused by circumstance must be made against economics and program demands.

Wellhead Sizes

Marine wellhead systems generally come in one of four basic sizes: 21¼-, 18¾-, 16¾-, and 13⅝-in. The 13⅝-in. is used only in a two-stack configuration. The wellhead system, which is usually designed to one of three working pressures (15,000, 10,000 or 5,000 psi), can perform underwater jobs that conventional wellhead equipment performs on land, namely:

- Suspension of tubular strings
- Annulus packing-off
- Attachment of control equipment (Christmas trees)

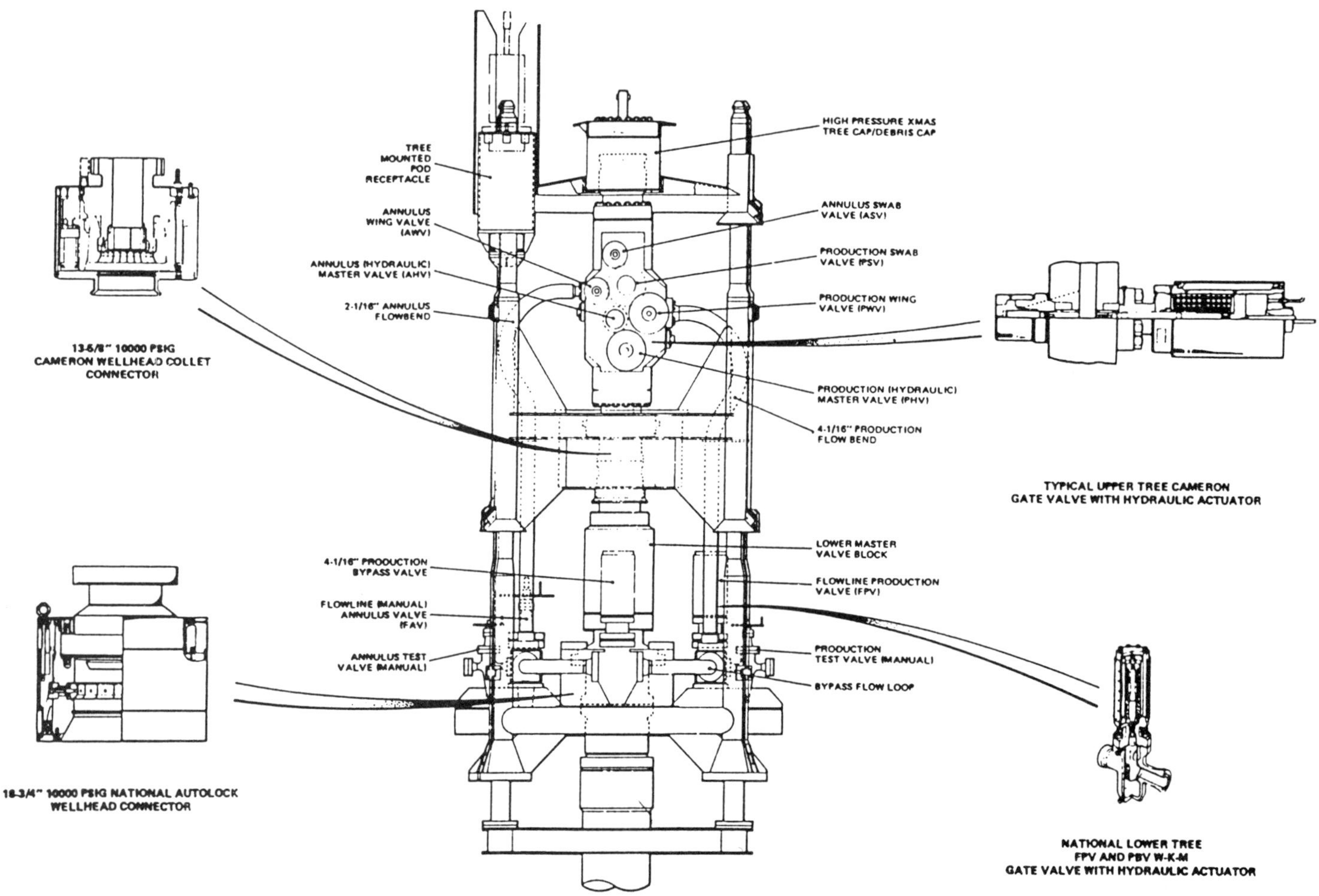

Fig. 7–2 Buchan Satellite Tree

Additionally, the wellhead system must be simple to install and come in standardized engineered packages. This will provide the assurance of stock shelf items being held at the manufacturer's service base.

Type Selection

Before a tree can be designed, it is important to know the planned configuration of the subsea wellhead system. Whether the tree is landing on a single- or two-stack system (Fig. 7–3) has a major impact on the arrangement of such tree components as the framework, wing valves, flowline connections, and interconnecting pipework when used on a template. It is generally accepted that production trees can be built for either application, although several differences exist, such as:

- With the dual wellhead system, the tree/template interface requries extensions due to the added stack-up height; this extension could be in the tree, in the template piping, or in both.
- The single 16¾- or 18¾-in. wellhead systems can generally accept larger outer-diameter completion strings (with free vertical access) than the dual 13⅝-in. wellhead systems.
- Rigs with BOP equipment for the dual wellhead system may be less frequently available in the future, due to a trend to convert most rigs into a single-stack system.

The use of the two-stack system can increase the tree elevation above the template, by as much as 30 in. Therefore, it is prudent to use a single-stack system to reduce the overall template completion height.

The single-stack wellhead system typically uses either a 16¾- or 18¾-in. wellhead for the high-pressure service, whereas a dual system uses first a 21¼-in. or 20¾-in. low-pressure wellhead and then a 13⅝-in. high-pressure wellhead. Bore size is generally not a limiting factor, as most 3-in. × 2-in. or 4-in. × 2-in.

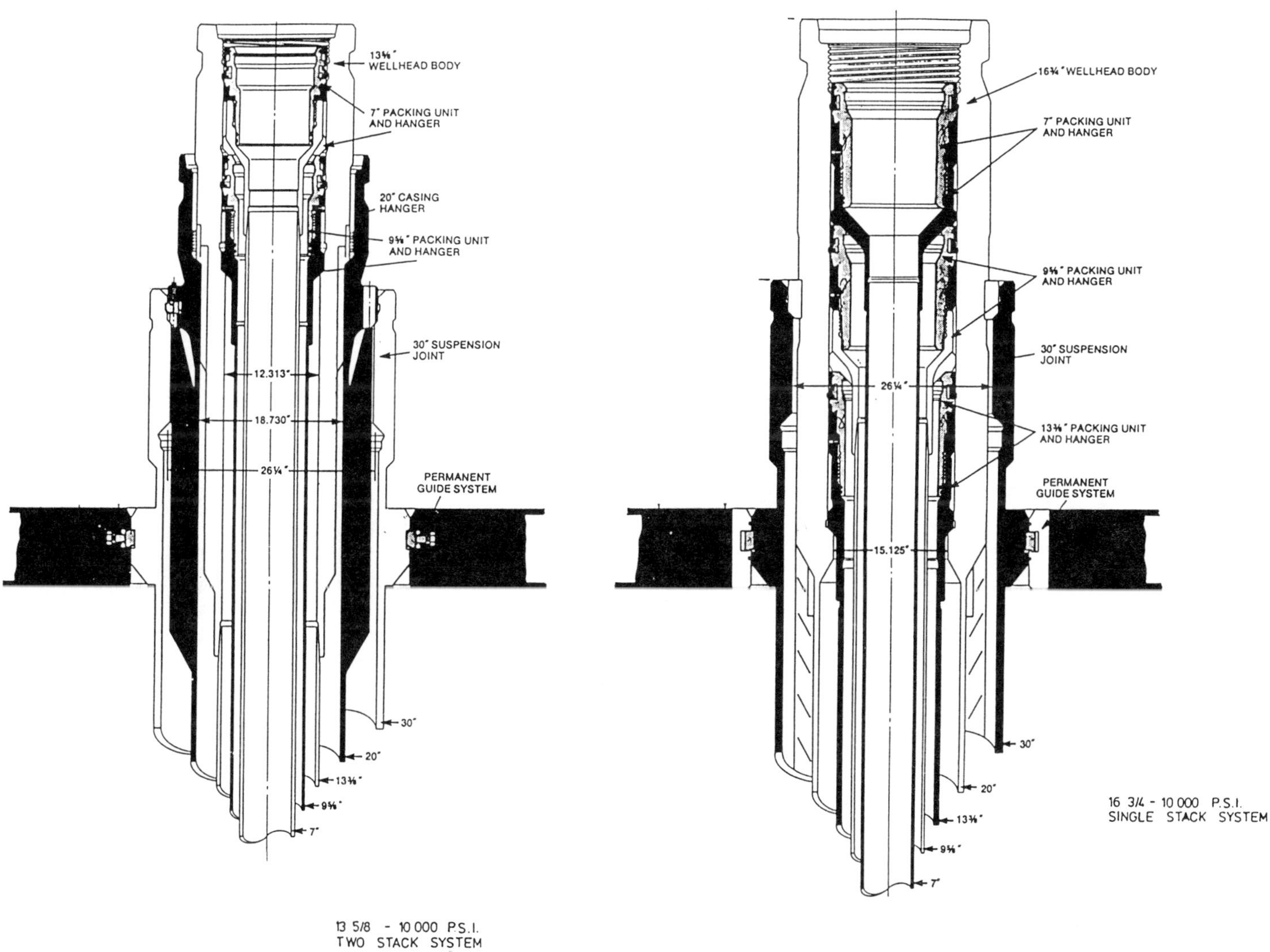

Fig. 7–3 Stack Wellhead Systems

bore profiles fit adequately inside the common wellhead sizes.

A key disadvantage of the two-stack system is the requirement for two risers of a different size, two stacks of a different size, running tools of a different size, and added storage requirements on the vessel when the entire system is being used. Two-stack systems also mean twice as much handling of heavy equipment, with twice the risk of damage and/or dropped objects.

Dual wellhead systems are an earlier generation than the present single-stack high-pressure equipment. The current trend is for drilling rigs to update their BOP systems with 16¾- or 18¾-in. 10,000 or even 15,000 psi equipment. For this reason, floating drilling rigs with single-wellhead systems may be more readily available in the future as the older dual-wellhead system becomes obsolete.

The only advantages with a dual system are that weight is minimized when handling BOP equipment, and a slightly lighter tree can be used because of its smaller connector. The smaller-stack system normally is preferred because of availability, the resultant shorter and simpler tree designs, and the ease of handling operations required for running and retrieving only one stack. The choice of whether to use an 18¾- or 16¾-in. system is an exclusive option for the operator based on economic, drilling, and completion requirements together with vessel/BOP availability.

Wellhead Types

Three of the most commonly used wellheads are shown in Figure 7–4, illustrating each manufacturer's connector profile. The internal bores on each of the wellheads show the approach each has to locking the casing hanger to the wellhead interior.

Many operators prefer to have the casing hangers locked into the bore, thus providing reassurance that the stack-up is correct. Without individual registers for each string, the final location of the tubing hanger is unknown until an impression block is run, at which point recertification of any errors becomes difficult. Again, the operator's preference should rule, as certain advantages exist with an elimination of the risk of latching into a locking groove that does not have a preferential profile.

Type preference may be influenced by the number of trips required to complete a casing program, and for this reason, an increasing number of casing hangers are run with their pack-offs providing one-trip installation and testing. Here, operators may prefer the reassurance of having only a pack-off to retrieve in case of seal damage, in preference to running special repair tools and pack-offs.

The basic requirements for a subsea wellhead system are that it should be compatible with commonly available equipment on board contract rigs, and that it is suitable for use with a single size of BOP. Secondly, that operator's preferences are respected with regard to hanger lock-down, pack-offs, and acceptable numbers of trips to complete a program.

Tree Configuration

At its simplest, a Christmas tree can be defined as an assembly of valves used for production or injection-well control. Generally, the main functions are that it can;

- Shut in production at the wellhead
- Facilitate chemical injection into the production bore
- Monitor pressures in annulus and production bore
- Provide a circulation path between annulus and production bores
- Facilitate downhole maintenance
- Facilitate well killing

A typical Christmas tree is made up of the following main components.

- Wellhead Connector: A mechnical or a hydraulic connector (see Fig. 7–5) with mechanical override provides a sealed connection between the production and annulus bores in the Christmas tree and in the wellhead.
- Valveblock: A valveblock (see Fig. 7–6), monoblock, or split contains manual and/or automatic master valves for the production and annulus bores, as well as providing access to bores for wireline intervention and chemical injection. It also may contain a production wing valve and swab and/or crown plugs. The valveblock can be bolted to, or made an integral part of, the wellhead connector. At the top, it has a mandrel preparation for the tree cap and running/re-entry tool. This also may be an integral part of the valveblock.
- Tree Cap: The tree cap is a mechanical or hydraulic cap that protects the valveblock mandrel from debris and corrosion and sometimes serves as a secondary pressure boundary. The Christmas tree control lines can be routed through the cap, thus enabling control of the tree functions in a workover situation. This is also achievable via a separate control and pod system.

Valving

Tree valving refers to the valves required to obtain certain tree functions not achieved solely by the vales contained in the valveblock.

They are necessary because the Christmas tree system must interface with other systems, namely:

- Wellhead system
- Overhead protection system, if provided
- Flowline pull-in or connection system when used
- Control system
- Installation and re-entry workover systems

Caisson Trees and Maintenance

Subsea trees may be either satellite or template mounted. The subsea trees should be kept as simple as possible for a marginal

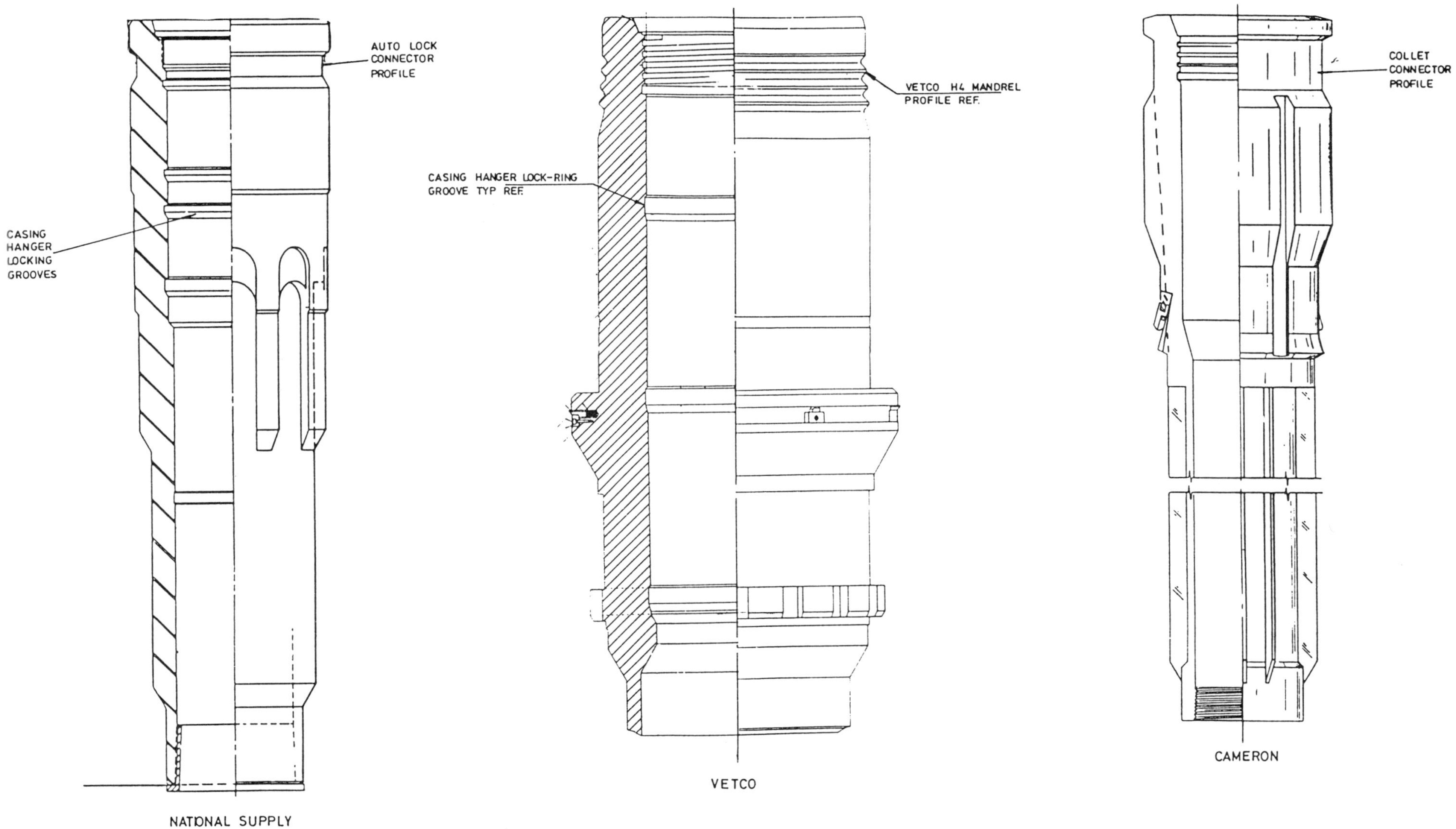

Fig. 7–4 Three Commonly Used Wellheads

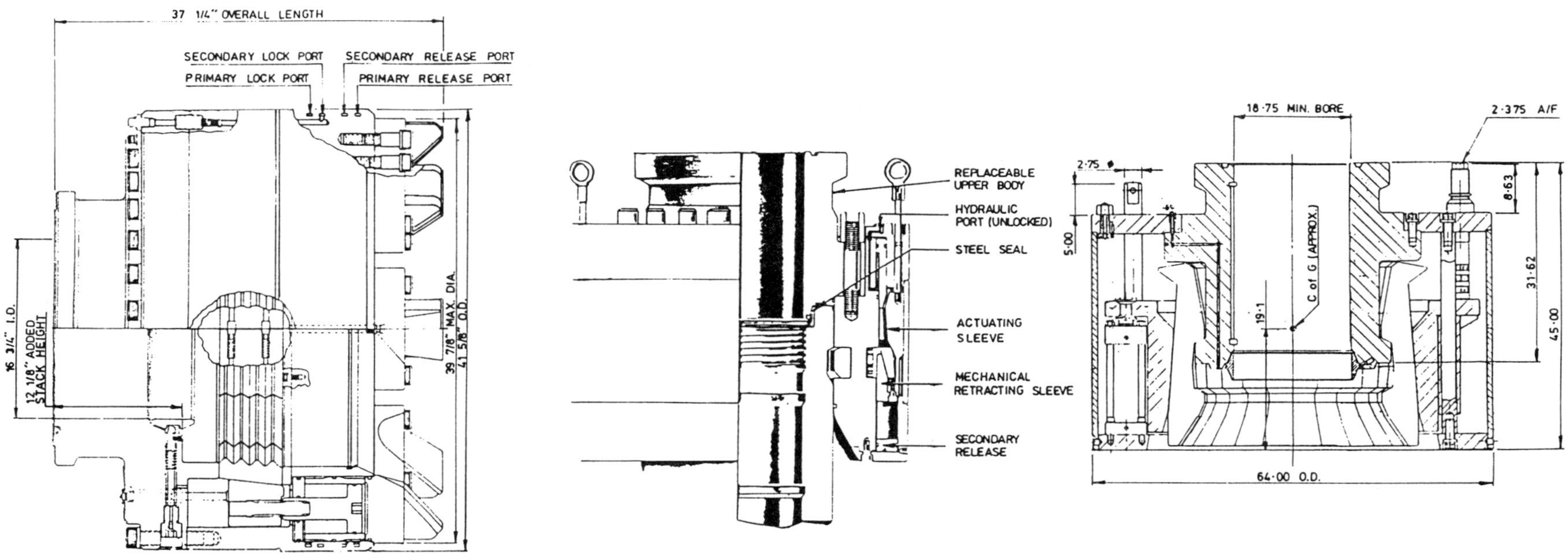

H4 CONNECTOR BY VETCO

HYDRAULIC CONNECTOR BY NATIONAL

COLLET CONNECTORS BY CAMERON

Fig. 7–5 Typical 15,000 PSI Wellhead Connectors

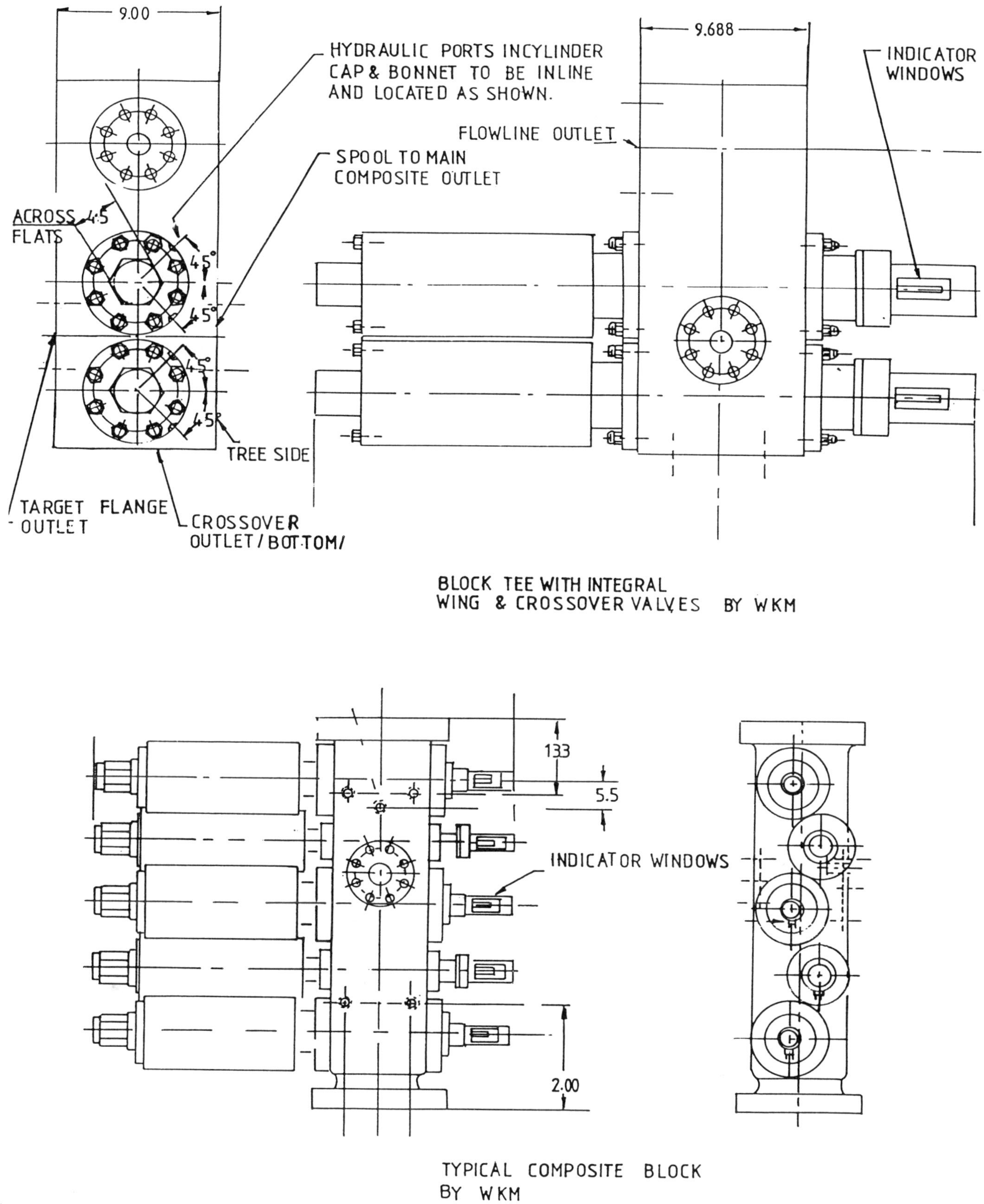

Fig. 7–6 Tree Valve Block Assembly

field development; therefore, caisson trees or dry-chamber systems are not at present considered suitable for marginal developments. Well maintenance should utilize wireline techniques. Thus TFL systems, due to their complexity, are not yet considered economic. The DHSV (downhole safety valve, sometimes referred to as a SSSV (subsurface safety valve), should be tubing retrievable, but nipples should be included to allow for use of wireline retrievable DHSVs to avoid delay during workovers due to DHSV failures.

Design Features

The required subsea well life must be established at an early date, allowing the selection of suitable materials and corrosion protection systems. The physical configuration of the designed tree should respect the requirements of simplicity. It has been found that in practice the use of standard pipe fittings, and the avoidance of the use of double off-sets in pipe spools, eases replacement problems. This aspect should be considered when evaluating tree-piping layout.

With the use of subsea chokes on trees, configuration has become more difficult. By using compatible configurations, the changing out of the chokes can be simplified if a tree needs to be reconfigured to a gas-lift arrangement. The exercise must be to arrive at a layout so that only one or two fittings are disturbed.

In an attempt to reduce the number of pipe fittings and to provide a less cluttered configuration, block tees and integral wing valves are commonly used. The inclusion of all the ancillary valves within one composite block is a feature of one of the major equipment suppliers (see Fig. 7–6). However, this should be avoided on the more marginal or smaller developments for two reasons.

- Manufacturing complexity is increased, leading to a high scrap rate during production and delays in obtaining replacement forgings which can have long delivery lead times.
- Damage to internal tappings, seats, or valve fixing arrangments during servicing or repair may result in the scrapping of the complete block, even though the damage may be of a minor nature.

While this approach is perfectly acceptable for a large field development with a long development program period, it could present some difficulties for a smaller project with a tight schedule.

Re-Entry for Workover

There are two classes of technique for performing minor well servicing tasks: direct vertical access via wireline or coiled tubing and horizontal access via the flowlines by using the TFL method.

The majority of the existing subsea completions have been designed for vertical re-entry into the well bore, because of its simplicity and it being an extension of proven land workover practices. The TFL method is less proven and requires the added complexity of subsea equipment and larger surface equipment. TFL is better suited to applications where vertical access to wells is relatively unavailable.

A decision should be made early in the system design of a marginal field development that the workover of wells will be through vertical entry by wireline or by coiled-tubing methods. Generally, TFL methods have not become accepted means for the various workover tasks. Apart from the continued efforts of Shell, Esso, and BP to pursue remote subsea maintenance, the industry has generally accepted that tree and control system maintenance should be by replacement techniques from the surface.

TFL has not yet achieved the market penetration that was thought likely by its sponsors in the early 1970s. This is based on a number of factors, namely:

- Flowline wear on the locomotive seals
- More complex wellhead control system requirements
- Problems associated with paraffin removal
- High reliability being achieved by tubing-installed downhole safety valves in relation to TFL-installed valves

Wireline Tasks

Minor well servicing carried out by wireline methods is achieved by attaching tools, designed for specific tasks, to a small diameter steel wire spooled onto a powered drum. The wire is used to lower the tool into the well to the point where the work is to be carried out and then, after the task is completed, retrieved to the surface. For safety and pressure control, a surface valve assembly is normally used to control well bore pressure while these operations are in progress. A wireline lubricator is attached to the top of the valve assembly. This consists of a length of pipe longer than the work tool, for parking purposes, with a stuffing box to seal around the wireline as it is lowered into the well or reeled back up.

For subsea wells, a riser can be used to provide a pressure-containing connection between a surface valve assembly and the subsea tree. The lower riser package should contain a subsea wireline BOP, and the tree swab valve should be capable of shearing the wireline in case of an emergency pull-off. Instead of a surface lubricator and riser, a subsea lubricator may be mounted on the subsea wireline BOP. Then a number of well servicing tasks can be performed by wireline techniques. The majority of wireline operations involve pulling and replacing downhole tools, such as subsurface safety valves and gas-lift valves. Wireline techniques are commonly used to set and retrieve plugs in the tubing hanger before trees are pulled for repair or before lowering instrumentation into the well to measure tubing bore size or bottom oil pressure and temperature. Wireline tools also may be used to shift sliding sleeves, cut paraffin, bail sand, and carry out minor fishing jobs.

TFL Entry

Entry for workover on a TFL system normally requires that the tree is provided with flow loops of sufficiently large radius to allow the train to traverse the change from horizontal travel along the flowline to vertical travel into the wellhead. The flow loop will normally enter the tree through a diverter spool. This spool provides for the insertion or removal of a tool. Therefore, the tree configuration is complicated by the large flow loops. Additional complications and extra leak paths are introduced by having a split tree which normally contains a lower master block, a separate spool for pump-down tool entry, and an upper swab valve block which can additionally serve as the means of vertical re-entry via a workover riser. A typical TFL tree schematic is shown in Figure 7–7.

As has been previously mentioned, the TFL system introduces extreme complexity into any design and should be avoided for marginal field developments.

Wireline Re-Entry

Wireline re-entry is achieved by the use of a re-entry hub or mandrel, on top of the tree located above the swab valves. To gain entry, it is necessary to pull the tree cap which can serve additionally as the control crossover during normal production. The tree running tool/riser connector is located on the re-entry hub or mandrel of the tree. This workover riser then provides the fluid path for circulation and the means by which the wireline tools are run down the riser into the tree bores to provide a variety of workover operations.

When using the workover riser, it is necessary to isolate the control system of the tree and close all valves before taking over command of the tree from the service vessel via the workover umbilical and cap.

In summary, to obtain means of re-entry for workover, it is necessary to provide a system. This sytem can be either a TFL system or a direct access system with a workover riser. The design philosophy should be toward a simple tree that incorporates a means of gaining a vertical entry via some form of workover riser. This workover riser must be of simple design and serve a dual purpose, that is:

- Of running trees and gaining re-entry to trees
- To be able to be used for running the tubing hanger in conjunction with a tubing hanger running tool (In this mode it is necessary to use the workover riser inside a marine/drilling riser connected to a BOP.)

The number of workover risers in service at present is very small. Those that are in service, or currently being designed, vary according to the manufacturer or operator controlling the development. When it is considered necessary to have up to 25 functions on a tree system, it is realistic to assume that the control of the tree during workover is through a control umbilical strapped to the outside of the workover riser. This is run off of a workover umbilical reel placed on the service vessel or on the floating production vessel.

Attempts have been made, and are currently under way, to incorporate all the tree control functions within conduits contained within the workover riser. Here problems may arise from the requirement to maintain very close linear and geometric tolerances at each workover riser joint. Obviously it is very difficult to make up each section of the workover riser when there are two main conduits, namely annulus and the production bore, plus up to 25 individual control lines. Therefore, a constant liaison with the manufacturers is essential to ensure that these problems are resolved.

DRILLING ASPECTS FOR WELLHEADS

Wellhead Connectors

Of all the components of a subsea completion for the production or the drilling of a well, the marine hydraulic connector is the most complex and critical.

It offers the only reliable method of remotely connecting a BOP stack to a wellhead or the lower marine riser package to the BOP. In addition, the connector forms part of the subsea tree unit and is used to connect the tree to the wellhead. The hydraulic connector also has been used extensively to attach rigid riser systems to subsea manifolds. In a miniaturized form, it has been found that many applications exist for their use as tree flowline stab connectors.

Each manufacturer approaches the design of a connector in a different manner. What is common, however, is that they each use sliding inclined planes to generate the clamping force required to secure the connector to the wellhead. This clamping force is required to energize a metal ring to seal cavity pressure between the wellhead and tree or wellhead to BOP.

Operating Principle

The operating principle uses hydraulic force to drive a sleeve, having a gradual inner-diameter (I.D.) taper, downward to drive dogs or collets radially inward. A taper on the dogs then engages matching tapers on the wellhead hub outside diameter (O.D.). The hub faces are pulled together by this action, deforming a metal ring to seal cavity pressure and generating a clamping force between them.

This movement involves the sliding of steel-to-steel surfaces under load. Therefore, the major factor influencing the connec-

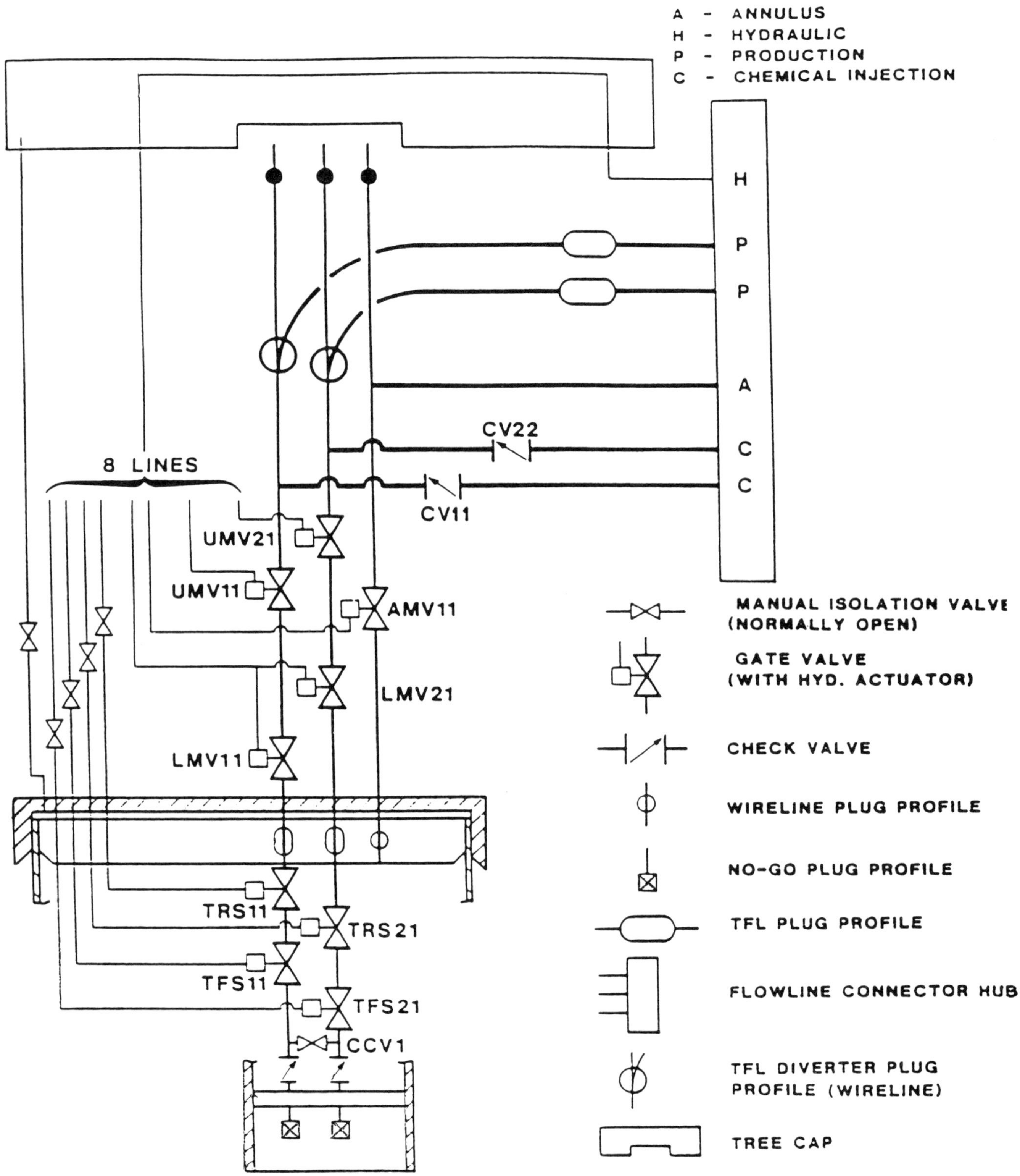

Fig. 7–7 Typical TFL Tree Schematic

tors mechanical advantage is the co-efficient of friction between the sliding surfaces.

For evaluation purposes therefore, it is advantageous to use a simple expression to assess a connector's suitability for any drilling or production application.

Mechanical Advantage

The connector produces a mechanical advantage through the sliding tapers, which equals the ratio of the clamping to that of the hydraulic force. Dure to the large friction losses however, it is necessary to assess the connector in terms of its efficiency. This has been defined as the ratio between the actual clamping force and the theoretical clamping force obtainable. The theoretical mechanical advantage then can be taken as the ratio between contact angles of the sliding surfaces.

Figure 7–8 shows the location and contact angles of the two mating parts on a typical wellhead connector.

Where α is the *sleeve to segment* contact angle and β the *segment to wellhead* contact angle, then the theoretical mechanical advantage (MA_1) is given by;

$$MA_1 = \text{Cot}\beta \ \text{Cot}\alpha$$

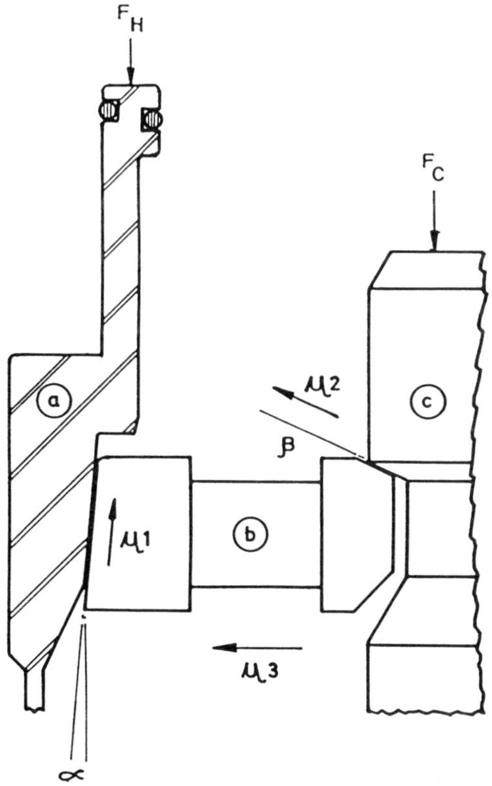

Fig. 7–8 Connector Operating Principle

A typical plot of coefficient of friction against actual mechanical advantage (MA) for a connector is shown in Figure 7–9(1). The curve is based on the following expression:

$$MA = \frac{\left[\dfrac{\text{Cot}\alpha - \mu_1}{1 + \mu_1 \text{Cot}\alpha} - \mu_3\right]}{\left[\dfrac{\mu_2 \text{Cot}\beta + 1}{\text{Cot}\beta - \mu_2} + \mu_3\right]}$$

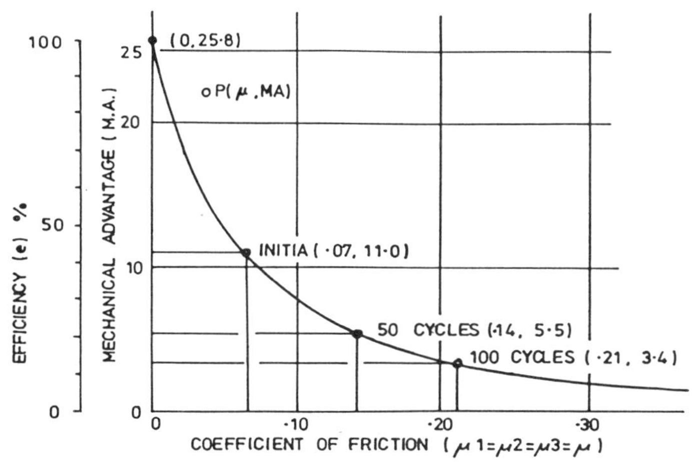

FRICTION COEFFICIENTS PREDICTED BY THE THEORY ARE SHOWN BESIDE MEASURED VALUES OF MECHANICAL ADVANTAGE

(1) FRICTION COEFFICIENT μ VS MA

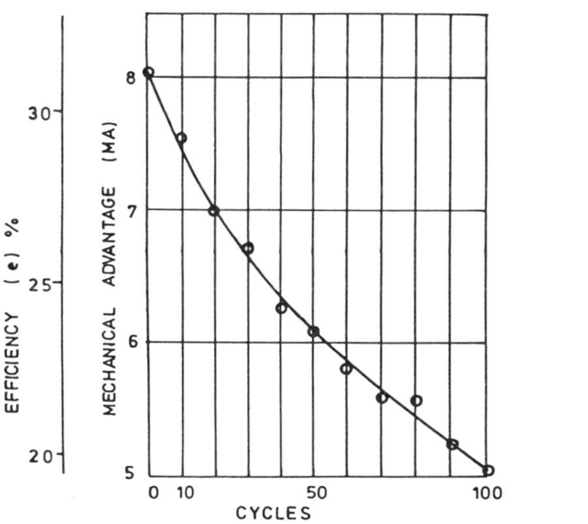

DATA SHOWS HOW MECHANICAL ADVANTAGE DIMINISHES WITH NUMBER OF CYCLES FOR GREASED 16 RMS FINISH

(2) CYCLES VS MA (ACTUAL)

Fig. 7–9 Connector Friction Coefficient Plots

This plot dramatically shows the effects of a change in the coefficient of friction (μ). During service, this increase in μ occurs as the mating surfaces squeeze out any lubrication, resulting in a dry surface coefficient value although the connector components may be totally immersed in the lubricating medium. In many recorded cases, this has prevented the disconnection of the connector. Effectively, this makes the provision of a secondary release boost mandatory. Actual tests verify this assumption, and a plot of an actual test is given in Figure 7.9(2). This applies to all connector manufacturers and must be considered when evaluating the design of a connector.

The efficiency of the connection would be given by:

$$\frac{MA}{MA_1} \times 100\%$$

Clamping Force

The marine hydraulic connector must resist the following loads without leaking cavity pressure:

- Pressure-separation load between the hub faces
- Bending moments generated by riser connection
- Tensile or comprehensive forces

When an external moment is applied to a clamped connection, it may be considered to fail when the seal ring pre-load is lost. At this point, the primary force through the connection changes from a direct line through the hub faces to a route through the clamp body. This failure might occur at a moment less than that required to structurally fail the connection.

Separation

Separation can be assured to begin when the load at the circumference of the hub face contact area approaches zero. Let F be the total compressive load exerted by one hub on another when the joint is in service, under pressure, and with axial forces but without external moments. Let A be the hub area actually in contact. Then the value of compressive stress, C, assumed uniform across the contact face is given by:

$$C = \frac{-F}{A}$$

From simple beam theory, the tensile stress, t, is:

$$t = \frac{Mr}{I}$$

where M is the applied moment, r is the outside radius of the contact area, and I is the second moment of area of the contact faces. Then, if $t + C = 0$

$$M_s = \frac{IF}{rA}$$

Since, I, r, and A are determined by the geometry of the interface, they can be assumed constant. Thus the applied moment to start hub separation (M_s) is proportional to the total load through the hub faces (F). Since F is the clamping force, adjusted for tensile forces and separating forces generated by cavity pressure, the connector clamping force will affect the moment carrying capability of the connector.

The above methods are approximate, although they do correspond closely to actual test data and provide a simple basis for assessment.

15,000 psi Connectors

The current trend toward 15,000 psi wellhead systems demands a new generation of connectors with extreme stiffness; two such connectors are shown in Figure 7–10. Usually, these are not required for tree service, and in these cases a lighter connector can normally be provided. The heavier drilling connectors are increasingly making use of annular pistons in preference to individual cylinders. Both annular and piston methods have their advantages and drawbacks. However, what is of primary importance is that the following features are included in the selected connector:

- A secondary unlock system
- Lock-down indicators
- Optional mechanical lock down

Manual Override

In many instances, a manual override is provided, but this feature is of dubious benefit. To manually release a connector after a lengthy period of being clamped to a wellhead or BOP, a force in the region of 250,000 lb (1 lb = 0.4536 kg) or more is required. This feature may be included if it provides a last ditch reassurance to the operators that, should the piston jam after initial breaking of the clamping surfaces, release may still be accomplished.

Tubing Hanger Orientation

The production or water injection tubing hanger is orientated to facilitate the correct landing of the Christmas tree and to locate the tree and DHSV stabs in the correct manner. This is important only for dual or triple bore tubing hangers and does not apply to concentric hangers.

The tubing hanger is run on a tubing hanger running tool. This running tool is attached to a BOP adaptor or rotation joint and run on a workover riser or tubing string within the marine drilling riser. The running string provides a hydrocarbon path for circulation and the path for the setting of plugs after the hanger is landed.

TYPICAL 15000 PSI COLLECT
CONNECTOR BY CAMERON

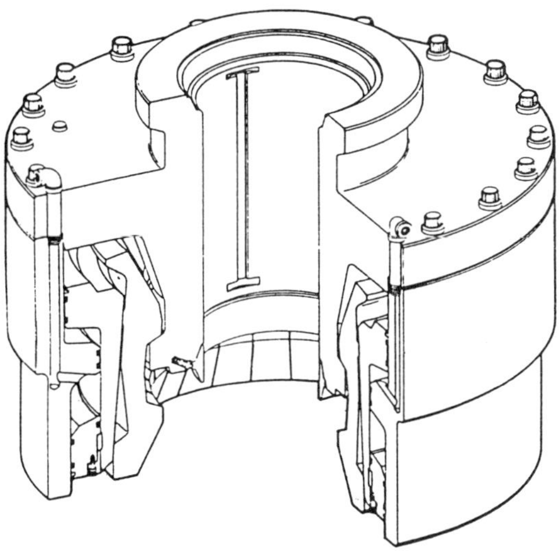

TYPICAL 15000 PSI WELLHEAD CONNECTOR
BY NATIONAL

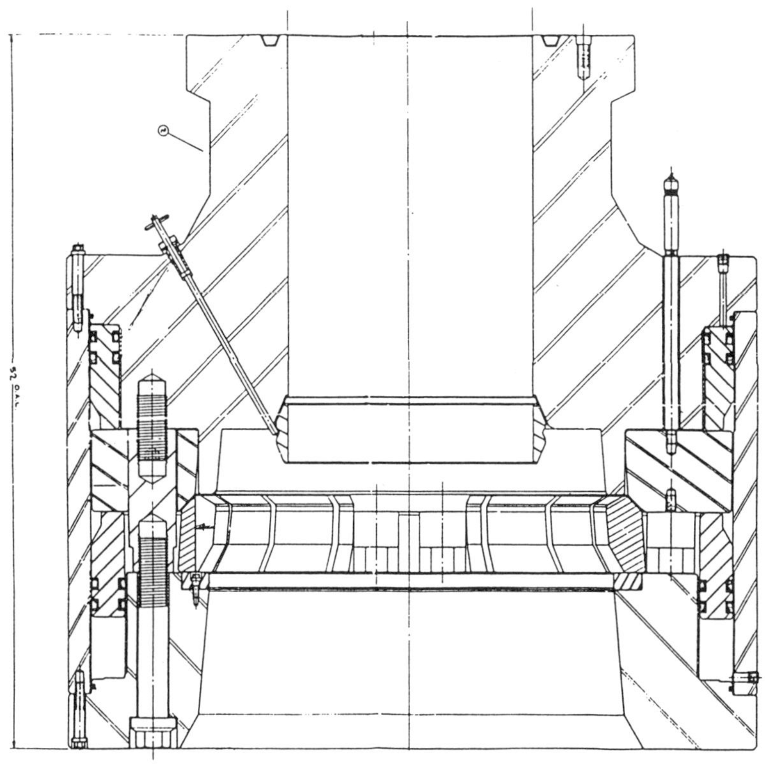

Fig. 7–10 Typical 15,000 PSI Wellhead Connectors

Two methods of orientation are commonly used. The first uses a bushing to orientate the hanger. This bushing, in turn, receives its orientation from the BOP stack connector. The orientation bushing has an internal helical profile which contacts a key in the tubing hanger. Subsequent installation of the running tool on the hanger is achieved by means of a slotted guide sleeve attached to the running tool. External hydraulic pressure, applied through the BOP stack kill line after closing the casing ram around the upper portion of the running tool, locks the tubing hanger to the wellhead. The hanger may be unlocked hydraulically through the running tool or by mechanical means.

The second method provides a vertical slot within the upper body of the wellhead connector. The tubing hanger is run and landed in the casing hanger, and the BOP ram is closed around the reduced diameter section of an upper sub. Then the tubing string is picked up and over-pulled. Rotation of the string is carried out until the running tool alignment key locks into this wellhead connector's vertical alignment slot. The torque is increased slightly to verify that proper alignment has been achieved and then eased off. The BOP ram is opened and the tubing hanger landed in the casing hanger.

An additional and desirable feature makes use of a return line from the tool to the surface control panel. This line provides a hydraulic indication which is read on the rig when a key springs into the wellhead connector slot.

In all cases, the running tool provides the means to additionally lock the hanger into the production casing.

A required feature that must be provided is a mechanical release to cover hydraulic power loss. This is not presently incorporated in all tools, and it is important that it is requested during any early discussions with equipment manufacturers.

Running Equipment

Irrespective of whether a well is to be drilled as a template or satellite completion, it is normal practice to use guidelines for running tools and BOP equipment. The method of attachment varies between manufacturers, as do the tools used for running the various casing programs.

Temporary Guide Base

Although the practice of using a temporary guide base is neglected by many operators, this particular item of equipment can provide reassurance that the wellhead is set at a predetermined elevation from the seabed. Before ordering this particular item, the subsea engineer must ascertain the seabed characteristics from the available soil reports and ensure that the seabed is able to support the weight of the base.

The primary tasks of the temporary guide base are to compensate for possible ocean bottom slope and to provide guidance for drilling the pilot hole. Guidelines are installed at the surface before running the guide base and, when desired, they can be retrieved after running. If additional weight is required on the temporary guide base, it can be filled with weight material before running.

Drilling of Outer Conductor

The first stage of running the temporary guide base is followed by the running of a bit/drill assembly on a two-arm frame. This is attached by using a bushing to the bit sub in the drill-string. After the drill-string is retrieved, the breakaway guide arm is attached with shear bolts to a 30-in. conductor 3–5 ft above the cementing shoe to facilitate initial conduction guidance (see Fig. 7–11).

Permanent Guide Base (PGB)

The last section of the 30-in. conductor is run with a 30-in. suspension joint and J-type running tool or 30-in. latch connection. At the same time, the permanent guide base (PGB) is latched and run together with the 30-in. suspension joint and landed onto the temporary guide base. The PGB post tops and posts are slotted, which allows the PGB to be used as the fixing point for the guidelines, either when lines are re-established or when a temporary guide base is not used. Posts and tops for template wells are not normally required in a slotted form.

Drilling the Inner-Conductor

Although the drilling of the inner-conductor may be accomplished without the need for any form of BOP, in many circumstances a 30-in. latch is used together with the lower marine riser package (LMRP). Then the lower riser package assembly is run on the riser and latched to the 30-in. conduction housing. Next, drilling of the 20-in. casing hole is performed through the riser, which allows directional control to bleed off any shallow pressured sands penetrated. The lower riser assembly is retrieved on completion of the 20-in. casing hole.

The last section of the inner-conductor has the wellhead attached and is run with a wellhead running tool, again attached to the two-arm unit. It is normal for the manufacturer to provide the option of a hydraulic running tool for this operation.

Landing of Casing Strings

After the inner-conductor is run, the main BOP stack, which is connected to the marine riser, is brought into use and the running of all equipment, casing, and tubing is made through it. Then the intermediate and inner casing strings are run in a similar manner.

The final stack-up is capped off with a low pressure corrosion cap. This cap may be of the high pressure type, and it acts as a pressure containment barrier. A cap is used to prevent silt ingress into the wellhead and as a protection against marine growth.

The selection and use of wellhead running tools is now routine, and the type used is obviously dictated by the selection of a particular manufacturer's product line. It should be remembered, however, that a manufacturer will provide these items on hire or loan. This is particularly attractive since the purchase can sometimes be expensive when it is considered that after the completion of the drilling program little, if any, use is made of them.

WELLHEADS AND WELL CONVERSION ASSEMBLIES

Marine Wellhead Systems

Marine wellhead systems have been developed for specific requirements and have not evolved from standard land and platform wellhead designs.

The marine wellhead is machined from a large one-piece forging. Materials used are normally ANSI 4140 or 4130, ASTM A668 (Class K), or EN19 (709M40). The forging is heat treated (as are the other components exposed to the hydrocarbon fluids) to accommodate hydrogen sulphide service; many manufacturers do not supply non-hydrogen sulphide wellheads.

The wellhead body provides a method of hanging the 20-in. casing string inside a 30-in. suspension joint, a location from which to suspend other casing strings, and high pressure protection when the BOP stack is connected. The wellhead body incorporates a male hub with a seal pocket and a generally large diameter left-hand Acme running thread, flow port areas for cement returns, and segments for locking the wellhead body in the 30-in. suspension joint. An extension to the wellhead body consisting of 20-in. diameter casing with a buttweld-prepared end normally is provided for attaching the 20-in. casing string.

The interior of the housing contains locking grooves for each casing string and lock-down grooves for the tubing hanger. The casing hanger is locked down together with or separate from the annulus pack-offs, the main load bearing shoulder being located at the bottom of the housing. In some of the new generation of 15,000 psi wellheads, each casing string has its own load bearing shoulder, resulting in a decrease in hoop and bearing stress within the housing.

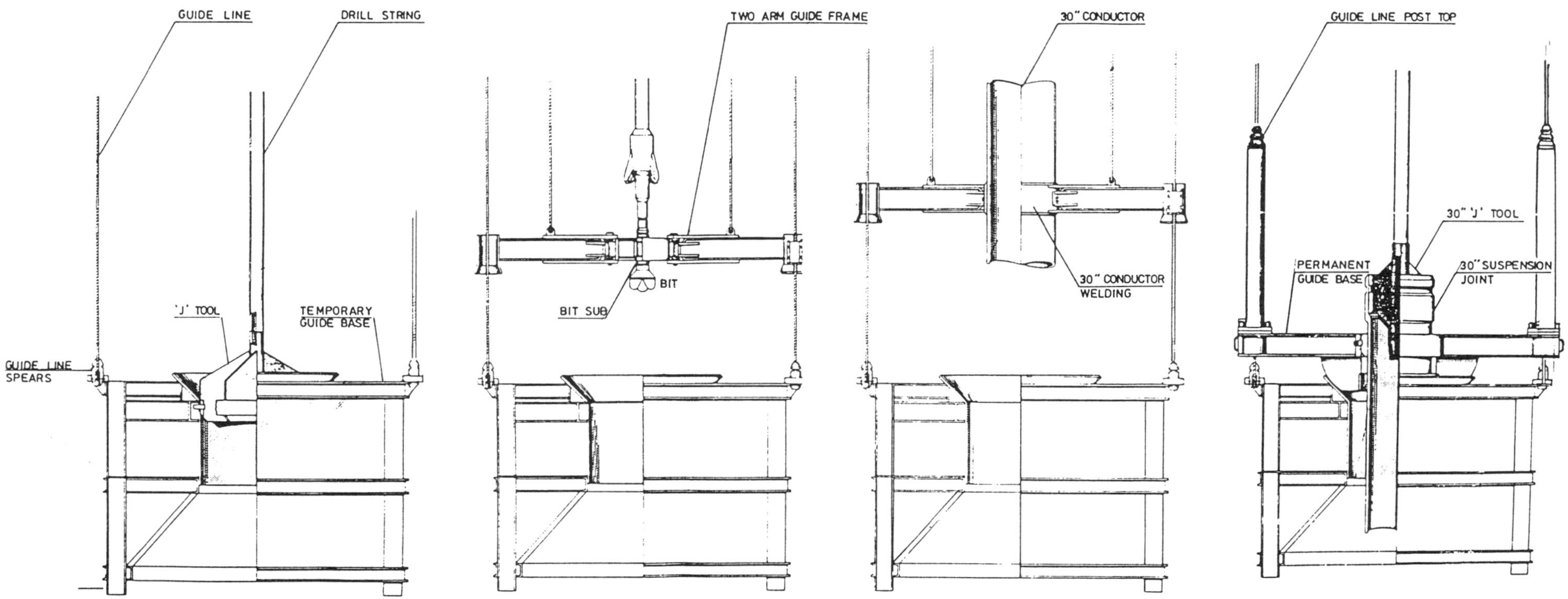

Fig. 7–11 Drilling for the Outer Conductor

Intermediate Casing Hanger

The intermediate casing hanger is the only hanger in the 10,000 psi wellhead systems that is supported off of a wellhead shoulder. Typical one- and two-trip casing hangers are shown in Figure 7–12. The 13⅜-in. casing hanger shown in Figure 7–13(1) is a combination mandrel and segmented-type hanger with buttress suspension thread. The hanger lands in the wellhead body and has a flow port area for cement returns. Run the hanger and cement the casing. A casing hanger installation tool is used to both run the hanger and cement the casing. After the cementing, a pack-off is run that seats on the upper surface of the casing hanger and locks into the wellhead. Refer to Figure 7–13(2).

Production Casing Hanger

The first production casing hangers, run subsequent to the intermediate casing hanger, land and seat above the pack-off body. Subsequent hangers are run onto the previously installed hanger. Each hanger has a landing shoulder and an internal sealing area for pressure test purposes. The preparation for the landing shoulder is normally specified by the operator. Refer to Figure 7–13(3).

A production casing hanger with the corrosion cap landed is shown in Figure 7–13(4).

Annulus Seal

The annulus seal between the casing hangers and the wellhead body is formed by the packing unit. The packing unit is run and retrieved on a packing tool using a drill-pipe, after the casing is landed in the wellhead body and cemented with the casing hanger installation tool. A pressure energized molded packing element affects the seal with the inside diameter of the wellhead body while O-ring seals are utilized in the inside diameter of the previously landed casing hanger. If the locking segments do not snap out, or seals require replacing, the pack-off is retrieved using a retrieving tool which depresses a tapered sleeve at the top of the pack-off to retract the locking segments.

Each manufacturer uses similar components, and it is the detailed design features that make one manufacturer's equipment more attractive than the others. The most important feature that should be considered is the sealing method. The types of pack-off used vary from a true metal-to-metal seal to a simple elastomeric ring.

Annular Seals

All the current wellhead seals incorporate elastomers to a greater or lesser extent, and it is worthwhile to consider three types currently offered.

1. The pack-off seal in Figure 7–14(1) is used on a one-trip casing hanger/pack-off unit. The construction is a mix of elastomers and metal seals. This Z-type construction is very effective and is field-proven on a number of North Sea installations.
2. A similar type of construction is used for the seal shown in Figure 7–14(2). This seal provides a very reliable solution with the attributes of both metal and elastomeric seals included.
3. The third seal type is a basic elastomeric seal and has no metal back-up. Although very simple, it is used in many existing wells and is found to be very effective. Refer to Figure 7–14(3).

Mudline Suspension Systems

Although not a marine wellhead, the functional design of mudline suspension systems accommodates a subsea completion, and they are extremely interesting. This stems from the fact that many southerly North Sea fields used them from jack-ups to provide early completions. The limitations of the mudline suspension system is the maximum depth to which they could be used. At present this is limited by the jack-up restriction of around 106 m (350 ft).

Mudline equipment is used when drilling from a bottom-supported platform, and it offers special economics for evaluation and development drilling, as wells can be drilled and abandoned at minimum cost. With mudline suspension equipment, the weight of the casing strings is suspended from hangers at or below the ocean floor. Concentric tubular risers reach out of the mudline caisson unit towards the surface. Topside, a conventional platform wellhead unit and BOP stack provide pressure control and containment. With the preventers at the surface, there is no need for an underwater control system or kill and choke line valves.

Setting the Drive Pipe

The 30-in. casing is driven, drilled, or jetted into place, and a suspension unit supports the casing at or below the ocean floor, connecting it to the riser. This connection can be made by using a weld ring between the casing and riser and, if this method is used, the connection may be driven into the ocean floor to any depth desired. As an alternative, a clamp hub is welded to the casing with the mating-half welded to the riser. If this system is used, the connection must remain far enough above the ocean floor to be easily accessible. Either way, casing is suspended at or near the ocean floor, and the riser is supported at the rig above the water line. Both weld rings and clamp hubs are normally available for most sizes of drive pipe.

Hydraulically or mechanically operated drive connectors are

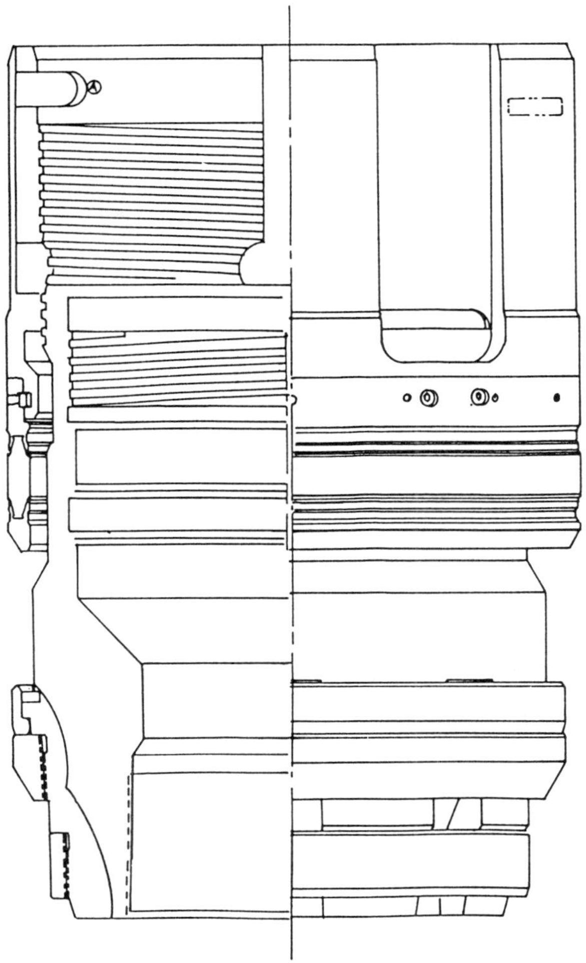

ONE TRIP CASING HANGER WITH PACK OFF.
BY VETCO OFFSHORE.

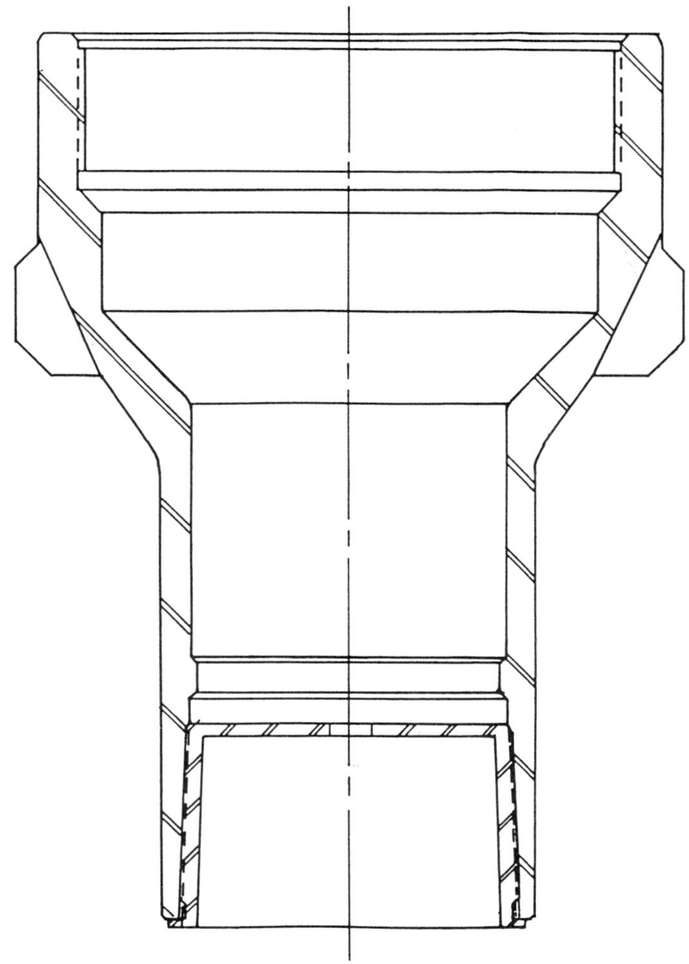

TWO TRIP CASING HANGER.(SEPARATE PACK OFF.)
BY NATIONAL SUPPLY.

Fig. 7–12 Casing Hangers—One and Two Trip

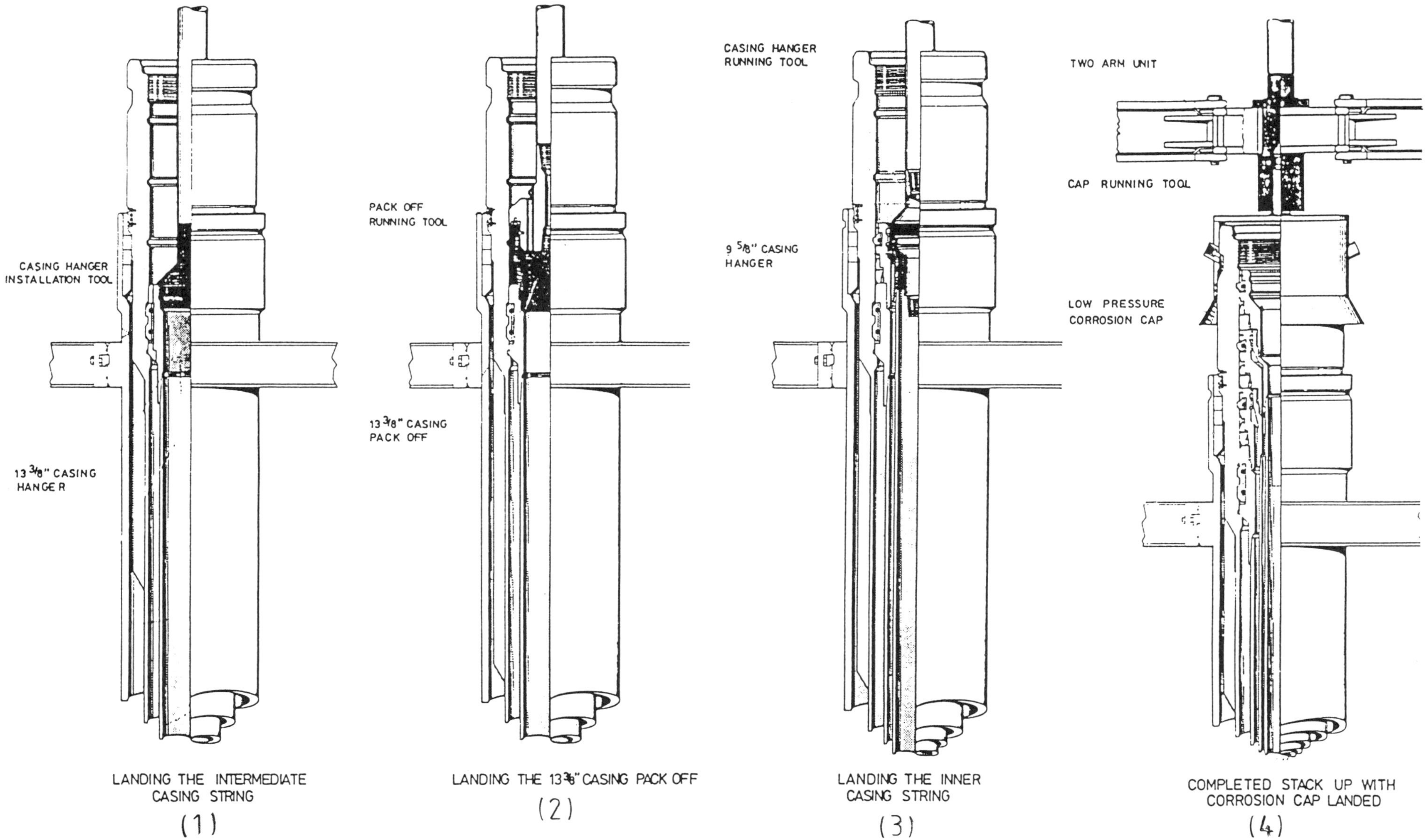

Fig. 7–13 Stacking up the Wellhead

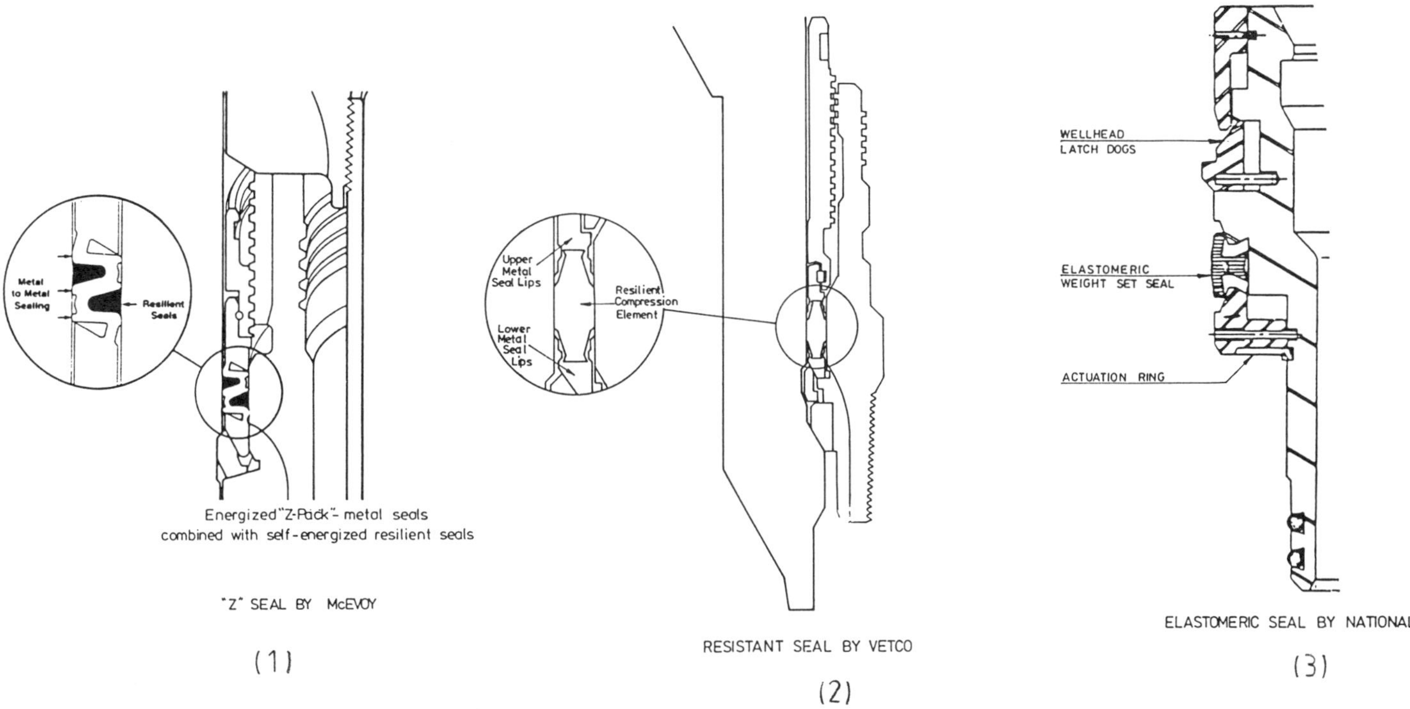

Fig. 7–14 Typical Casing Hanger Seal Types

available also. At this point, flow diverters or BOPs are sometimes fixed to the drive pipe for control at the surface.

Running Conductor and Casing

When the drive pipe has been set, the 30-in. casing is used to provide a means to drill through for the 20-in. conductor. This casing is run and landed, then a casing flange is installed at the surface to attach a BOP. All of the casing load is suspended at or below the ocean floor, and cement returns are taken through external flow ports. It is normal to provide a means to wash cement from above the hanger, and this is achieved in many systems by rotating the 20-in. casing so that wash ports are exposed. Immediately after washing, the ports are closed by rotating in the opposite direction.

After drilling through the 20-in. conductor casing, 13⅜-in. surface casing is run and landed. The 13⅜-in. casing is cut and packed off at the surface for installation of the drilling head and BOPs. The load of all three of the casing strings is being suspended at the ocean floor on the 30-in. suspension shoulder.

The drilling of the casing program continues with the 9⅝-in. intermediate string and the 7-in. oil string. Each of the strings is cut and packed off at the surface. Preventers can be installed and evaluation work on the well commences. A stack-up of a mudline suspension system is shown in Figure 7–15.

Capping the Well

After drilling and testing, the well may be plugged and temporarily abandoned. In doing so, risers and handling connection tools are retrieved. By using the remaining 30-in. drive pipe as a guide, the corrosion cap is run and landed. Its purpose is to provide a protective cover to exclude silt and salt water and prevent corrosion or marine growth at critical areas. It does not provide a high pressure seal although special designs with this capability are available.

If a clamp hub has been used above the ocean floor on the drive pipe, it is removed by divers. However, if the choice had been to use a weld-ring connection, the riser is cut wherever desired.

Returning to a Temporarily Abandoned Well

The attraction of the mudline suspension system is that it provides a simple and efficient method of re-attaching risers when the operator returns to a well after temporary abandonment. If desired, the operator can be selective in choosing which strings to tie back to the surface. After the corrosion cap has been removed by the retrieving tool, risers are run back in. They are attached to the hangers by latches with weight-set mechanisms. Rotation of the riser is commonly used or just weight-set latching. At the same time, connections are automatically packed-off making a seal between the risers and the casing

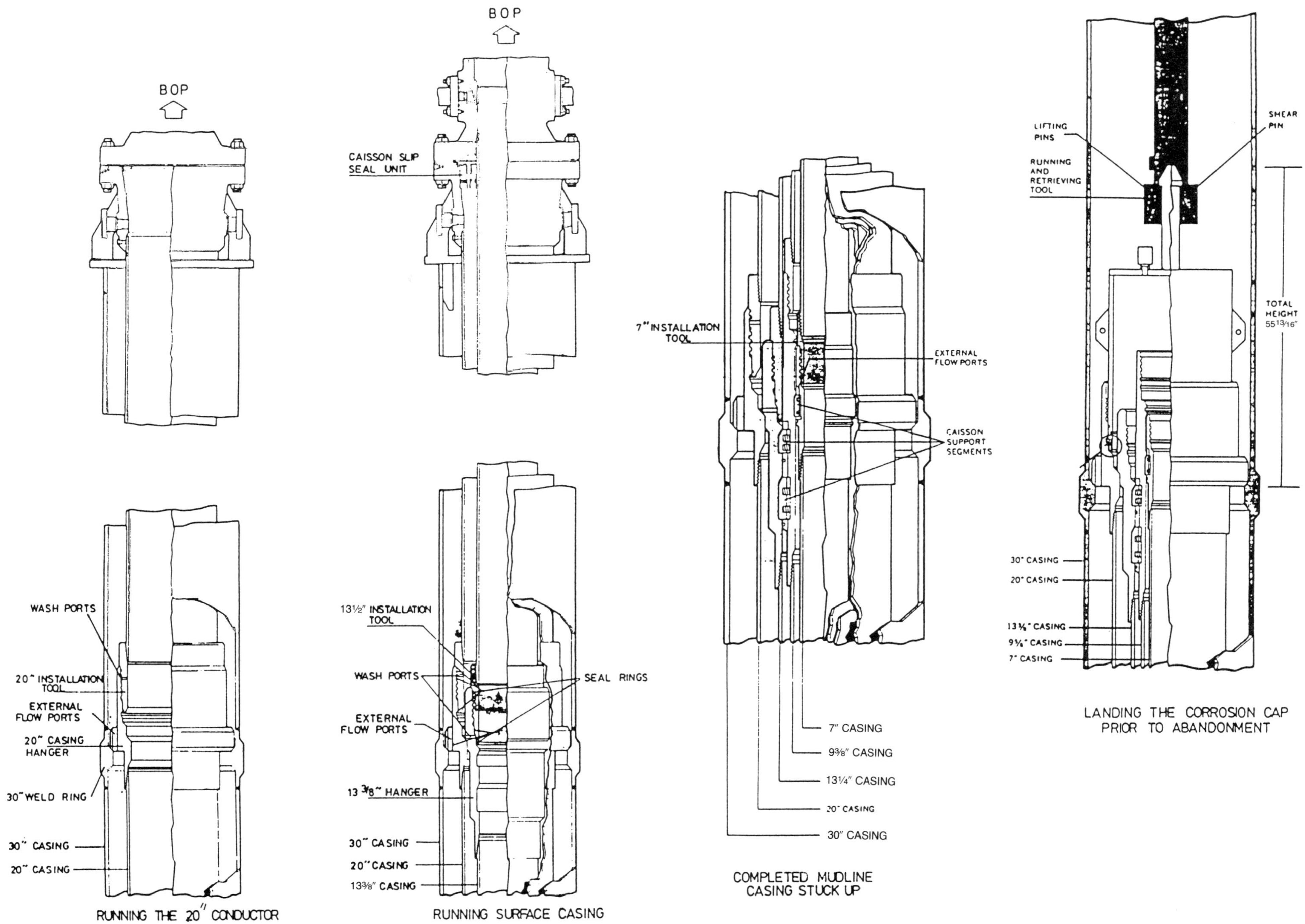

Fig. 7–15 Mudline Suspension System

hangers. If it is desirable to remove the riser, it can be unlatched by rotating or some other means. Once the risers are extended to the platform, a conventional wellhead unit can be installed. The tubing is suspended at the surface, and the standard surface Christmas tree assembly is used.

Obviously, re-entry to a suspended well can sometimes prove difficult, and normally the wellhead manufacturer will provide two methods of re-entering: the first uses the same handling tool connections as for the initial installation; the second, if minor positional misalignment is found, riser latch mechanisms are utilized instead of the handling tool connectors.

Marine Wellhead Conversions

In many instances, exploration or appraisal wells are drilled and the wellheads run by using different manufacturers' wellhead systems or of a size different from that of the final development wellhead selection. In these cases, it is more economical to design a wellhead conversion assembly than mixing equipment types. Such an assembly provides the means of running the tubing hanger and tree connector in a size that is compatible with the development wells.

The wellhead conversion assembly (WCA), Figure 7–16, generally consists of a spool which latches to the existing wellhead, a tree connector, and a mechanical back-up locked and run on a standard wellhead running tool. The well is completed after the wellhead conversion spool is in place.

The connector's upper body, with the conversion spool, provides a reduced size mandrel for latching the drilling BOP stack and the production tree. The mandrel has an internal profile to accept the tubing hanger. Mounted around the connector is a heavy guide frame with four guideposts to fit over the existing permanent guide frame and to provide guidance for the BOP stack and tree.

Particular attention must be paid to the method of testing the tubing hanger seals after running. This often means modification of standard cap testers or the specification of a suitable purpose-built tool.

The method described above has recently become standard practice and enables one manufacturer's wellhead to be converted into another's. The major penalties to be paid however, are an increase in the height of the completed well, increases in leak paths, and the necessity of having two wellhead connectors.

The wellhead conversion assembly connector must be provided with a secure mechanical lock-down system to prevent any accidental delatching. Additionally, it must be remembered that the connector may never be operated during most of the well's life. This may lead to problems if and when the connector is removed, and it is necessary to ensure that an adequate secondary release system is provided.

Mudline Suspension Conversions

Although mudline suspension equipment is not primarily designed for use with subsea production units, it is possible to provide a subsea production head for use with standard components (see Fig. 7–17). After running and suspending the tubing string on the ocean floor, a Christmas tree unit and flowlines can be installed. It is not necessary to install the production head during the initial installation of the mudline suspension wellhead.

The attraction of this method of conversion prompted many operators (Shell, Arco, Phillips) to consider it as a means of returning to previously abandoned wells in the southern North Sea.

The general method is to run the mudline suspension system with the normal surface tie-back on each casing size, up to the packing-off of the production casing. The tie-backs are removed, the 30-in. conductor cut as required, and the well temporarily abandoned. When the decision is made to return to the well to complete subsea, a spool is run onto the intermediate casing, the tubing and hanger installed, and the tree run to provide the subsea completion. The type of spool used, to provide the crossover from a tie-back system to a subsea completion, may be a standard production head or a purpose-built wellhead adaptor with a tree connector to land the tree.

The completion may use either a single or a dual-string tubing hanger. Conversion of a well as a single string dictates the use of a sliding-sleeve tubing hanger to permit annulus monitoring.

The ability of the wellhead conversion of a mudline suspension system to resist bending moments is less than that for marine wellheads. For this reason, any flowline base or anchor point must tie-back to the 30-in. conductor for increased strength.

TEMPLATE TREES

Tree Connectors

The tree wellhead connection is a hydraulically operated unit used to lock and seal the tree directly to the wellhead housing. Generally, it is remotely operated and incorporates design features which enable it to provide lock and seal integrity with a release and retrieve capability.

Early subsea trees did not always use hydraulically operated connectors. A typical example is the Molino field trees installed by Phillips in 1963, which used manual locking dogs to preload and secure the connector to the wellhead. The second generation of connectors were modified marine riser drilling connectors; though recently, it has been recognized that tree connectors do not require such a substantial connector. A new generation

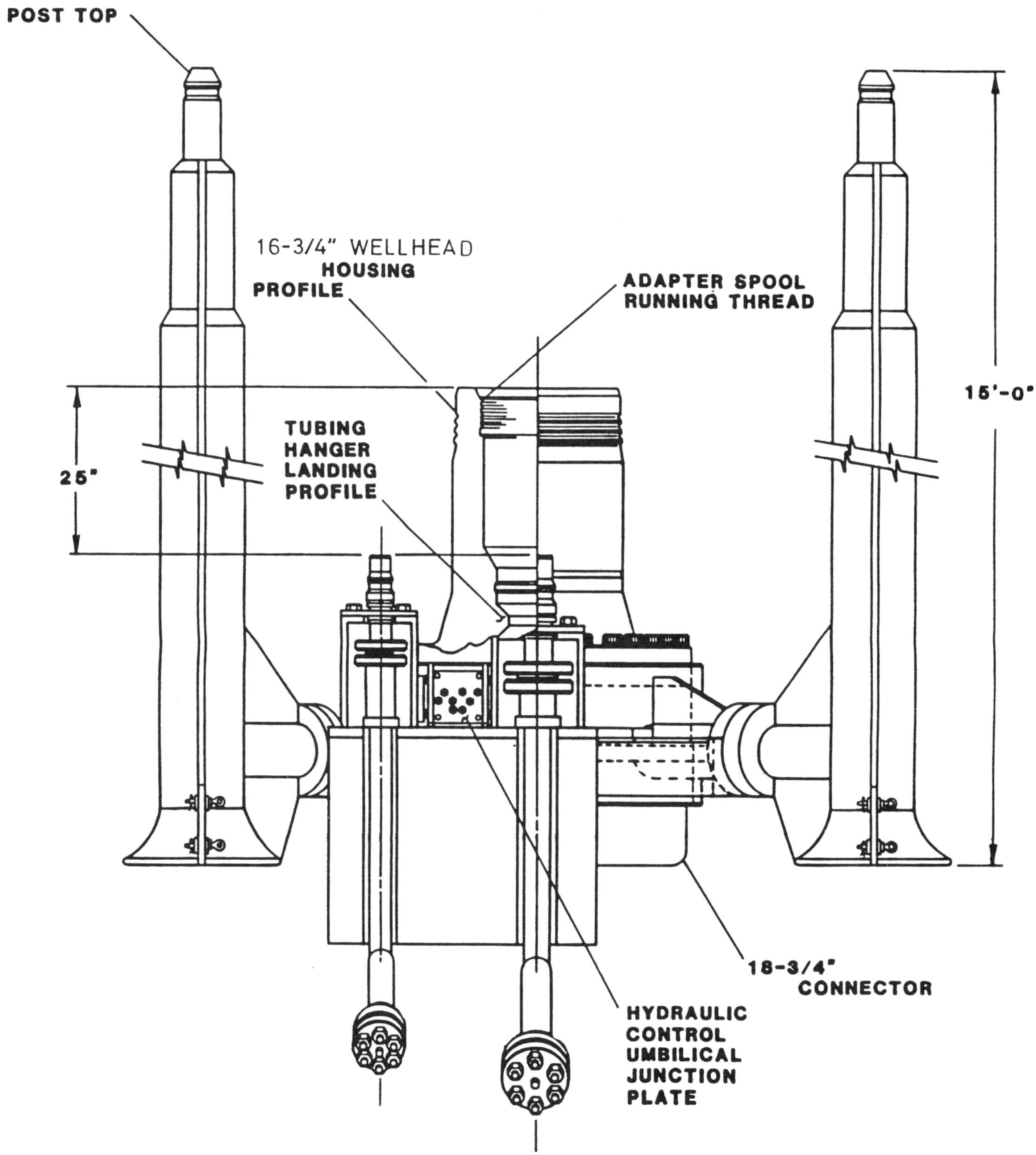

Fig. 7–16 Marine Wellhead Conversion Assembly

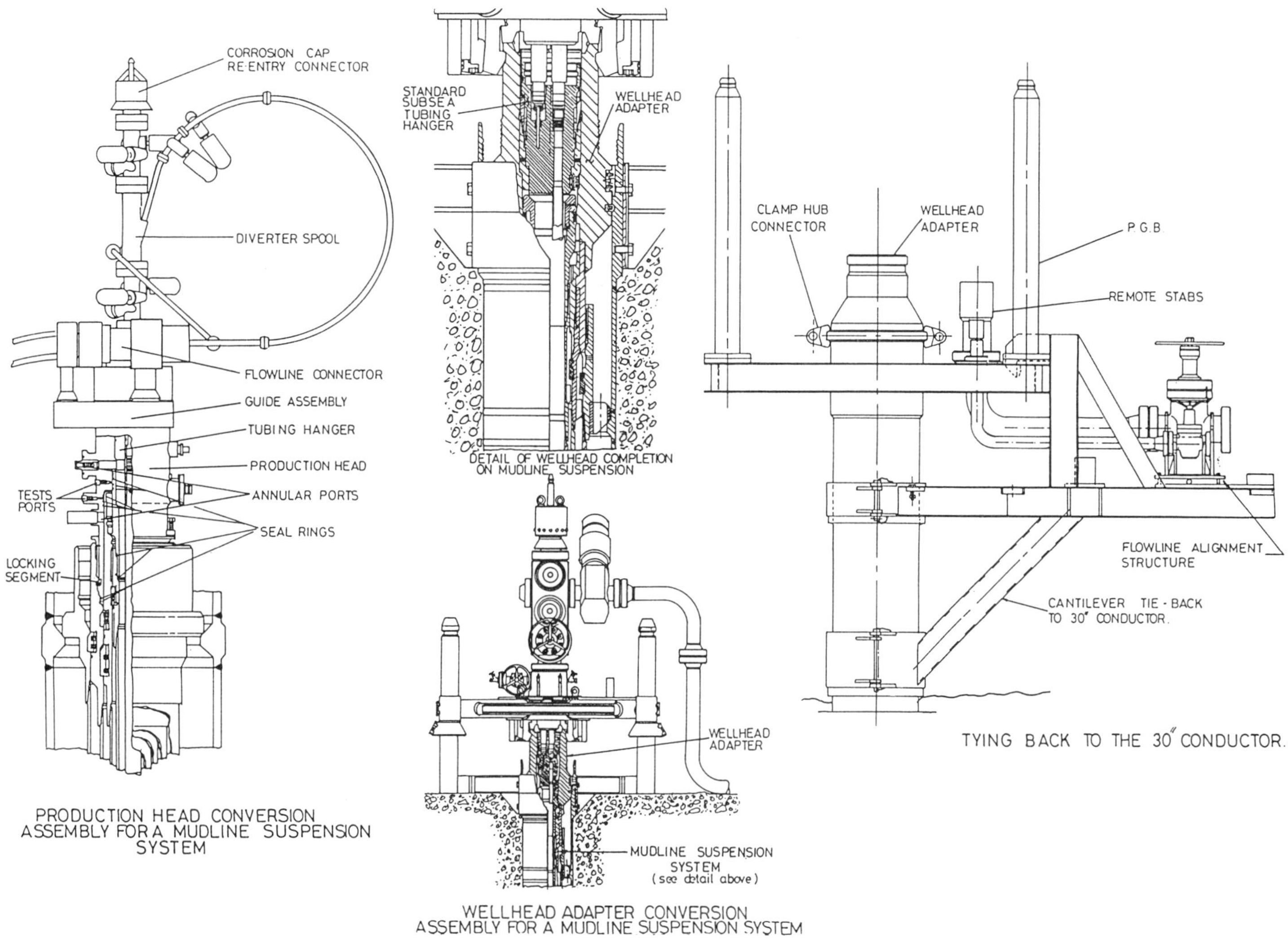

Fig. 7–17 Conversion of Mudline Suspension Systems

of connectors has been designed by one manufacturer to provide lighter, less rigid components. These are more economical and simpler in construction and operation.

The connector illustrated in Figure 7–18 features a single piece main body of relatively small ruling section, with six windows through which the connector dogs are fitted. This is similar to the drilling marine riser connector and provides a direct load path between the connector dogs and the body. The upper body may be supplied studded or with a clamp hub and has a ring gasket for sealing the tree valve block connection.

The connector is operated by applying hydraulic pressure to the rod side of the six hydraulic cylinders. This moves the cam ring down behind the connector locking dogs, in the same manner as the standard connectors. Release pressure is applied to the piston side of the cylinder, providing greater unlocking than locking force due to the increased area involved. All operating cylinders are plumbed together from separate lock and unlock ports.

As with the standard connectors an AX-type ring gasket, held in place by three retainers, seals the connector to the wellhead. The connector is protected by a tree guide frame center ring, and a protective cover encloses the cylinders and hydraulic piping.

Two large mechanical release rods are fitted to the cam ring, for:

- Mechanical release of the connector by pulling directly upwards.
- Mechanical lock down of the connector once the tree is set and locked in place. The lock collar carried on each of the release rods is screwed down into the mating female threads. This mechansim also incorporates an override release feature.

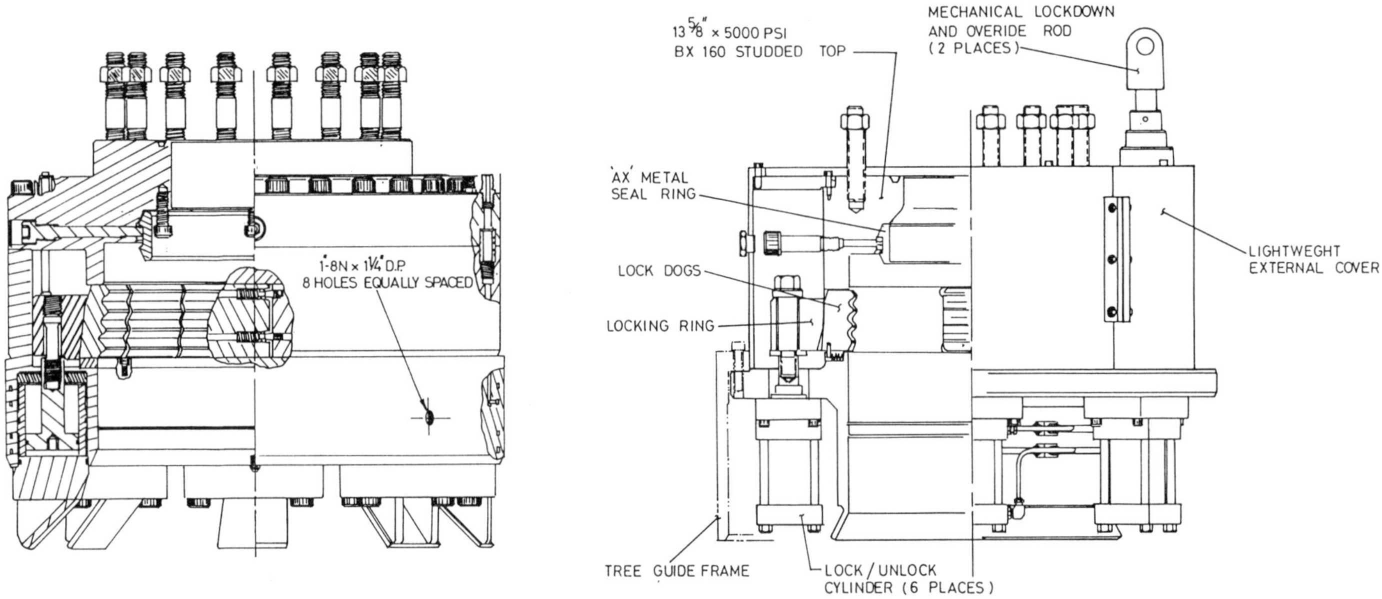

HEAVY DRILLING CONNECTOR
ADAPTED FOR TREE USE (VETCO)

LIGHTWEIGHT TREE CONNECTOR (VETCO)

Fig. 7–18 Tree Connectors

- Visual verification of the locked or unlocked condition of the connector. When the connector is in the unlocked position, the lock collar stands proud of its mating female thread receptacle.

In many instances, it is necessary to provide an entry to the wellhead for downhole pressure transducers in addition to the mandatory DHSV. These provisions require that certain connectors must contain through-bores for this purpose. The advantages of a separate upper body type connector are the lower costs involved in having a common connector with a multiple choice of upper connections and the shorter stab lengths of the various stab stubs.

When specifying a connector, it is prudent to consider the following check list:

1. Wellhead locking profile, size, and rating
2. Top profile of connector
3. Provision for remote operation
4. Positive locking of clamping ring or collets
5. Need for diver lock down
6. Independent lock and unlock parts
7. Secondary release requirements hydraulic or mechanical
8. Wellhead seal type
9. Lock down indication
10. Operating pressure
11. Orientation requirements
12. Downhole instrumentation

In addition, construction strength should be considered in the light of the workover system used and the tensioning requirements needed to operate the system.

Tree Valves

The general layout of the valves for subsea trees is shown in Figure 7–19. These schematics provide for test/kill crossover, choke installation, annulus crossover, and annulus monitor. However, recent attention has been given to the number of leak paths, questioning the inclusion of a lower master valve. The reliability and safety of trees is a function of the number of leak paths, and the deletion of the lower master valve currently is being considered by Shell, Arco, and Sun Oil among others.

The nomenclature and layout of the valves on a tree generally has arisen from early practice on land wells. Traditionally, a land well was completed with a casing contractor by running the casing for the drilling contractor. The well control was contracted out also. This resulted in the need for the casing contractor to provide a closure at the casing head, and it was this that gave rise to the use of a lower master valve. In addition, the lower valve was used by local and national bodies as a means of sealing off the well independently of the operator.

The well control contractor generally supplied and mounted the tree on the valve left in place by the casing contractor. The tree in its simplest form was a stack-up of valves with a lower valve to close off the hydrocarbon line, this valve being

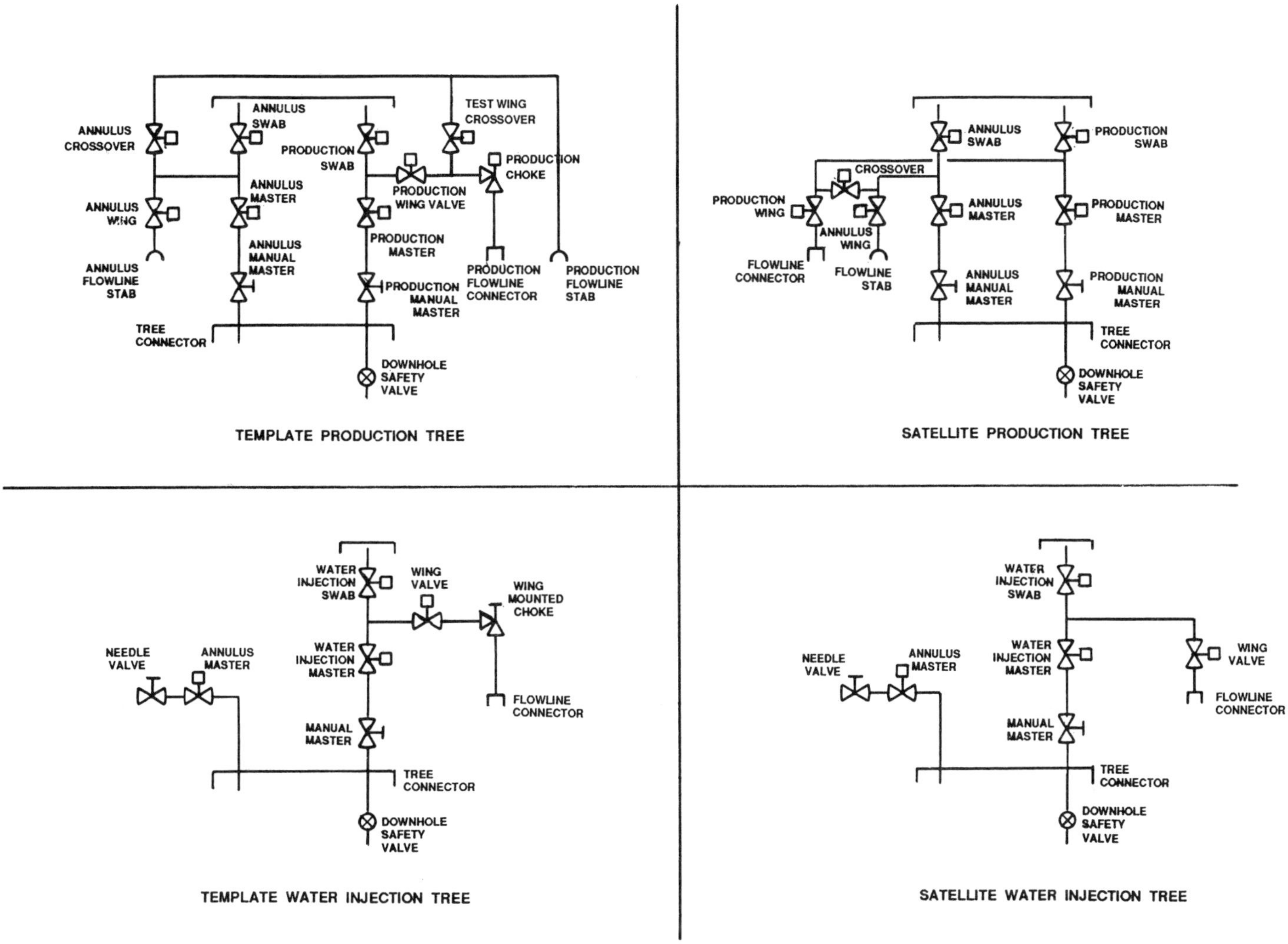

Fig. 7–19 Valve Layout Schemes

known as the ''master valve.'' Mounted directly above the master valve was generally a tree with a gate valve, where the horizontal branch was used for hydrocarbon discharge. The gate valve was most frequently used and literally took its name from the wing-like appearance created when mounted on the tree. The valve mounted above the tree and in line with the master valve provided a means of entering a swab into the tree to encourage start up and consequently was referred to as the swab valve.

The above is a very simplistic view of the history behind the use and nomenclature of Christmas tree valves—variations of the tree having been designed to suit various applications. However, the custom of using two valves, one on the casing head and a second in the tree acting as master valves, was established. It must be added, that this double master valve arrangement was extremely functional and allowed the Christmas tree valves to be serviced in-situ, thus alleviating the need to remove the tree.

It was natural that the first subsea trees were stacked-up valve arrangements very similar in appearance to the early surface trees, and the custom of having two master valves was carried over to these early designs. As the water depths and currents increased on subsequent developments, it was obvious from a strength point of view that the stack-ups were inadequate (see Fig. 7–20), and various attempts were made to compensate for these increased bending loads. This led to the design of the first composite block tree.

An early example was the Molino field trees, which did not use lower master valves as the need for them had effectively disappeared. Subsequent trees in the North Sea continued to use the lower master valve in an attempt to provide a means to service the tree and as an added back-up. Recently, failure mode and effect analysis (FMEA) has shown that this is not the case and that lower master valves do not provide any substantial gains. Care when designing trees must, however, be taken

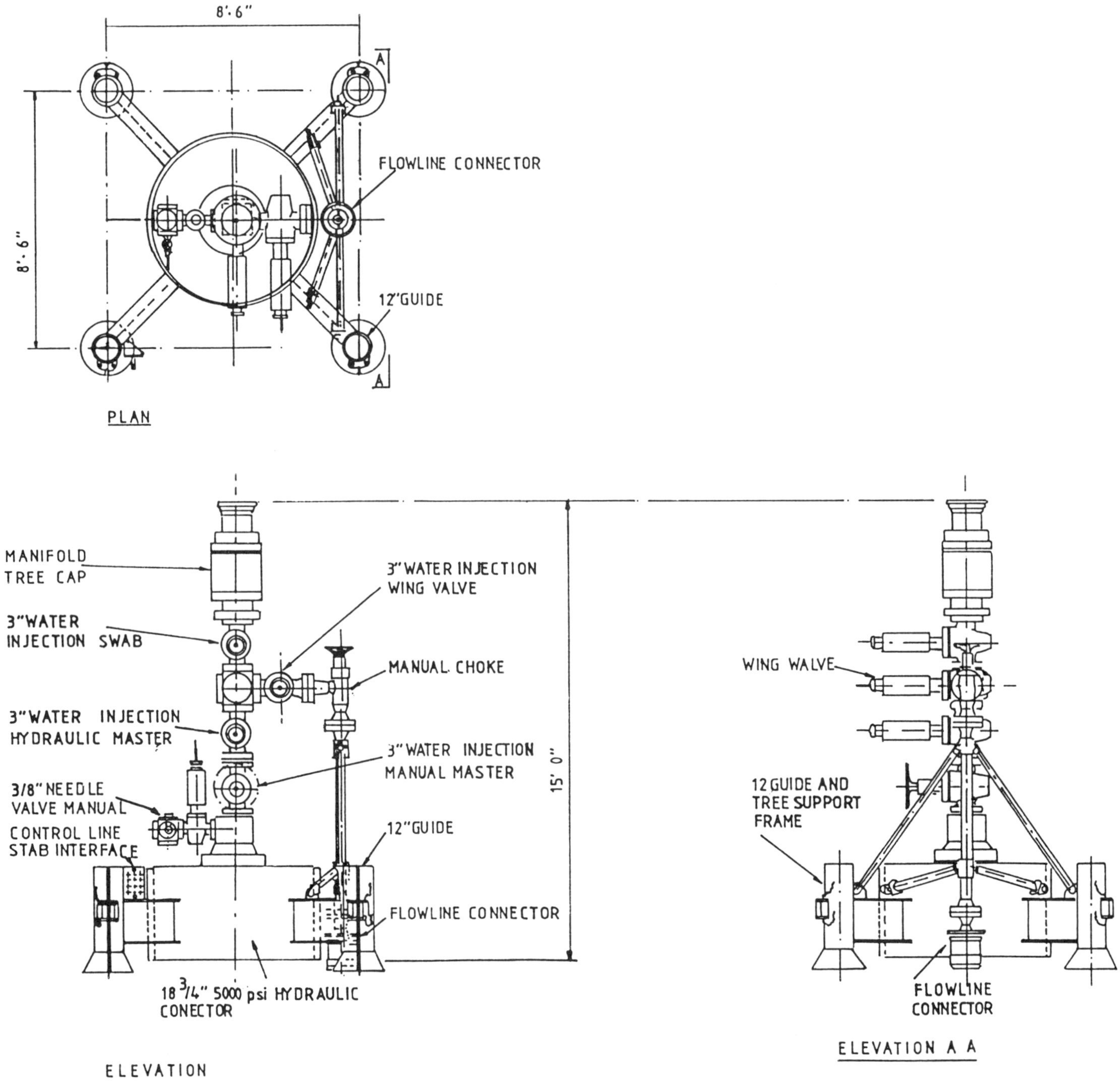

Fig. 7–20 Stacked Tree Configuration

to assure the regulatory authorities of the adequate safety of the proposed system. However, Det Norske Vertias' tentative rules do not exclude trees without lower master valves.

In summary, therefore, any tree should be subjected to an FMEA at the design stage.

The comments and details given in Chapter 2 apply equally to the construction of tree valves. All composite valve blocks on trees are constructed using standard subsea valve gates, carriers, and bonnets. The main forged body of the composite is machined to accept the valve trim and gate carrier. It is becoming increasingly common to incorporate the wing and crossover valves into their own forged body which then is attached to the main composite. However, the combination of all the valves into one block has been used on one particular development in the Norwegian sector.

Actuation of tree values may be hydraulic or manual. Hydraulically actuated valves also may be used for the lower master valve. This depends upon the philosophy adopted for a particular field development. However, it is important to recognize that the choice of actuator may result in the need for local accumulation units on the tree. These prevent ingress of seawater and contaminants into the spring cartridge.

Subsea Connections and Control Interfaces

Template trees require interfacing with both the production piping system and the control system. In addition, provision is required to allow the tree to be connected to the workover control system, simultaneoulsy breaking the connections into the main production control system.

The main interfaces and connections are:

- Tree flowline connections to the template pipework
- Production hydraulic control connections
- Workover hydraulic control connections
- Downhole pressure transducers
- Control system module signalization and power

The methods adopted for these interfaces may vary from system to system depending on the control and workover philosophies chosen. However, the general overview given is pitched towards the use of a workover riser and an electro-hyrdaulic multiplexed control system.

Tree Flowline Connections

The connection of the tree flowlines to the template pipework must be the subject of an early appraisal which should attempt to arrive at the most suitable method consistent with the template structural design and layout. Operationally, the most attractive method must be the remotely actuated hydraulic connector (see Fig. 7–21). This requires that the template upward locking stabs are accurately located to prevent seal and stab damage. It is achieved by using a stab base plate accurately machined and jib-bored to provide an anchor point for the template piping beneath and precise stab location above. The base plate should be initially installed on the template when locating the well-bay landing rings, using a common jig for both purposes. Subsequent to template fabrication, the jig may be used for the tree vendor's works to facilitate the laying out of the tree flowlines.

The use of flexible tree jumper spools may be considered if space requirements allow. Bear in mind that the minimum bend radii of flexible hose can take up valuable room.

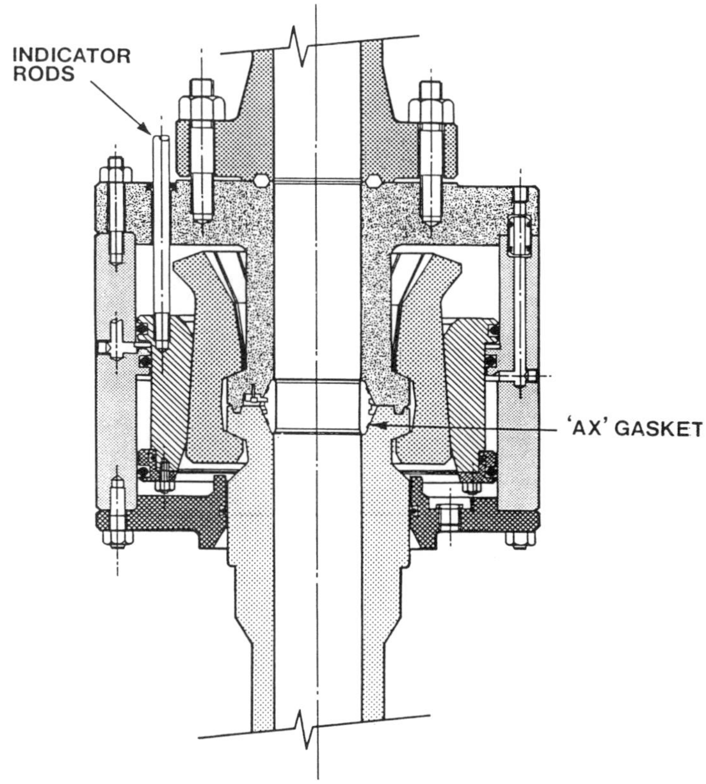

Fig. 7–21 Hydraulic Flowline Connector (Source: Cameron)

Production Hydraulic Control Connection

These are normally of the quick connect and disconnect type. The two main requirements for a diver connection system are: 1) that the connection force to make-up is not excessive with full hydraulic line pressure applied and 2) that the connector can be made up when unbalanced forces occur across the stab plate. This originates when various lines are under pressure and others are not. The stab units may be proprietary-brand stabs or purpose-made. However, they must be capable of making and breaking under pressure and still maintain cleanliness. A typical stab plate and a connection are shown in Figure 7–22.

Workover Hydraulic Control Connections

The workover control connection may be made by using a separate pod arrangement (see Fig. 7–23). Alternatively, the tree cap may be used to provide crossover for the control function. In this case, a running tool is required to remake the connection to the workover control panel when tree cap is retrieved. The tree crossover cap arrangement uses a ring dressed out with cartridge receptacles and mates with a tree-running tool cartridge.

The umbilical stab plate is similar in construction to that used for the production hydraulic control connection.

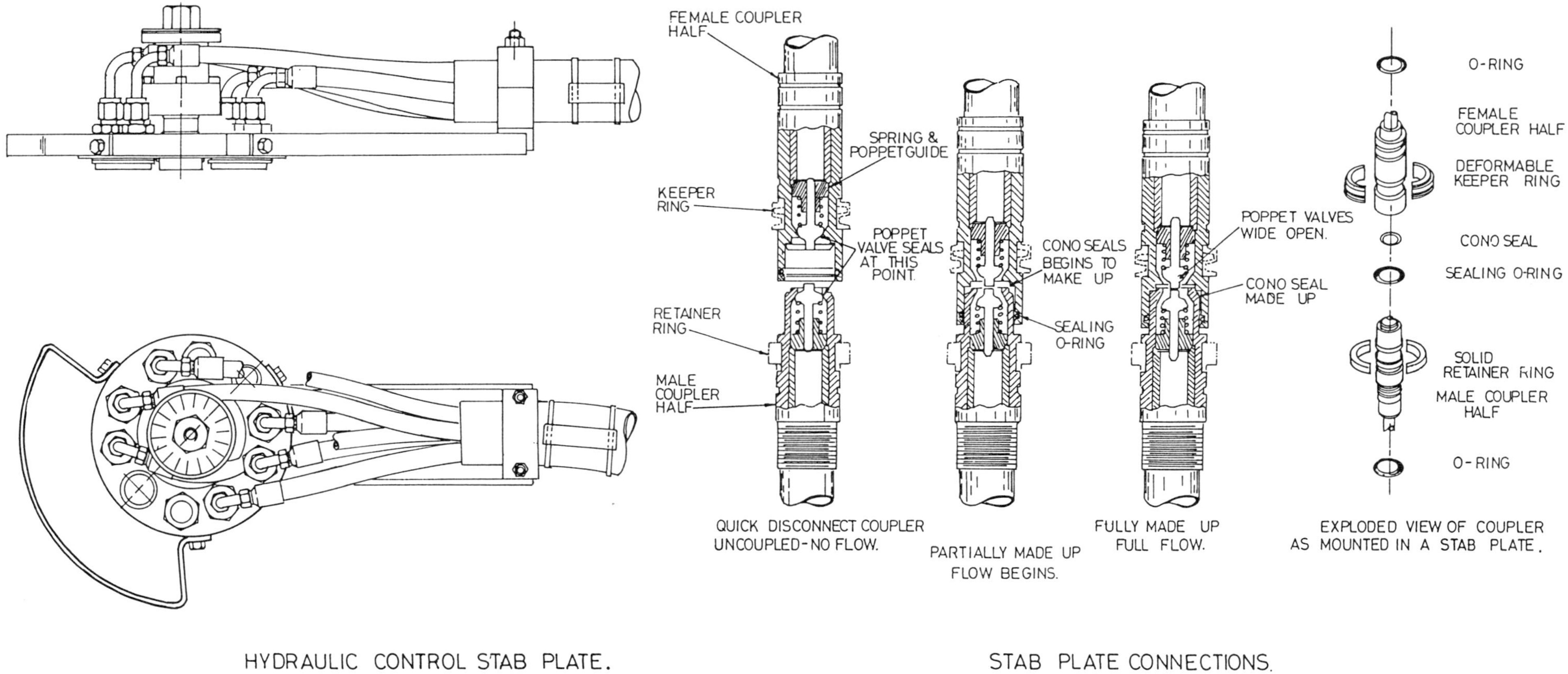

Fig. 7–22 Tree Hydraulic Stab Connection

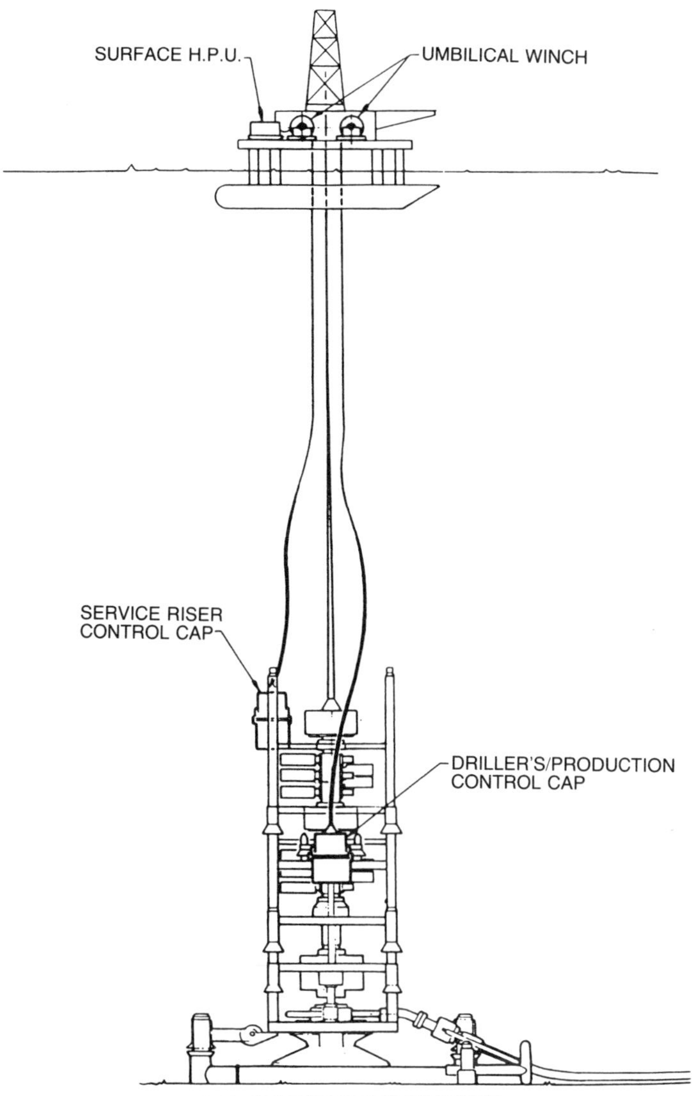

Fig. 7–23 Workover Control Using Separate Pod Units

The alternative method of interfacing the production and workover control system uses a separate pod and receptacle. The pod provides a crossover path between the production control system and the tree functions. It is removed and replaced for workover by a workover pod which is landed in the same female receptacle. The main disadvantage of this method of interfacing is the need to provide a separate guidepost system to guide the running tool. When used with a subsea wireline BOP arrangement, the stack-up height is considerable. The attractive feature, however, is the decrease in the risk of damage associated with the making and breaking of multiple cartridge connectors on a tree cap.

Downhole Pressure Transducers (DHPT)

The DHPT interfaces with the tree by using an electrical wet connector, the male portion secured in the tubing hanger (see Fig. 7–24) and the female stab sealed into the tree valve block. The second interface between the tree block and the control pod is mounted either on the tree or on the template, local to the tree. Particular attention must be paid to the sealing of each break point through the tree block to the outlet connector.

This should always be carried out by the manufacturer's representatives to ensure a satisfactory seal by the epoxy sealing plug usually poured on site.

Control System Module Connections

Connections to the control modules include those for power and signal. It is these which interface tree instrumentation signals with the multiplexed modules. Connections to the modules are made by using inductive couplers, which provide a simple, safe method. The method of attachment will vary with the size of the coupler used; heavy power couplers requiring a complex pull-in arrangement are three times the size of a single coupler. Lighter sized couplers however, can make use of a twist action with final lock down achieved with the tightening of bolts on clamp lugs. The signal couplers are used for each of the signal paths from choke position indicators, differential pressure cells, and flowline pressure taps.

Water Injection Trees

The operation of template mounted water injectors may vary from that of satellite trees. Operationally, it is preferable for the choking of the flow to be carried out on a template tree. Satellite trees on the other hand, due to their remoteness, would preferably be choked at a manifold located on a template or manifold base.

Owing to the less demanding workover system used for a single bore access tree, the water injection tree valve block need not necessarily be of composite construction and can consist of a stacked-up valve configuration (see Fig. 7–20). However, consideration should be given to bending moments induced by the marine riser before selecting the configuration.

Although a single through-bore design may be used, provision normally is made for annulus monitoring. The dual bore side outlet spool mounted on the wellhead connector will provide this capability. As the tree is landed, a stinger sub in the spool opens a check valve in the annulus bore of the tubing hanger (see Fig. 7–25). This action opens the well annulus to a 2-in. annulus master valve. To monitor annular pressure, a diver attaches a monitor line downstream of a needle valve provided for this operation.

If a need to kill the annulus of the water injector arises, diver intervention is required again. This is carried out by closing the annulus master valve (verified by monitoring pressure on the needle valve), connecting a kill line, and opening the master valve to commence killing.

The tree running tool does not require the complexity associ-

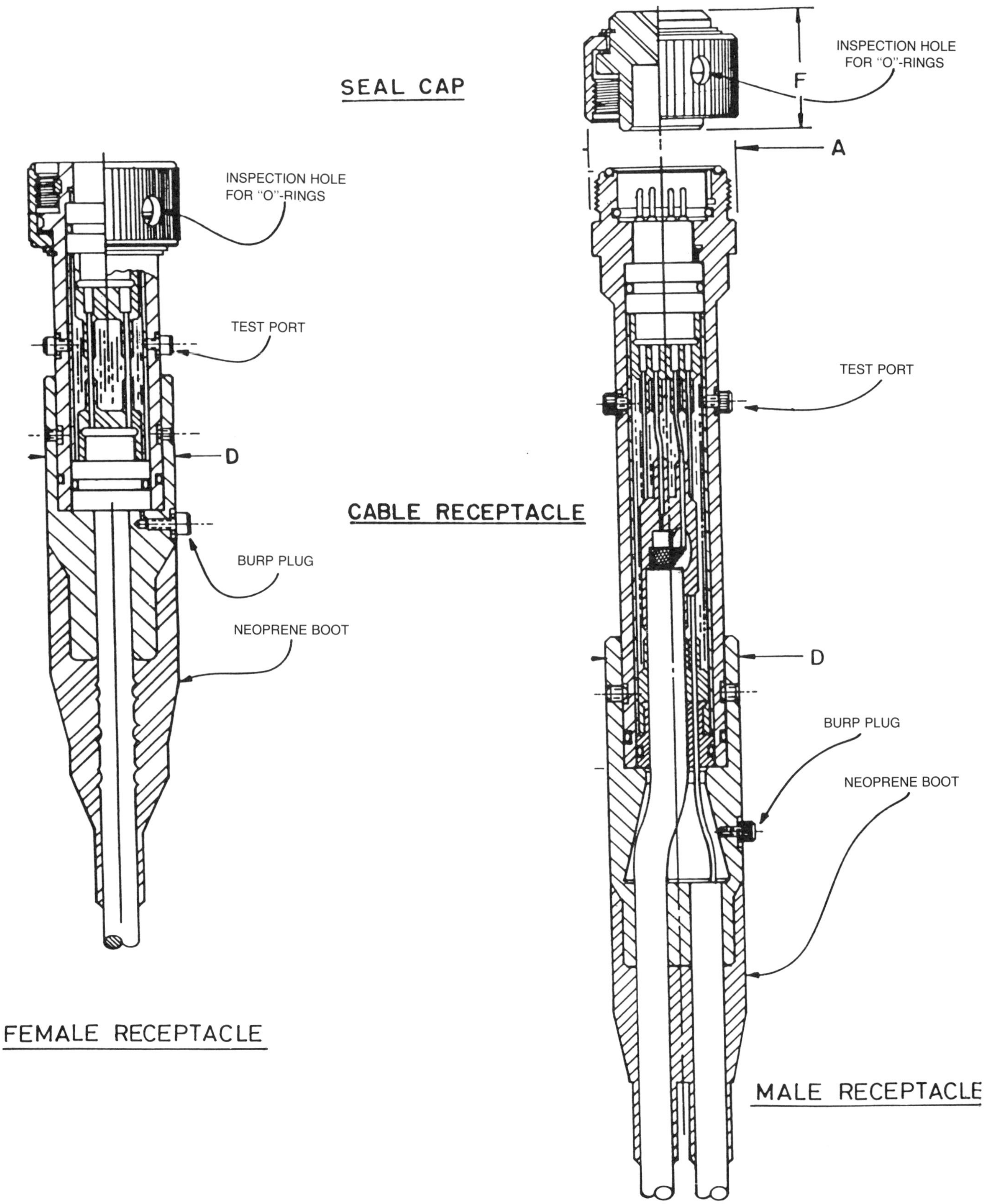

Fig. 7–24 Downhole Pressure Transducer

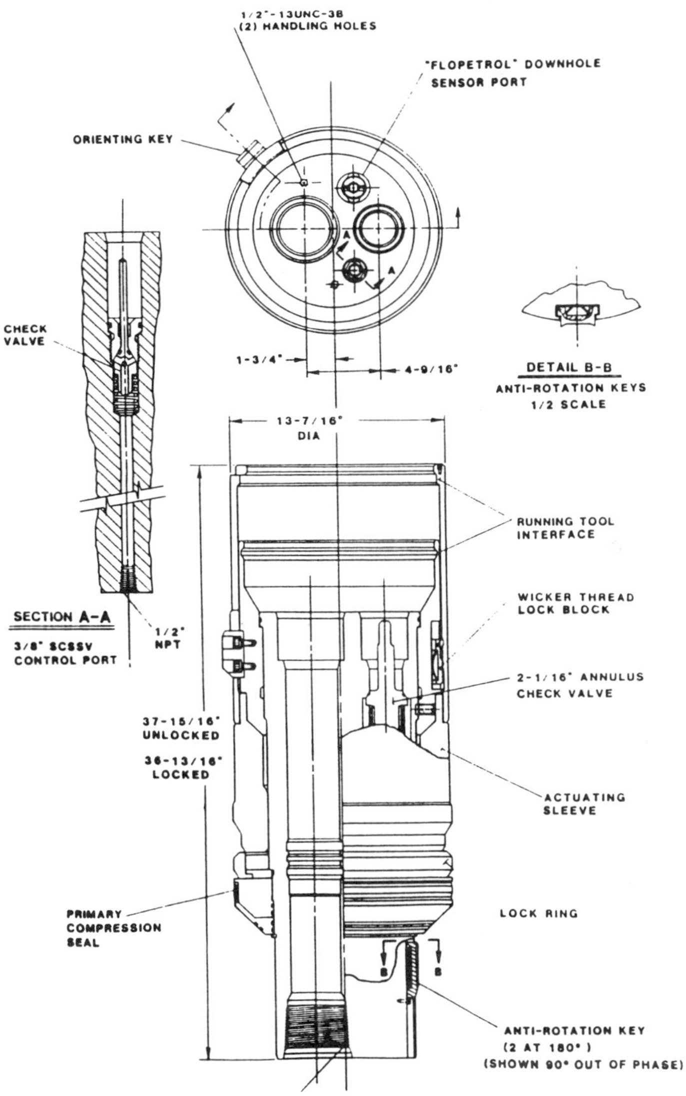

Fig. 7–25 Water Injection Tree Tubing Hanger

ated with production trees due to the relatively low number of control functions. The locking on of the tool and the tree cap may be achieved by pressuring up the drill string used for running the tool.

Additionally, the number of workover umbilical connectors are decreased. This enables a small stab plate to be used, which may be connected to the production control system junction plate after disconnecting from the template control system.

However, water injection trees are subject to reasonably hard service, and it is prudent to ensure that the valve trim is suitable for hydrogen sulphide service, even where the production crude is designated as sweet. Additional wall thickness requirements should be specified for the tree flow loops to ameliorate the effects of the higher erosion rates associated with water rejection trees.

The location of the choke in the tree configuration should be such that manual operation by divers, although infrequent, can be easily achieved. In addition, if manual chokes are fitted, access for maintenance must be considered even though subsea maintenance may not be important in the general maintenance philosophy.

Field development may require both satellite and template trees. Therefore, it is necessary to ensure that the choke flange center distances match standard elbow dimensions. This will allow an interchangeability between trees both with and without chokes. Changeover then involves the removal of the choke and replacing it with a flanged elbow spool.

For remotely operated chokes, a hydraulic connection interface will be required for choke actuation. The actual connection should be of the quick-connect type, and the number of connections may vary between manufacturers.

An interface is required in the signal line to establish communication between the remote position indicator sensor and the electro-hydraulic multiplexed (EHMUX) pod. The interface between the linear variable differential transformer (LVDT) and the pod will require up to four pin connections ganged into one line.

Flow indication may be obtained by the use of differential pressure transducers. The interfacing between the choke and the control system is a twofold operation: 1) the tapping points upstream and downstream are located according to the DP-cells suppliers' requirements, and 2) the signal lines from the cell to the inductive coupler are connected to the control pod.

Tree Choke Interfaces

Choke interfaces can be mechanical and electrical, to suit the requirements of manual or remote operation. Maintenance and instrumentation philosophies also will dictate the number and type of interfaces required. As a base case, consideration should be given to the following:

- Choke location in the tree flow-loop layout
- Deletion of choke from tree configuration
- Hydraulic connections to remote chokes
- Monitor lines
- Differential pressure transducer requirements

Choke Location

Whether or not the choke is to be maintained subsea, careful consideration must be given to the location of the choke within the tree layout. The ease with which the choke may be removed is important as it is considered the weakest system component. With manual water injection chokes, it must be expected that severe erosion problems may arise. If so, subsea maintenance of the choke may become mandatory if excessive downtime is to be avoided. Therefore, the choke should be placed in a position so that the actuator and trim are accessible, and the flange make-up should allow the choke to be removed subsea.

In addition, tapping points upstream and downstream of the choke should be available on straight lengths of pipework. The location of these points would be specified by the DP-cell supplier.

Tree Operation and Maintenance

The operation and maintenance of template trees is similar to that for satellite trees, being simple and avoiding excessive use and disturbance of the tree valves. General operations should ensure that the tree is shut in at the wing valve for most purposes. Only when a well is shut in for a long term or when moving off station with the floating production vessel should the master valve be operated.

Provided that the tree design is simple and the construction conventional, there is no reason for a heavy maintenance program for trees. This has been borne out by retrieval of the Molino trees after 20 years in operation. Very little would be required apart from seal changes to obtain many further years of safe, reliable service from these trees. Since those early days, many material and component advances have been made which should be a very reassuring trend for operators.

Maintenance, when required on trees, should be carried out when the trees are retrieved to the surface for repairs or refurbishment. In-situ repair of components, other than control plumbing, is not recommended. The maintenance of trees should be scheduled for the period when major workover of wells is carried out, as this generally necessitates the retrieval of the tree.

Routine visual inspection by divers or ROV is required only with the functional and pressure testing. Ideally, this is carried out during workover periods, if these coincide with annual testing requirements.

Subsea components such as valves and chokes, which are integrally linked with the production and annulus bores of the trees, should be repaired, replaced, or serviced on the surface. Diver repairs should be performed only on items not critical to the recovery of pressure integrity of a tree (i.e., control system connections).

However, access must be provided for divers to work on components which could fail and put the tree recovery operation in jeopardy. For instance, subsea chokes may be made diver removable, with replacement by a blind flange. This procedure could be used if the choke should erode through the choke body, preventing the operator from setting plugs before tree removal.

Periodic maintenance should be planned for all tree components once the tree has been retrieved to the surface. The only exception to this would be diver inspection of bolt torques and a visual search for hydraulic leaks.

Workover operations and tree replacement should be planned for periods of optimum, summer weather window conditions. Considerable time must be allowed for suspension of workover operations in the event of heavy weather forecasts. Special maintenance for satellite trees should be conducted by using a separate diving support vessel (DSV). Periodic surveys must be conducted which may be carried out on an annual basis or as dictated by statutory bodies or operations.

Extraordinary maintenance of subsea systems requires that tools and facilities be available to immediately effect retrieval and repair of major components. This requires the following:

- An adequately equipped workshop with tools, work space, and furnishings needed to repair valves, chokes, and other components. This area also must include utilities and space for testing and check-out of components and control modules.
- A restricted access storage area adjacent to the workshop with sufficient storage space and access for handling all onboard spares.
- The facilities should be located adjacent to the vessel moonpool. The workshop should have access to the moonpool area, and the storage area should have crane access for handling of large components.
- To facilitate major repairs of subsea trees and manifolds, crane capacity should be sufficient to handle these components between the moonpool and a supply vessel.

Spare Parts

Spare parts availability plays an important role in any maintenance philosophy, and, as a general guideline, a normal onboard spares inventory should include:

- Template valve spares for each size valve with additional sets of gates, seats, and packings
- Spare manifold valves and repair parts
- Complete choke assemblies and parts for repair or refurbishment of trim and actuators
- Wellhead and workover riser connector repair kits
- Hydraulic controls piping and fittings
- Replacement jumper spools and seals
- Miscellaneous studs, nuts, and ring gaskets
- Spare electro-hydraulic jumper umbilicals
- Spare control modules (minimum one onboard and one onshore)
- Control module electrical and hydraulic interconnection cables
- Transducers with umbilicals and couplers
- Spare guidelines
- Guideline installation and retrieving tool plus spares

Dependent on the number of trees in-service, major capital spare items such as tree replacements, workover risers, tree running tools, connectors, etc., should be held at the base stores for immediate trans-shipment when required.

SATELLITE TREES

Production and Water Injection

Satellite production trees differ from template located trees in a number of ways. The principal difference is the need to land onto a permanent guide base and provide the means to interface satisfactorily with the intrafield flowlines. Additionally, for operational purposes, it is more attractive on marginal fields to provide remote retrieval of the satellite trees, to reduce the time spent by a service vessel in heavy workovers and tree replacement.

Although this requires a more complex tree in terms of flowline alignment and remote workover interfaces, the actual piping configuration and crossover valve arrangements may be of a more basic nature. This is especially true if the tree produces to a manifold base or to template mounted manifolds. Commingling of the flow in a field development may be controlled from the satellite tree by using a tree mounted choke; this, however, places what may be considered the weakest subsea component farthest from the main work area. The use of chokes in the North Sea therefore has been confined to satellite manifolds rather than the tree itself. Three basic concepts are now considered.

Single String With Vertical Access

The simple single string assembly with verbal access to the tubing string is shown in Figure 7–26. This tree has an annulus monitoring line and a production line each with a T-branch to a wing valve. The wing valves are connected with a crossover. This arrangement allows production to be carried out down both lines, production and annulus, with reduced back pressure. In addition, the crossover can be used to facilitate ball pigging of both lines from one entry point.

The valve block arrangement has a hand-wheel operated lower master valve, a fail-safe hydraulic upper master, and a fail-safe hydraulic swab valve. The lower end of the valve block is attached to the upper body of a connector. The top end terminates with a re-entry hub and high or low pressure corrosion cap. Both the wing valves and the crossover valve are fail-safe hydraulic as is the valve between the annulus outlet and flowing tee.

The flow lines are attached with diver assistance and are normally secured back to the permanent guide base (PGB) frame. Production systems of this design have been operating in the North Sea since 1970.

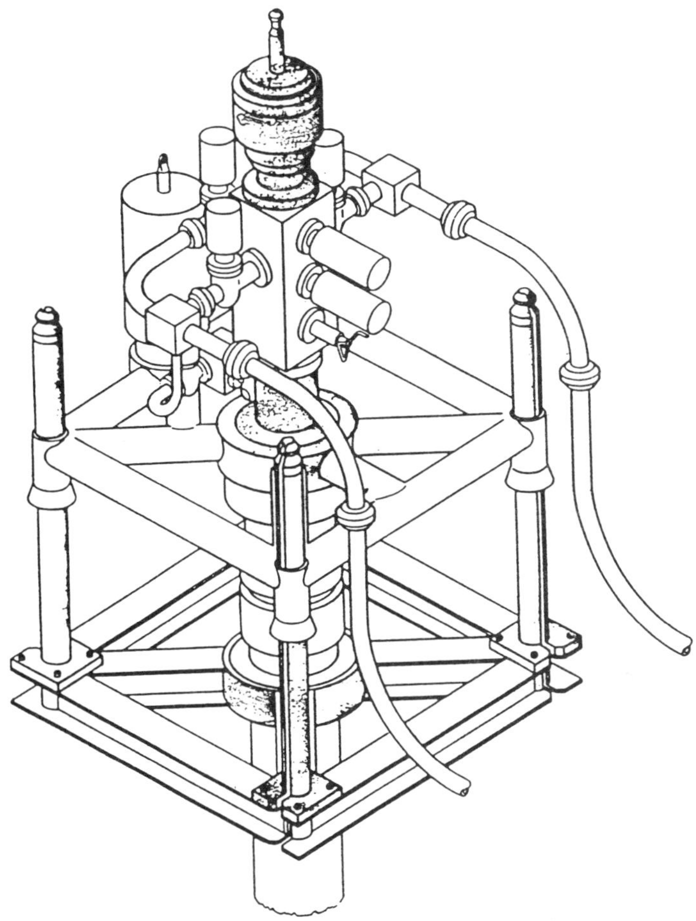

Fig. 7–26 Single String, Diver-Assist Flowline, Non-TFL, Satellite Tree

Multiple Tubing Strings with Hybrid Flowline Connector

Another typical satellite design is one with multiple tubing strings and a hybrid type of flowline connection (see Fig. 7–27). This tree has a valve block arrangement which contains hand-wheel actuated lower masters. If desired, these could be supplied as hydraulically operated fail-safe valves, as are the upper masters and swab valves. The lower end of the valve block is attached to the connector, and the top end terminates in a re-entry hub which is ported to allow vertical access to all tubing bores. It also provides a point of attachment for BOPs, workover risers, or a high or low pressure corrosion cap.

The tree is manifolded with two flowlines, wing valves, and a crossover valve, all fail-safe hydraulic actuated. This arrangement allows the flowlines to be ball pigged as well as providing one as an annulus or monitor line. The flowlines require diver assistance during initial installation but are diverless for all subsequent running or retrieving of the tree. This is accomplished by flowline stingers on the tree which stab into receptacles on

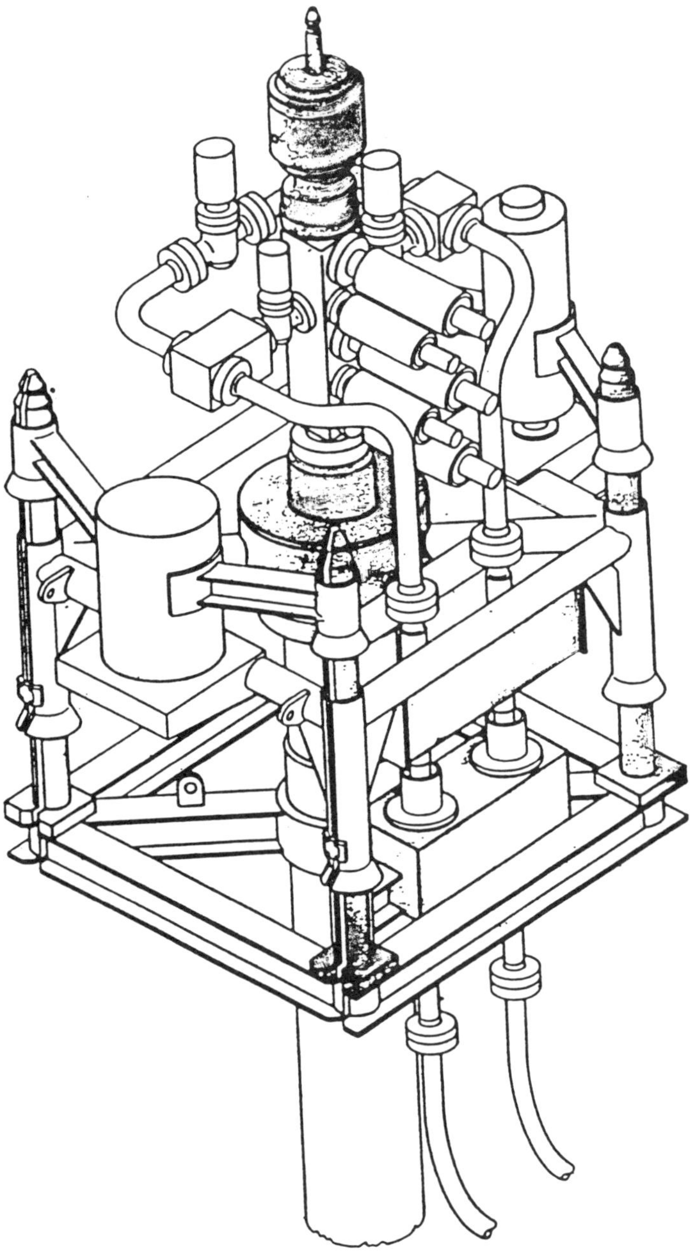

Fig. 7–27 Dual String, Diver and Diverless Flowline, Non-TFL Satellite Tree

the flowline and base frame end. The flowline stab connection normally utilizes a metal-to-metal seal.

Satellite Tree with Flowloops for TFL Tools

Alternatively, the third basic satellite type tree incorporates flow loops to allow the use of TFL or pump-down tools (see Fig. 7–28). Normally, the flow loops will enter the tree through a diverter spool. This spool provides for the insertion or the removal of a tool to change the internal flow path from vertical to the flowline 5-ft radius flow loops. The tree arrangement shown has a dual tubing string with fail-safe hydraulic master valves, diverter spool, fail-safe hydraulic swab valves, and a re-entry hub.

The layout provides a hydraulic crossover valve between the flow loops, hydraulic wing valves at the diverter outlets, and

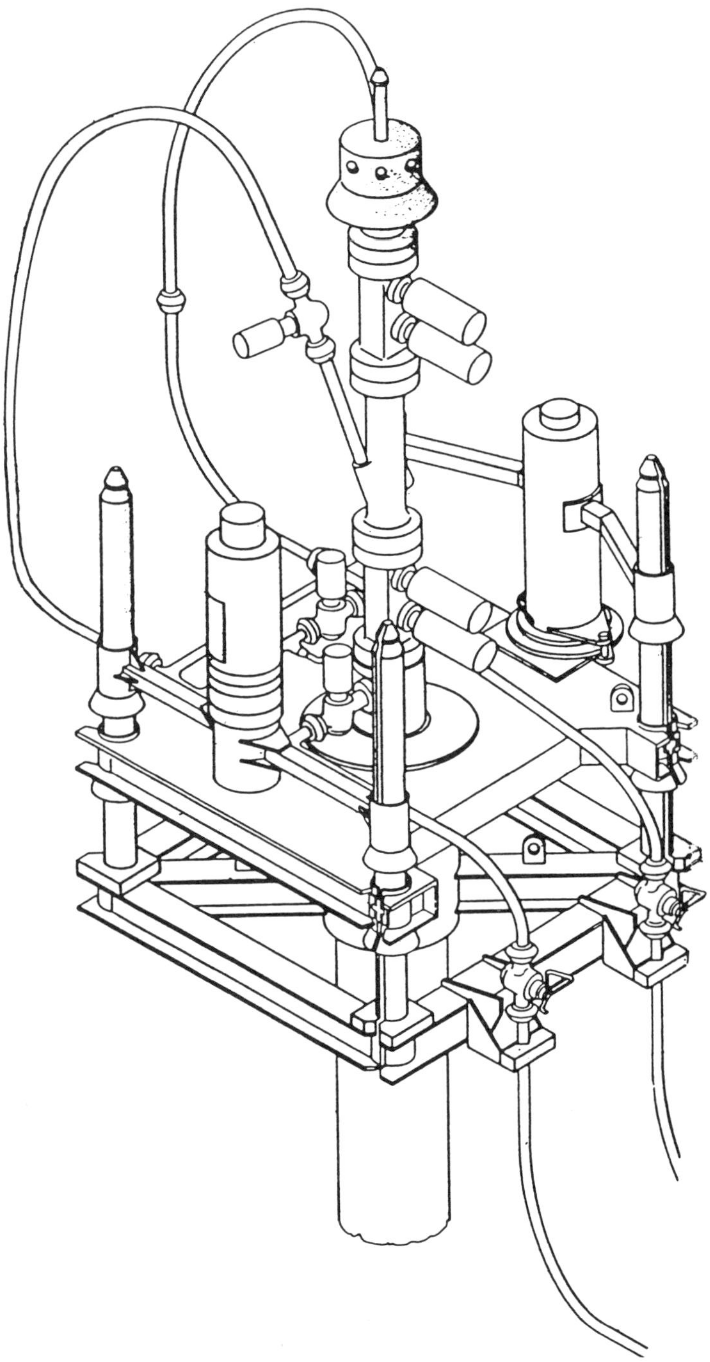

Fig. 7–28 Dual String, Diver-Assist Flowlines, TFL, Satellite Tree

a pair of hydraulic valves which link an injection line into either of the flow loops. The annulus access line also has a hydraulic operated valve tied back to the flow loop. Flowlines are diver-installed, but the stab type or diverless types may be available also.

Water Injection Trees

Satellite water injectors do not differ greatly from template mounted trees. What does differ is the requirement for an annulus monitor flowline. In general, the annulus may be monitored via the control umbilical connected to the upper body of the connector just above the master valve. As physical annular access is not required, the tree composite block may be of the single string type. This both reduces the number of valves, and hence the control requirements, and enhances the reliability of the tree due to the reduction in moving parts and seals.

A series of tree layouts have been shown previously and illustrate the different styles adopted for production trees. It also can be seen that the lower master may be housed in a separate lower master block. This type of split tree design finds favor with some operators. It does, however, introduce a number of extra leak paths. This splitting of the tree provides the ability to mix equipment from a number of manufacturers as in the case of the Buchan trees.

Flowline Alignment Structures (FAS)

FAS provide a landing base for the tree connections and, in certain cases, the structure for locating and securing the flowline pull-in tools. The primary purpose of the FAS is to ensure alignment of the flowline and its stab mandrels with the tree flow loop hydraulic connectors.

The system of orientation of the tubing hanger has been outlined earlier in its own section. Additionally, it is important as it enables the tree flow loop connectors to land closely over the upward facing flowline mandrels.

The FAS shown in Figure 7–29 is landed on a previously installed permanent guide base and allows the mounting of the production and annulus flowline elbows, which in turn interface with the tree flowline connectors and the subsea flowlines. The flowline elbow provides a flange for connection to the production

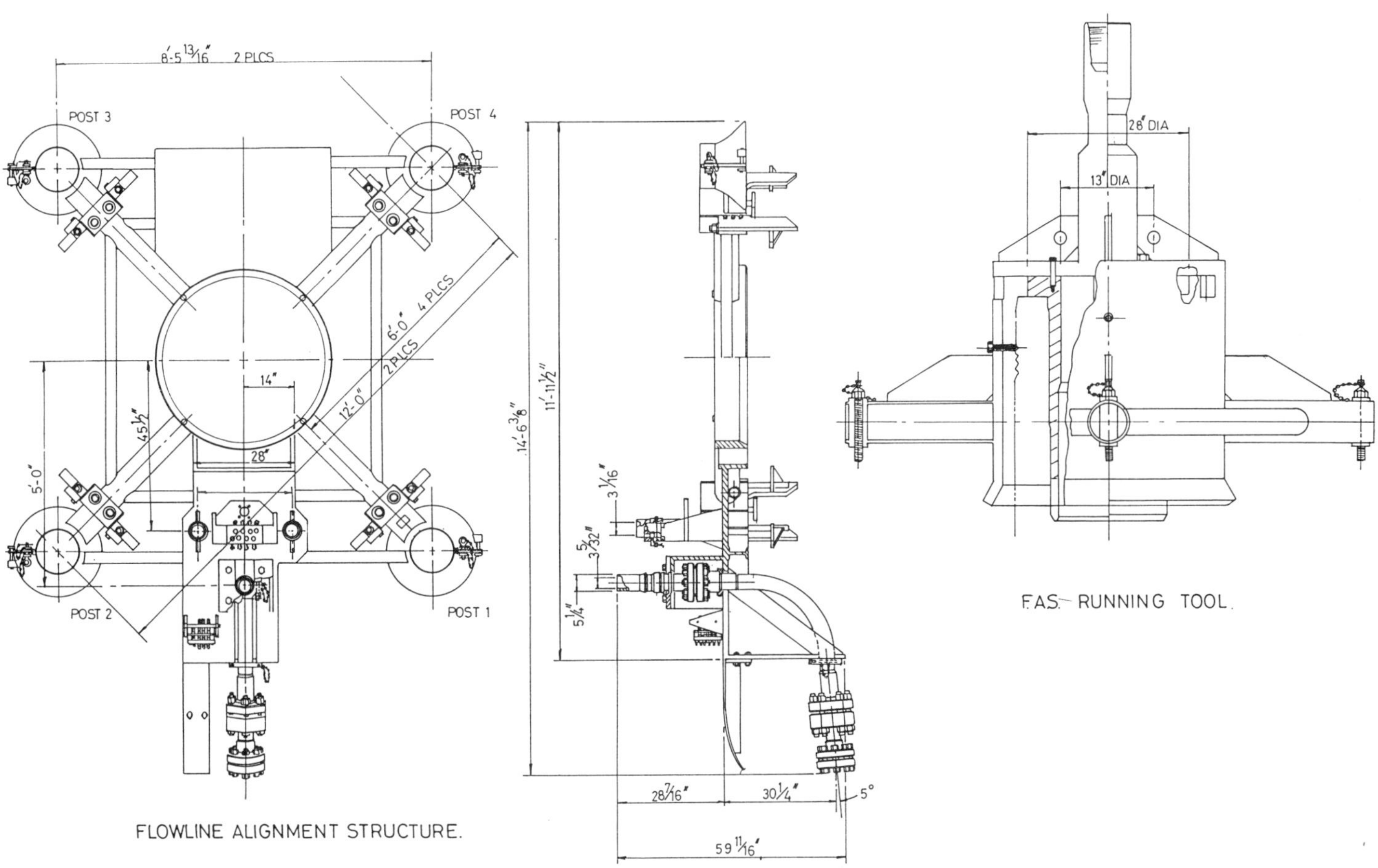

Fig. 7–29 Flowline Alignment Structure and Running Tool

flowline, and it is strongly recommended that misalignment flanges are integrated into this arrangement. The flowline elbow has a flanged connection to attach the flowline connector mandrel. Mounted between the two flowlines is a hydraulic stab plate which interfaces with the intrafield hydraulic control umbilical. This plate is plumbed to a second hydraulic stab box which will mate with its counterpart on the tree.

After the tree is landed, future retrieval is totally remote, without the need for diver intervention. Although the arrangement appears cumbersome, it is field-proven and has been used in Brazil and in the Balmoral field development.

Other methods are available but are normally specified for the more complex pull-in designs. Irrespective of the type of FAS employed, it is necessary to ensure that the arrangement is designed strong enough to withstand the loads imposed by snagging of the flowlines. In this respect, the design load must exceed the rating of the flowline breakaway joint if the FAS is not to become the weak link.

Workover Interfaces

Workover interfaces for satellite trees are identical to the methods already outlined (in the "Re-entry for Workover" and "Subsea Connections and Control Interfaces" sections)—that is by means of a composite workover riser or a dual string tubing riser run using dual slips. The control interfaces also are identical to the methods already described. It is intended here to highlight the aspects of a tree-cap system which is becoming the most common method of providing control crossover during production and the control interface during workover. Figure 7–30 shows a typical tree-cap design using cartridge type couplers.

The main areas of concern when establishing a workover

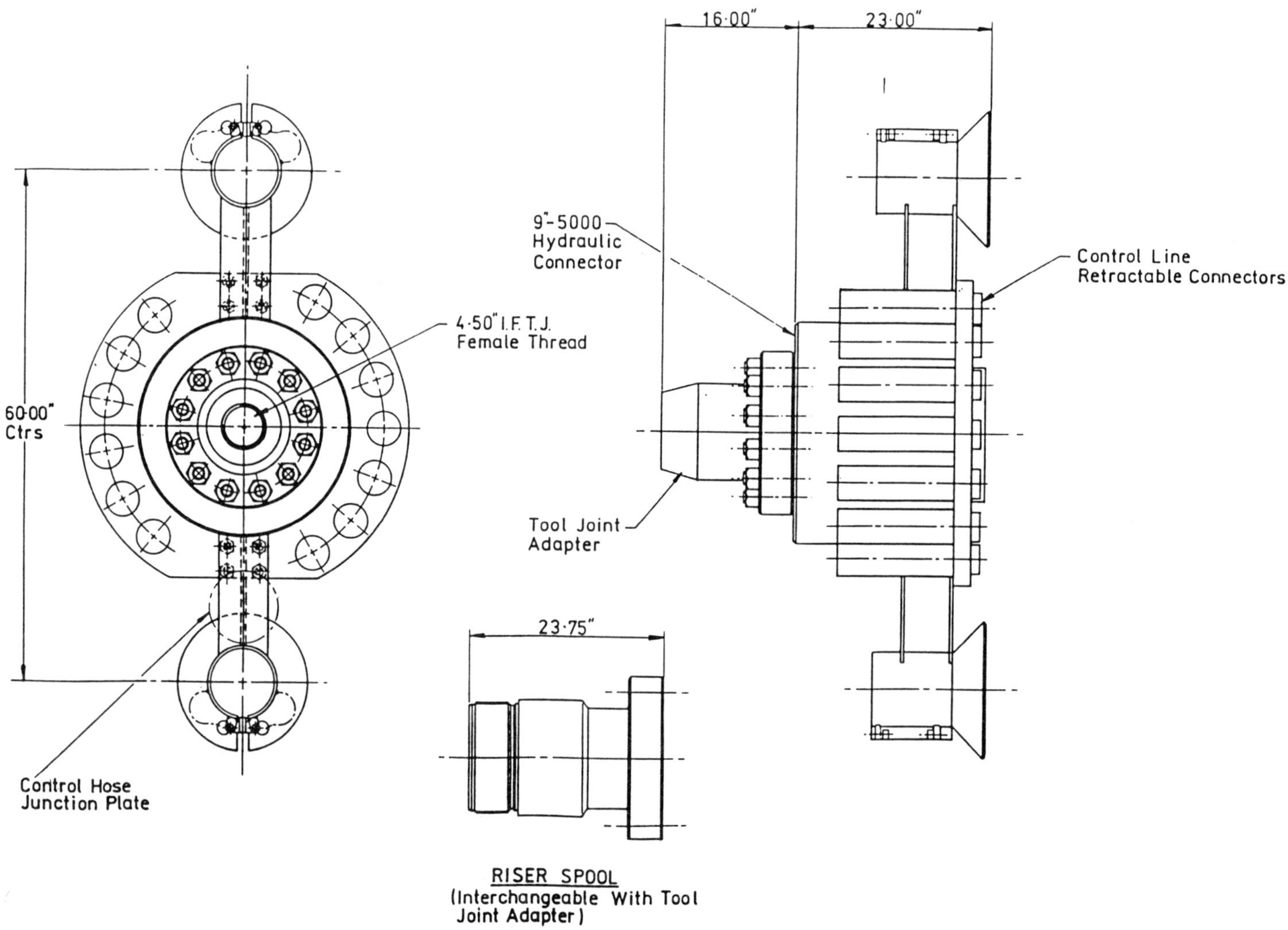

Fig. 7–30 Workover Control Connection With Crossover Cap

interface on a tree mandrel equipped with stab cartridges are the maintenance of hydraulic cleanliness and the safe connect and disconnect of the workover system. The seven problems to overcome are:

1. Stabbing the tree mandrel and workover umbilical couplers together subsea positively displaces a small amount of seawater into the control lines each time the coupler is made up. Therefore, the design must eliminate this deficiency.
2. Since a male check valve coupler requires a check valve in the mating coupler, the tree running tool/workover interface must have check valves also. Therefore, when running the tree running tool, if there is any air trapped in the hose bundle, hydrostatic pressure will force the check valve open and allow seawater to enter the hose. It will be almost impossible to completely bleed all of the air out of the hose. In addition, after the hose bundle is run the first time, it will be impossible to flush the seawater or old fluid out of the lines for storage. This will undoubtedly lead to additional contaminants being forced directly into the control system.
3. The couplers must not be dependent on a single O-ring to effect a seal. Further back-ups must be provided.
4. Remote disconnect of the couplers with any pressure differentail may cause a single O-ring to be blown out of the coupler. Therefore, the stab operation can be performed only once without tripping the running tool or tree cap back to the surface to replace the O-rings. Due to the small size of the couplers and the number normally used, it would be virtually impossible for a diver to replace the O-ring subsea.
5. Accidental disconnect of the tree cap or tree running tool may cause the check valves to trap pressure in the control lines holding the valve operators open, defeating the fail-safe close function to the tree valves. Also note that, without a tree cap in place, a production control system is inoperative; therefore, all control of the tree would be lost.
6. High vertical tolerances are required to mate the couplers properly. It is important that the designer verifies that the tree supplier is fully aware of this in preliminary discussions.
7. Pressure separation loads are placed on the tree cap by the control system pressure. Consequently, it is necessary that these loads are considered when finalizing the method of tree cap lock down.

A system should be developed with the tree vendor that eliminates the above problems and limits the intrusion of seawater into the control system. A tree cap system block diagram is shown in Figure 7–31. Each of the components has been described more fully in previous sections.

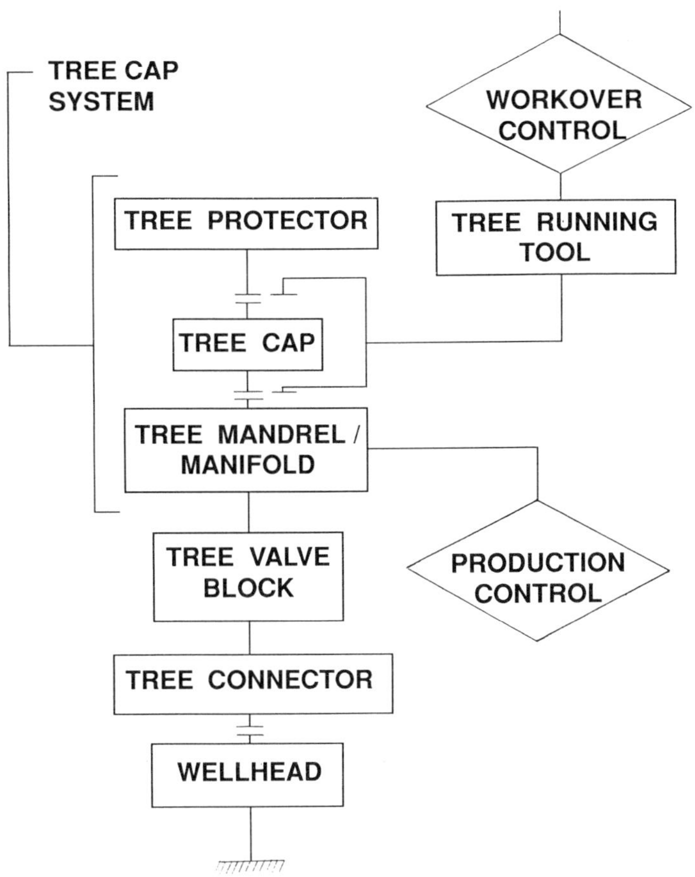

Fig. 7–31 Schematic of Tree Block System and Production Tree

TFL Operations

TFL (through flowline) or pump-down refers to a method of accomplishing simple downhole maintenance. It is similar to wireline in that all operations utilize a tool string to provide simple push or pull forces required to manipulate various devices downhole. Where wireline methods use gravity for downward push and the wire itself for upward pull, TFL relies upon hydraulics for both push and pull. That is, hydraulic pressure provided by a pump on the surface acts upon the cross-sectional area of the tool string to create the necessary force, no wire being needed.

By establishing a circulation path between two tubing strings in a well, TFL tools can be hydraulically transported in and out of the well. This circulation path is generally an H-member but can be a Y-block on the end of dual strings.

The use of hydraulics for servicing a well offer three distinct advantages:

1. The well can be shut in at any time during TFL operations.
2. The well is always under control, owing to the hydrostatic head of fluid being used.
3. Tremendous forces can be generated by the tool string during setting and pulling downhole tools.

All maintenance operations normally carried out by wireline can be achieved by TFL operations, which require five basic items:

1. Suitable pump
2. Fluid
3. Circulation path (H-member)
4. Tool string
5. Suitable conduit

The fluids used to circulate the tool string depend largely on availability but can be salt water, dead oil, diesel oil, calcium chloride, and jelled water (sand washing).

H-Member

The H-member (see Fig. 7–32) is designed as the communication path between the two completion strings. It is available in various configurations to allow more than one to be installed as deep in the well completion as possible.

Tool String

The tool string, also shown in the Figure 7–32, consists of a fish neck, piston units stem, a running or pulling tool, and the various downhole tools being run or pulled. The string is designed to negotiate a 5-ft radius on the TFL Christmas tree.

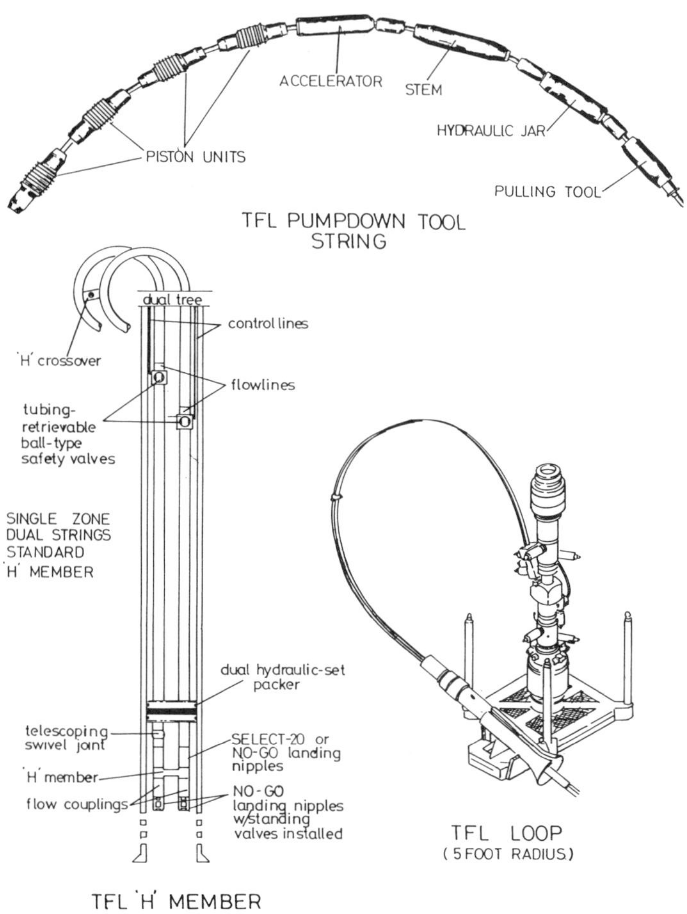

Fig. 7–32 TFL Pumpdown Components

Conduit

The conduits must have the correct internal diameter to ensure the passage to the TFL tools through long flowlines. Welding misalignments, welding slags, undersized or oversized pipe, and undrifted loops can present difficulties for a successful pump-down operation.

Tool String Position Monitoring

The piston units on the tool string are designed to bypass certain amounts of fluid. Tool string position can be monitored by:

- Pressure build-up as the packing stacks and piston units pass through the packing bores in the various downhole components
- Specially installed, reduced bore nipples approximately 10–15 ft long
- Monitoring of fluid pumped and returned on the barrel counters of the control console

Positive Tool Location

For positive location of a specific tool downhole, TFL utilizes a no-go system. This allows the use of a number of different nipple and key combinations to be run in one tool string. Prior to running in the hole, the specific key locator is attached to the pumpdown tool string. The string is pumped into the hole until the locator finds and locks into the matching profile.

History

Research into the application of hydraulic well servicing (pump-down) began in the early 1950s with an effort to adapt wireline procedures to specific functions rather than a total pump-down system. Its capabilities were limited. With the susbsequent oil and gas industry's increasing interest, the first pump-down equipment was developed (1956–1958). This system relied on a limited number of no-go tools for providing location points in the well. Though the system was still limited on application, the basic concepts for pump-down were developed at this time. With increased demand, previous concepts were expanded into a wider variety of functions and tools.

In 1962, developments occurred in selective locating (position) nipples and operating procedures. Most of the testing during this period was performed on land wells completed for pump-down servicing.

In 1965, the H crossover, a circulating mandrel with an integral landing nipple profile, was introduced. Also, an array of flow control devices, concentric gas-lift valves, paraffin scraper, etc., were developed to perform tasks previously done by wireline. Experience during this period was limited, but several operations previously not possible with wireline were successful by using pump-down techniques. Examples are: 1) retrieving stuck dummies and perforating guns which could not be fished successfully by wireline or the use of 1-in. tubing. 2) Sand washing jobs in the casing below the tubing were done using pumped-in wash pipe.

By late 1970, a program for a new system which employed eccentric side-pocket devices to move flow controls out of the flow path was initiated. Pump-down equipment development continued with simpler, sturdier, and more reliable tools as a result.

Recently developed equipment has virtually eliminated many of the operational problems encountered in the early wells, including fluid loss in pulling standing valves from wells with low bottom hole pressure, difficulty in determining location of the tool when pumping into long flowlines and/or deep wells, mutliple runs required for servicing gas-lift installations, and more stringent downhole safety valve requirements.

Pump-down capability has expanded from servicing storm chokes, standing valves, and plugs to an extensive flow-control service capability plus many remedial and completion operations.

Sliding Sleeve and Dual Bore Tubing Hangers

Non-orientating tubing hangers were initially developed to fulfill the simpler requirements of water injection wells. This initial requirement resulted in the forerunner of the sliding sleeve valve tubing hanger. The first non-orientating hangers were of a simple mechanical nature with the ability to open and close an annulus communication port. No physical annulus access was obtainable, only hydraulic communications. The annulus port was spring-operated with pressure assistance. The encouraging results from this design and the availability of suitable running tools led to further work on this principle.

The hanger was modified to allow for remote opening or closing of the annulus valve by using either the multi-purpose running tool when used through a BOP or with the tree in place. Testing also may be achieved from above or below the hanger to verify the annular production cavity. A sliding sleeve hanger is shown installed in Figure 7–33. This type of arrangement has been in use on the Hamilton Brothers Argyll field trees and is being considered for gas-lift operations. It should be pointed out, however, that for gas lift the sealing arrangements would have to be able to provide a bubble tight seal. The sliding sleeve valve arrangement is in use in the North Sea in both the water injection and production modes.

The latest development resulting from this approach is the through-the-tree tubing hanger. The sliding sleeve valve, due to its compact nature, lends itself to being pulled through the tree, thus obviating the need to retrieve the tree when changing out a tubing installed DHSV.

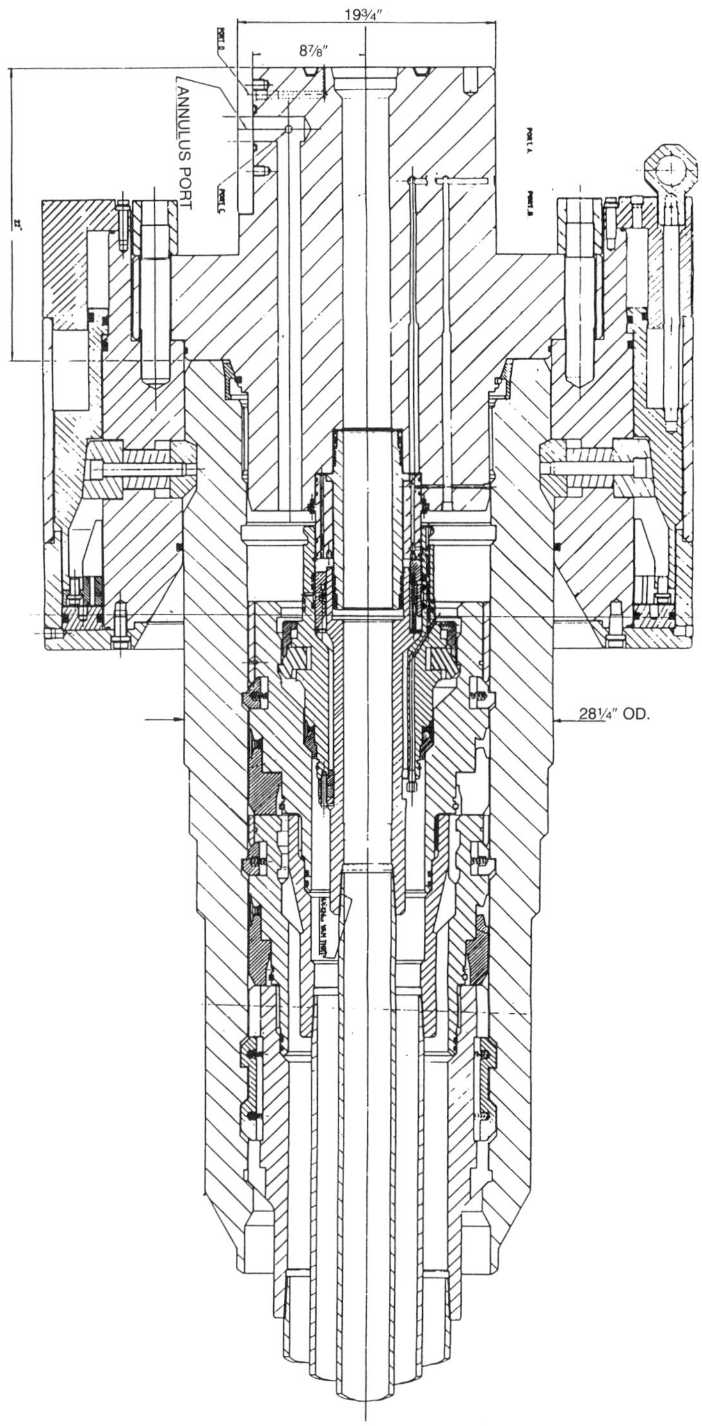

Fig. 7–33 Sliding Sleeve Tubing Hanger

These trees are currently in use on the Duncan field development and already have resulted in considerable operational savings. The tree size is obviously increased with the valves having a 7-in. nominal bore. This development by National Supply is worth some considerable attention.

Dual Bore Tubing Hangers

The orientation tubing hanger system is required when vertical access to the annulus or multiple string completions are necessary. Under normal situations it is only the dual string hanger which is orientated, although it is feasible to orientate the other wellhead components. A typical tubing hanger is shown in Figure 7–34.

The orientating tubing hanger may be run with two basic types of handling tool: mechanical or hydraulic. Mechanical tools are field-proven and quite adequate to land, orientate, test, and provide completion capabilities in the deep water depth range. The mechanical tool approach has the advantage that no hydraulic control system or hose reel is required to operate the tools. Normal rig floor operations are all that is required. Although there is a hardware cost advantage in the mechanical tools, this cost advantage is soon dissipated by the additional operating costs associated with deep water.

Orientating multi-port tubing hangers are successfully utilized on many wells with a hydraulically operated running tool. The tool performs the tubing hanger orientation, landing and testing, and vertical access into the production strings and annulus bore as well as downhole safety valve communication—all in one tool through one trip of the tool and tubing hanger. Important features of the tools include large orientation load-bearings to reduce torque required when being orientated. A hydraulic orientation position indicator has the ability to retrieve the tubing hanger and tubing at any time during the completion operation, if so required, without the necessity of a round trip.

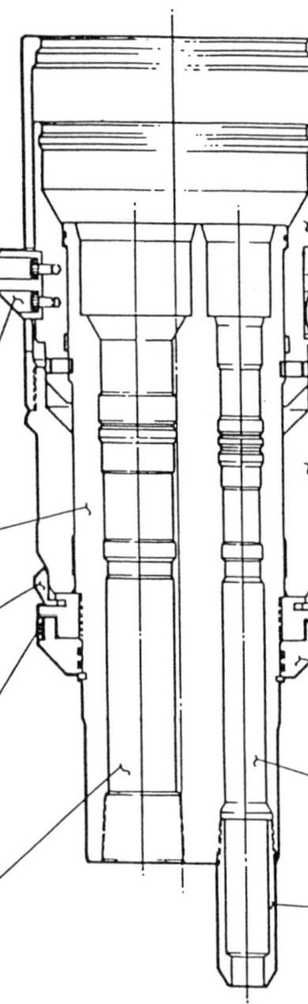

Fig. 7–34 Typical Dual Bore Orientated Tubing Hanger

Control Interfaces

The satellite tree control interfaces are common with those discussed in the "Subsea Connections and Interfaces Control" section, the main departure being the requirement to interface any downhole pressure transducer line with the intrafield umbilical. This is achieved by linking flying leads from the connection shown in Figure 7–22(1) to a diver make up pin connector from the intrafield umbilical. The main disadvantage with this particular method is the requirement for a miniature hand habitat when making the connection. Investigations to eliminate this type of connection should be made in the early stages of a design. Further discussions on control system interfaces are given in the "Control-system Interfaces" section (Chapter 5).

Chapter 8

Flowlines and Export Pipeline System

RANGE OF PIPELINES

A major item in the development of any field is the design, laying, and subsequent tie-in of the intra-field flowlines and the associated export system. the latter could be either an export pipeline or alternatively an offshore loading system. Typically, a small field development, as will be reviewed in this chapter, generally requires:

- One flowline set into each satellite production well. These are 3-in. to 10-in. nominal diameter lines for return of production fluids. An additional 2-in. to 3-in. line also may be required for gas lift, or annulus monitoring.
- One flowline for each satellite water injection well. These are 6-in to 8-in. diameter lines for injection of water into the reservoir.
- One flowline may serve a number of satellite wells when the products are commingled using a subsea manifold. Water injection, annular monitoring, gas lift, and chemical injection, if needed, would require additional lines.
- One pipeline for oil export (typically 10-in. to 16-in.).
- One pipeline for gas export (typically 6-in. to 12-in.).

These flowlines may be installed by a dynamically positioned reel ship, conventional lay barge, or in bundles by a submerged tow method. This chapter briefly examines each of these methods and their technical and economic relationship. Diver-assisted and remote pull-in systems are examined with respect to tie-in activities once the flowlines are in place.

PRELIMINARY ROUTING STUDY

At the start of a project many of the satellite wells and template and export systems are undefined. Assumptions have to be made regarding placement of structures and routing of flowlines. A preliminary routing diagram must be produced to be used as the basis for discussions with the potential pipeline and flowline contractors and with governmental organizations. The comparative economics of the various methods depend on the size and number of the lines and market conditions at any given time. The assumptions made in this chapter are based on recent cost rates and the condition that the contract for each line is "let" independently of the others. Timing and market conditions could therefore completely alter this situation.

If a floater is used as a production platform, the mooring pattern presents some constraints on placement of the flowlines. A twelve-point (or more) mooring system will create a complicated situation which can be alleviated by limiting this to eight points. This arrangement could reduce the cost of flowline installation approximately 10 to 20% by increasing the available corridor widths and hence reducing the complexity of the laying operations.

As further definition of the reservoir is obtained, revisions to the assumed layout will obviously be necessary. It is recommended that a more in-depth study of pipe routing, analysis of flowline placement, and tie-in method to be carried out once specific well and template locations are determined. These exact locations and pipe routing will have a bearing not only on the method of placement but on the cost of the operation.

The sizing of lines is important and must be made with sufficient accuracy. Broad assumptions are to be made on pipe wall thickness for this purpose, and the resultant overall pressure losses are subject to variation as the flow rates become more definitive.

Anticipated flow rates for each individual well and for the topside production system are required in the calculation process. Also, any insulation requirements for the flowlines and the method used for laying the lines are alleviated at the design stage; which allows stresses associated with the lay method and in-service conditions to be checked against allowables.

Allowable stresses for various loading conditions are specified as guidelines, for the design of pipelines and risers, by codes of practice issued by authorities such as Det Norsk Veritas (DnV) and the American Petroleum Institute (API).

VARIOUS PIPELINE/FLOWLINE LAYING METHODS, GENERAL REVIEW

Lay Barges

Lay barges generally provide the fastest and most economical means of installing pipelines. For instance, with single line installations, speed of one mile per day is possible.

For applications where small numbers of lines must be brought to a structure, the lay barge is ideal. Pipe sizes of 6-in. and over are commonly placed in this manner. However, the lay barge is well suited for placement of a few large lines; it is not recommended for use with multiple small diameter flowlines.

Each line to be placed requires one pass of the barge over the pipeline route. The repeated mooring operations dictated by this method extend the laying time and cost to unacceptably high levels.

A further constraint on this type of installation is mooring of the barge in areas of multiple flowline concentration. As each additional flowline is laid, the risk of rupture or snagging by a mooring line is increased.

Submerged Tow Bundles

This method has considerable merit for applications where multiple lines are to be laid to a subsea structure. Its main advantage is the relatively short offshore work time required, compared with the method of laying individual lines and the fact that multiple lines in bundles use the minimum possible space in congested seabed areas. Other advantages include: the capability to weld, inspect, and fully pressure test onshore, and that no mooring of work vessels is necessary during installation.

Submerged tow requires encasement of the flowlines in a carrier pipe for buoyancy and protection during launch and towing operations. Either end of this carrier is fitted with a tow head which provides a tow point and pressure seal for the flowline bundle. The assembled package is pressurized with nitrogen to the site seabed pressure. This pressure is periodically monitored during the tow to check the watertight integrity of the carrier.

Fine tuning of the bundle weight, and in turn its buoyancy, is accomplished by attaching chains to the carrier pipe. The flowline bundle of these chains is negatively buoyant, by 1.5 lb/ft during launch. This is achieved by strict weight control throughout the assembly operation. Once launched, the bundle is allowed to settle to the sea floor. As the chains touch bottom, their weight is supported by the seafloor and the pipe, which has positive buoyancy and floats above the sea floor. Once an equilibrium point is achieved, divers are used to remove excess chain to obtain a final weight trim before towing begins.

In a more recent innovation, the forward speed of the tow is used to control depth. The assembly is negatively buoyant, without chains, being towed at controlled depth tensioned between fore and aft tugs. After the final weight trim, the tow to location can begin. Tow depth is controlled by tow speed, 3.5–5 knots, and the hold back force applied by a trailing tow vessel plus the length of towline paid out. The major disadvantages of this method are the additional cost of carrier pipe and towing equipment together with the limitations in length for towing and the limitations in length of fabrication yards offering this kind of technique.

Although the method is relatively new, it has had substantial field exposure. Occidental and Shell have both had good results with submerged tows on Claymore field and on the underwater manifold center (UMC). Similarly, Placid used a submerged tow to their Green Canyon development in the Gulf of Mexico. In their case however, it was second time lucky, as the first attempt resulted in a number of miles of pipe having to be recovered from over 270 m (900 ft) of water when the tow unfortunately hit a submerged obstacle.

Other advantages of the towed bundle configuration include easy thermal insulation of flowlines (formed foam or gel type) and less expensive control/chemical injection umbilicals.

A major factor in its favor is the significant offshore time saving which is possible when laying multiple lines. It is difficult to generalize and accurately judge the total impact this could have on installation plans, other than to point out the reduced effect of weather downtime it offers when compared with individual line laying methods.

Reel Ships

Reel ships installation is judged to be the most suitable method of installation of single or dual satellite flowline. The use of either flexible or steel flowlines has been examined in detail, and both have been installed in the past using this method.

From an installation standpoint, the dynamically positioned reel-ship approach is ideal for this operation, as it avoids mooring complications and can provide extreme accuracy in flowline placement. Additionally, the lay time for single and dual flowlines is relatively low, with possible speeds of up to 1,000 ft per hour in good weather.

As with conventional pipe lay operations, the cost can vary with market conditions. An in-depth analysis of conventional steel flowline and flexible flowline lay costs is therefore to be recommended.

FLOWLINE CONNECTION METHODS—DIVER VERSUS REMOTE

Diver-made connections are relatively common and many operators have shown themselves to favor diver operations, water depth permitting. This includes the North Sea's Balmoral and Highlander field installations.

Remote Connections

Diverless connections have been developed with a view to reducing the cost of diver intervention and to provide a tie-in method that can be used at depths where diver intervention becomes either limited or impractical. Applications of this technology include Shell's UMC and Chevron's Montanaza field, well D2 installation. Diverless connection also have been used on the Ninian and Brent Fields.

From a functional standpoint, any remote pull-in tool has a limited tolerance for error and needs the presence of a drilling vessel to operate it. Use of this approach also prohibits the individual repair or replacement of flowlines due to the type of end connections used. If a drilling vessel is already on location, no major additional cost is incurred. Generally, this is not the case, and availability and mobilization costs of a suitable vessel for the remote pull-in operation tend to favor the use of alternative methods, using divers, if possible.

One of the major inconveniences of these remote connection or tie-in methods is that they require template space. For a small field with seven to eight satellite wells to be connected, the space taken will increase the template size by 30% to 40% with, obviously, a direct and similar impact on cost.

To overcome the problems associated with offsets and misalignments between flowlines and template interface points, use of special misaligning flanges on all lines is recommended. Also recommended are pressure balanced safety joints on all lines, to protect the template and minimize damage to flowlines in the case of dragged anchor incidents. These joints, which are called load limiting connectors, separate when a pre-set stress is reached. In addition, safety valves in either side of the joint can be installed which automatically close to prevent pollution; a major inconvenience is the complication of the mechanism. As an alternative, smaller lines, less than 6-in. could be fitted with machined bolt connections to provide a safety break.

Diver Connection and Flexible Spools

For connection between the end of the flowlines and the piping in the template structure, the use of rigid, made to measure, spool pieces is often recommended. An example of installation is shown in Figure 8–1. It is a good practice to use this technique for lines of 10-in. diameter and over.

With favorable soil conditions, this is a good solution, but it is diver intensive for two reasons:

1. The final dimensions of the spool must be measured by the diver.

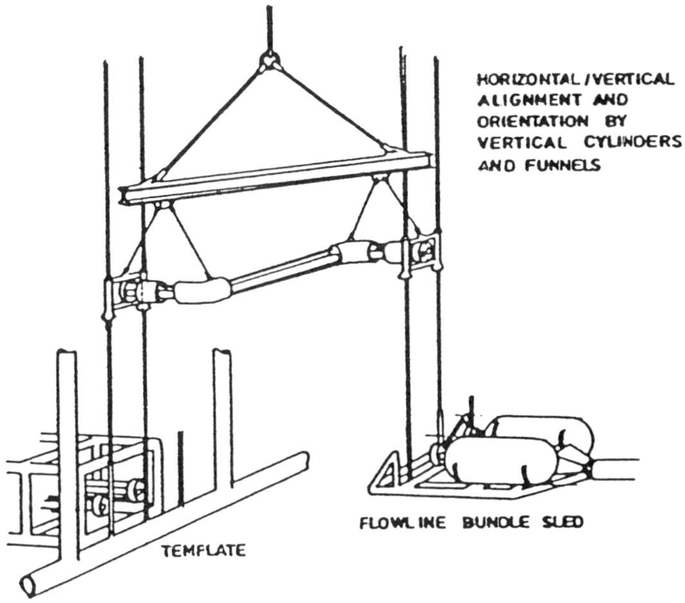

Fig. 8–1 Flowline Connection by Spool Piece Tie-In

2. After measurement, the spool piece must be cut and welded offshore, which requires more time, equipment and expertise.

In addition, the installation of large diameter spools and long spools necessitates complex and expensive handling techniques.

A very attractive solution is to use flexible spools or "jumpers," supplied by companies such as Coflexip, Dunlop, and Pagowag. Such pipes have been used extensively in the North Sea, recently by Texaco and Sun Oil in fairly large quantities. The flexible spool option can be compared with the design of a long flexible flowline later in this chapter. When soil conditions are poor and unstable, flexible spools are the most suitable recommended solution.

Dunlop ARMALINK is the trademark of a family of flexible pipes based on an elastomeric construction reinforced with either steel or polymeric layers. Its application includes jumpers, spool pieces, risers, and flowlines, but not of continuous lengths. Construction is customized to account for the type of fluid conveyed, the environment surrounding the pipe, pressure, and temperature. These parameters determine, to a large extent, the type and strength of the reinforcement and the choice of compounds used in the manufacture of such high pressure flowlines.

Range of Jumpers Available

Internal diameter (I.D.)	2-in. to 12-in.
Working pressure (W.P.)	1,000 psi to 10,000 psi
Test pressure (T.P.)	1.5 times W.P.
Internal working temperature	−20°C (−4°F) to 130°C (266°F)

Fluids conveyed include	Crude oil sweet or sour with or without gas; Drilling muds; Gas; Water; Acid.
Fire tests	Certification by Lloyd's Register of Shipping
Unit lengths	150 ft (46 m) (This limit is being extended.)

A typical structure is shown in Figure 8–2. A detailed description is as follows:

- 316L stainless steel interlocking flexible tube 6.4 mm thickness.
- Extruded Duralon lining—3.75 mm thickness; change over rubber layer—1.00 mm thickness.
- Two layers rayon fabric at alternative bias, encapsulated in rubber compound; plus nylon breaker fabric—total thickness 4.0 mm.
- Two piles high tensile steel cable wires applied at 55° to hose axis at alternative bias and encapsulated in rubber; cables are brass coated to achieve bond to rubber compounds—cable diameter 2.5 mm.
- Interply between cable wires to prevent fretting, consists of open weave nylon breaker fabric encapsulated in rubber plus a layer of rubber—total thickness 2.0 mm.
- Sub cover build-up consisting of nylon breaker ply and two plies of rayon fabric as used above to provide strength; fabrics are encapsulated in rubber. A rubber sub cover and nylon breaker complete the sub cover.
- Cover of blue hypalon 2.0 mm thick.

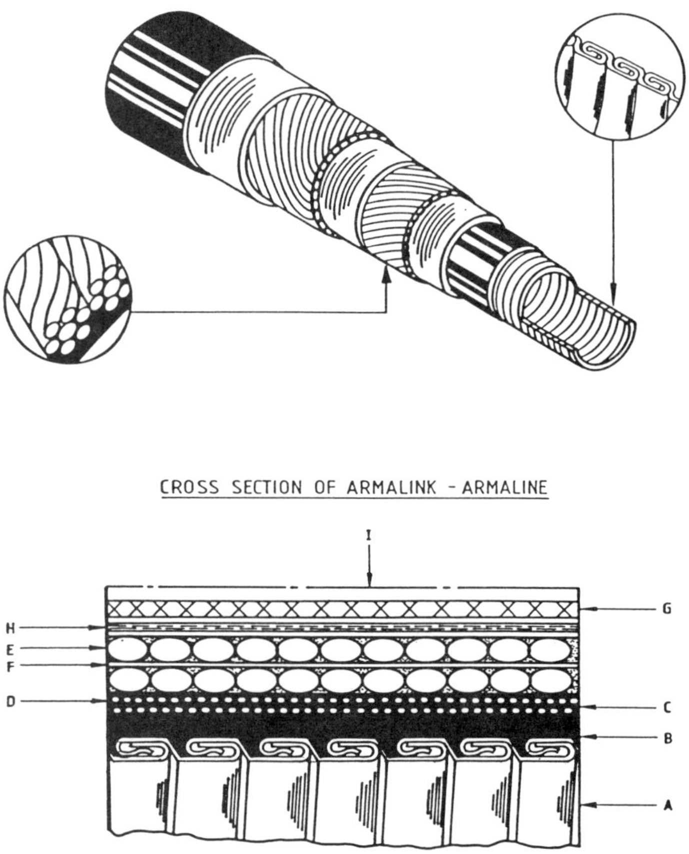

A. STAINLESS STEEL, STRIP WOUND, INTERLOCK TUBE.
B. DURALON ELASTOMERIC LINER.
C. TEXTILE, ANTI-EXTRUSION LAYER.
D. HYDRAULIC TRANSFER LAYER.
E. CLAD STEEL REINFORCEMENT.
F. ANTI-CHAFFING LAYER.
G. ABRASION RESISTANT COVER.
H. FIRE RESISTING LAYER (OPTIONAL)
I. ARMOURING (OPTIONAL)

Fig. 8–2 Armalink—Armaline High Pressure Flexible Pipes (Source: Dunlop)

A typical ARMALINK flexible spool data sheet is shown in Table 8–1 below, and typical specifications for 2-in., 4-in., 6-in. and 8-in. pipes are shown on Table 8–2.

End fitting or couplings can be designed to suit the specified requirements. A typical design is shown in Figure 8–3, and the various types of couplings are shown in Table 8–3.

STEEL FLOWLINES

Santa Fe offers the laying of steel flowlines from their dynamically positioned reel-ship Apache. The method is well proven, and its track record in the North Sea since 1979 is excellent.

The main advantages over conventional pipe-laying methods are:

- No anchor interference problem, as the vessel is dynamically positioned (DP).
- Pipe-laying speed of up to 500 m/hour is achieved.
- Ability to mobilize quickly and accomplish its mission between storms.
- Precision in the placement of flowlines and small diameter pipelines in deep water with its D.P. system.

The designated routes are pre-programmed into the ship's computerized position control system, which receives actual positioning data and transmits appropriate signals to the main D.P. control system.

The Apache reel capacity is:

Nominal Pipe Diameter		Approximate Reel Capacity	
Inches	Meters	Feet	Miles
4	81,380	267,000	50.5
6	48,760	160,000	30.3
8	31,690	104,000	19.7
10	22,250	73,000	13.8
12	16,450	54,000	10.2
14	13,700	45,000	8.5
16	9,140	30,000	5.7

Table 8–1 ''ARMALINK'' Technical Data Sheet

1) Spool piece location	At template end or tree end	
2) Description	4-in. I.D. production test/kill line	
3) Construction:		
Interlock carcass	Stainless steel (316L) tube	
Liner	Duralon	
Primary carcass	Textile reinforced	
Reinforcement	2 ply 2.5 mm steel cable	
Secondary carcass	Textile reinforced	
Cover	Specially formulated blue elastomeric cover to resist marine growth	
4) Technical:		
	English	Metric
Internal diameter I.D.	4.0 in.	101.6 mm
Outside diameter O.D.	6.68 in.	169.6 mm
Internal volume	0.087 ft^3/ft	8.107 l/m
External volume	0.24 ft^3/ft	22.59 l/m
Weight in air empty	21.52 lb/ft	32.02 kg/m
Weight in air full of seawater	27.11 lb/ft	40.34 kg/m
Weight in seawater empty	11.53 lb/ft	17.16 kg/m
Weight in seawater full of seawater	15.54 lb/ft	23.12 kg/m
Working pressure	1650 psi	112 bar
Test pressure	2475 psi	168 bar
Reinforcement stress at W.P.	352 lb	160 kg
Reinforcement stress at T.P.	529 lb	240 kg
Burst pressure	7200 psi	490 bar
Hydrostatic collapse pressure	5700 psi	401 bar
Impact collapse pressure	1770 lb/in	31.6 kg/mm
Damaging pull in straight line	152100 lb	68900 kg
Minimum Bend Radius (MBR)	3.3 ft	1.0 m
Bending stiffness at 20°C	5200 $lbft^2$	219 kgm^2
Maximum elongation at T.P.	±2%	±2%
Thermal conductivity		4.4 w/m°C

Table 8–2 Sample Specifications for Dunlop ''ARMALINK'' for Conveying Sweet or Sour Live Crude Oil

BORE (inch)	2	4	6	8
Working pressure psi	5,000	5,000	5,000	5,000
Factory test pressure psi	10,000	10,000	10,000	10,000
Outer diameter body (mm) couplings	124 160	178 225	240 280	325 375
Pipe weight (kg/m)	28	48	112	169
Weight in seawater full of seawater (kg/m)	15	23	70	93
Weight in seawater full of air (kg/m)	13	15	51	60
Coupling weight (kg/end)	30	42	60	90
Minimum band radius (m)	0.9	1.4	1.7	2.0
Number main reinforcing piles	2	2	4	4

The reel has an 82-ft. diameter, and it can haul shorter lengths of various sizes on one trip—enough for more than one project. When required, portable reels are mounted on deck so that multiple pipeline bundles or hydraulic control umbilicals can be laid simultaneously.

Steel could be of A106 grade B through and including 5LX65, according to API specifications or equivalent. All pipe joints supplied by the steel manufacturer shall be seamless, and all billets should be peeled or similarly treated before being formed.

Chemical composition, heat treatment, physical and mechanical properties, yield and tensile strength, hydrostatic test, dimension at tolerances, Non Destructive Testing (NDT), and traceability of manufacturing date and run are all of great importance for the manufacture of steel pipes.

Epoxy coating of insulation, if required, will be carried out after the pipe is received from the pipe mill and before the joints are welded at the spool base.

Pipeline and flowline coatings are usually:

- Export pipeline coating (not for reel ship laying) 450 micron of FBE corrosion protection, 2 mm of concrete impingement protection and 50 mm of concrete impact protection.

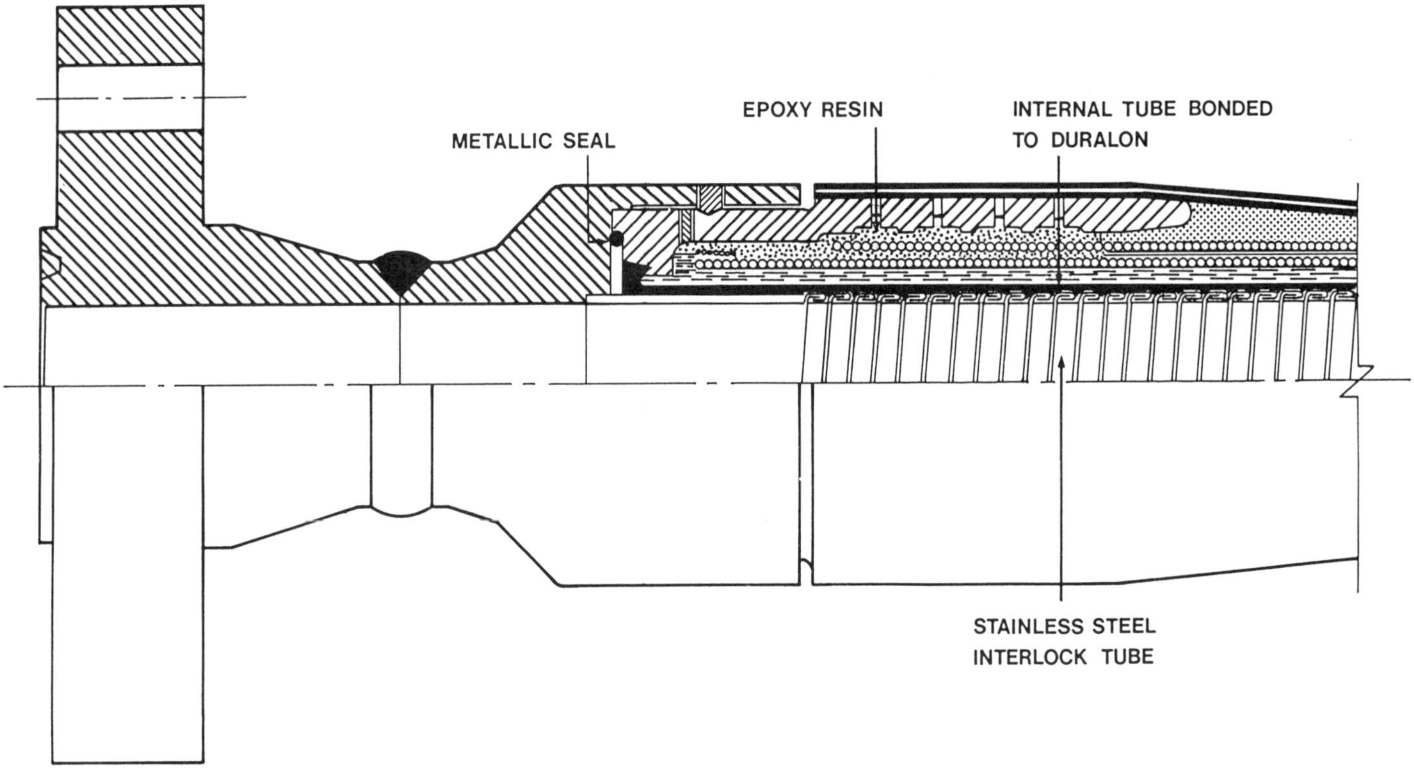

Fig. 8–3 Typical Two Piece Fitting/Coupling (Source: Dunlop)

- For export pipelines with epoxy coating, up to 420 microns of fusion bonded epoxy are usually applied.
- Insulation of lines is described in more detail later in this chapter.

Except for offshore tie-ins, all welding is carried out under controlled conditions onshore. Joints are welded into sections of 1,000 meters (3,280 ft.) or more. Welds are x-rayed and coated and the prewelded pipe stored on racks. Joints are coated also and insulated. Then strings of pipe are ready for spooling on the reel. At the end of the first spool, the operation is halted briefly so another length of pipe can be welded to the first string. While the vessel is at sea, other sections of pipe can be welded into stalks and stored at the base, ready for spooling to recommence as soon as the vessel returns. The Apache can sail under its own power to any area of the North Sea within 48 hours at speeds of up to 10 knots. The maximum capacity of Apache's reels is approximately 2,000 tons.

Table 8–3 Dunlops ''ARMALINK'' Couplings and Fittings

Dunlop's Oil and Marine Division at Grimsby designs its own fittings, which are presently built into the pipe during maintenance.

Coupling systems which can be incorporated into the fittings include:
- A.P.I. line pipe thread (up to 7,500 psi W.P.)
- Cameron hub and clamps
- Grayloc hubs
- Hammer Unions such as those made by WEBCO
- Flanges to suit particular requirements
- Cryongenic couplings

Typical Laying Operation

Figure 8–4 (1) shows the general arrangement of the Apache reel ship. Its application in a typical laying operation is best covered under a number of headings.

Navigation

Prior to operations, beacon stations are set up on a number of surrounding locations, i.e., platforms. They define a field which is calibrated and within which the vessel, or any suitably equipped vessel, can move by way of accurate location checks.

A back-up system, such as Artemis, has a beacon mounted on an adjacent platform. This provides a distance and direction indication.

Prior to the operation, the reel ship's navigation system would be programmed with the pre-determined route details.

Laying the Flowline

The reel ship is on location, loaded with steel pipe and with its stern ramp set at the correct angle according to water depth, pipe weight, etc. Refer to Figure 8–4 (2).

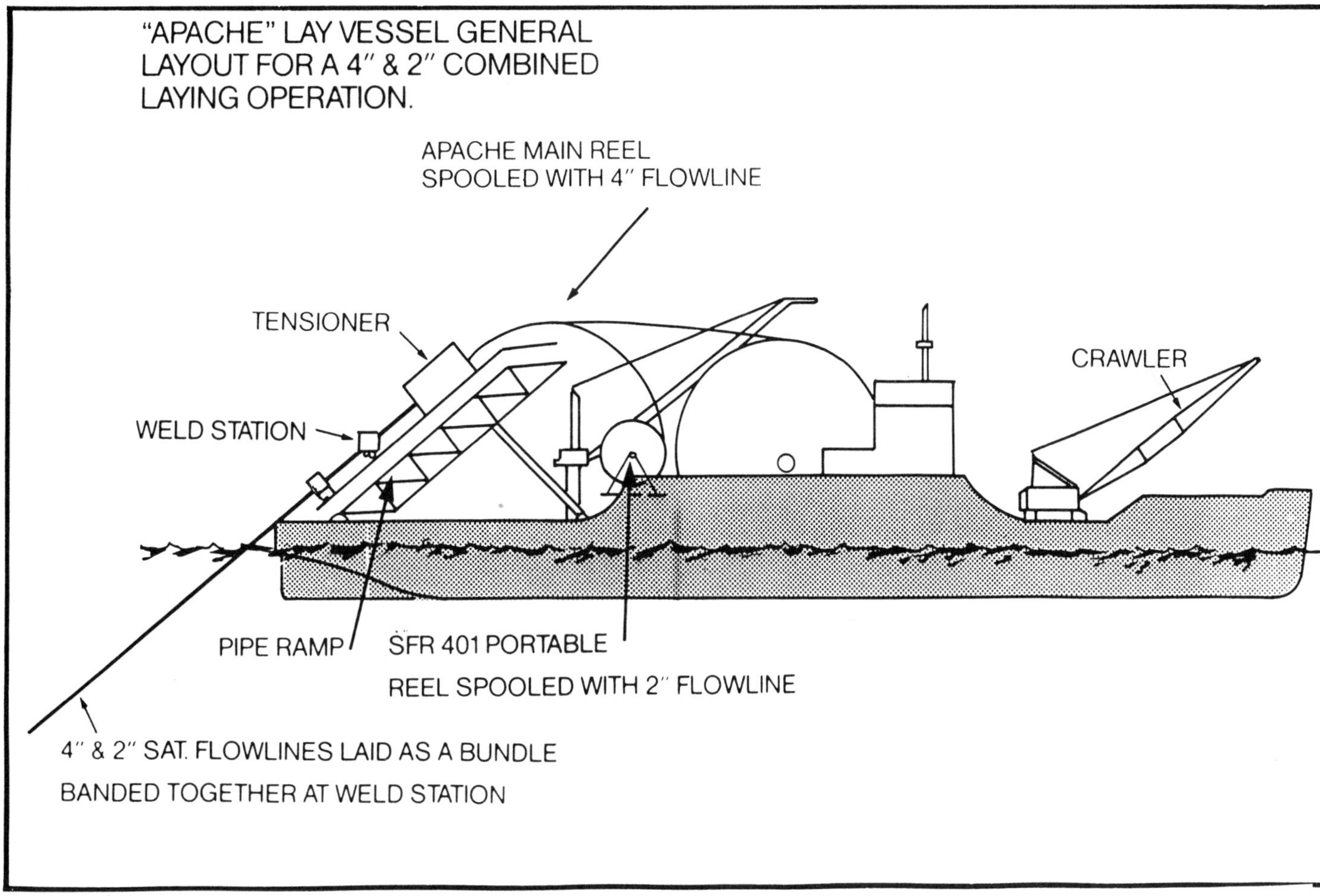

Fig. 8–4(1) Apache Lay Vessel General Layout (Source: Santa Fe)

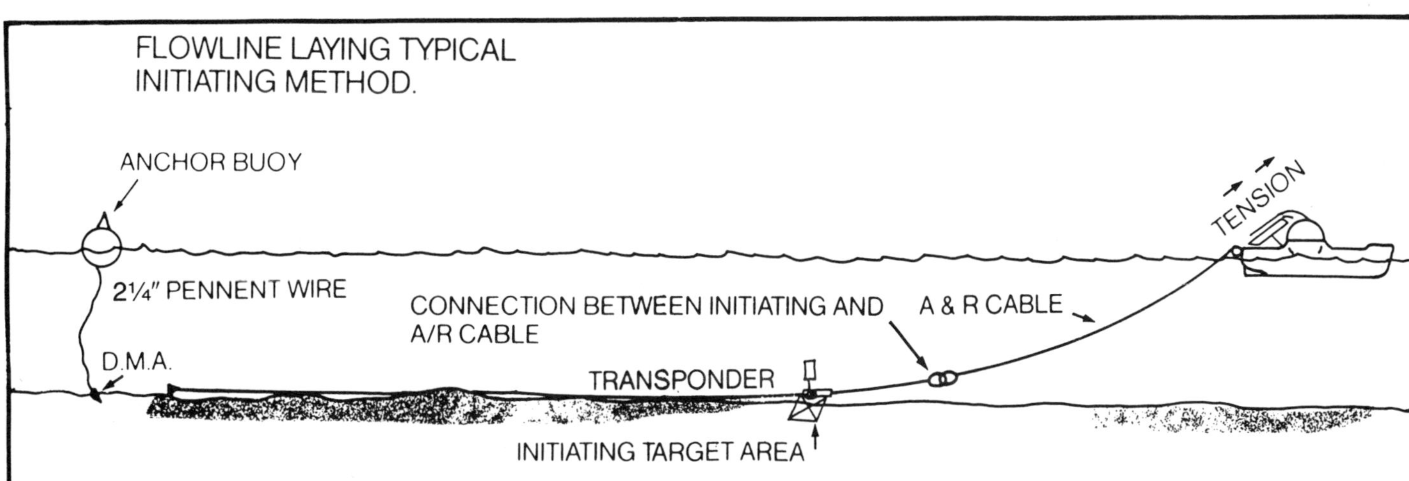

Fig. 8–4(2) Flowline Laying—Typical Initiation Method (Source: Stena Offshore)

A lay-down head is bolted into the end of the pipe. Attached to the head is a pennant/abandonment-recovery (A and R), on the end of which is a dead man's anchor (DMA).

The first action is to lower the buoyed DMA to the sea floor. Its positioning is critical as the vessel now moves off, keeping the A and R wire under tension, so that the pipe is finally fed from the ramp.

The pipe lay-down head, with a transponder attached, now can be lowered to the sea floor so that it lands within an Initiating Target Zone determined by the client's requirements.

The ship now lays the pipe within the predetermined corridor, pausing as necessary to attach anodes to the pipe from a station on the ramp. A typical anode attachment procedure would involve:

1. Pipelay halted as anode location reaches the weld station.
2. Anode pads cleaned with a power brush.
3. Hinge bracket anode lowered over a pipe using a chain fall and loosely drawn together.
4. Anode rotated to align bonding straps with anode pads. Bolts tightened and conductivity leads attached.
5. Bonding straps are welded to the anode pads in accordance with welding procedure.
6. Weld inspection carried out.
7. Exposed metal support wire brushed and coated.

Figure 8–4 (3) shows an initiation method for use near subsea structures.

About 300 m (980 ft) from the final Laydown Target Zone, a transponder is attached to the pipe. The vessel moves ahead and the transponder reference co-ordinates are taken at touch down.

The additional pipe length to be laid from transponder to the Target Zone is calculated, and the pipe is clamped and cut accordingly, once this point is reached as laying continues.

The pipeline's end is now bevelled and a pup-piece, consisting of 1 m of pipe welded to a W.N. flange assembled to a lay-down/test head, is welded to the pipe.

Now a transponder is attached to the lay-down head along with an abandonment and recovery (A and R) cable. Then the A and R cable is tensioned, the pipe-clamp released, and the vessel moves slowly ahead.

The lay-down head transponder allows it to touch within the Target Zone by way of vessel sideways movement.

With touch down, A and R cable tension is released while it is unshackled from the lay-down heads. Piggings, gauging, and pipe testing now can go ahead.

Abandonment and Recovery

All laying operations have a procedure to cover pipe abandonment and recovery should this be required during the lay for any reason. Refer to Figure 8–4 (4).

Essentially, it is the same as laid out above with an abandon-

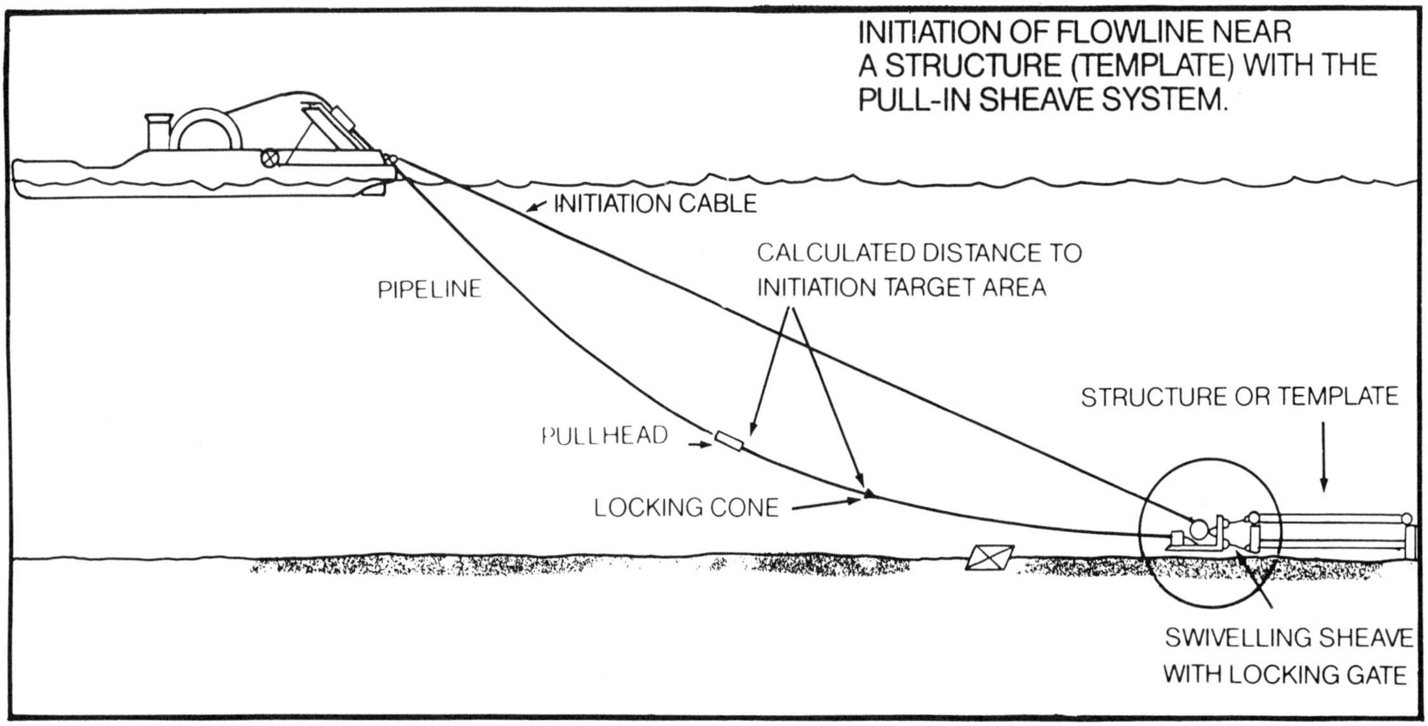

Fig. 8–4(3) Initiation of a Flowline Near a Structure (Source: Stena Offshore)

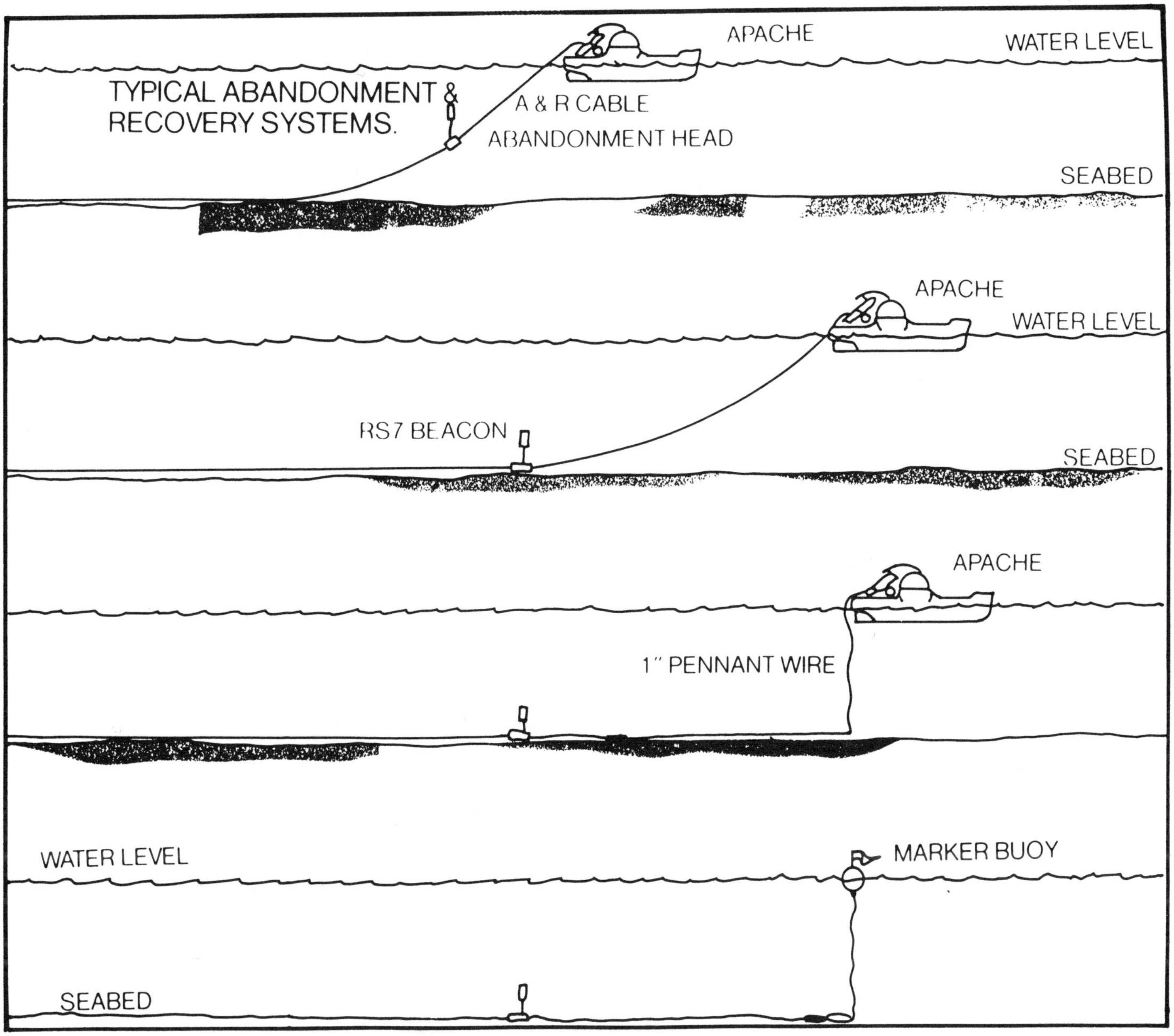

Fig. 8–4(4) Typical Abandonment Recovery Systems (Source: Stena Offshore)

ment head, with attached beacon, touched down on the sea floor attached to an (A and R) cable which is bouyed.

FLEXIBLE FLOWLINES

The laying of continuous flexible lines, such as those supplied by Coflexip, is a relatively easy operation. Various types of surface support vessels can be used, depending on the site conditions and extent of the job. Advantage can be taken of short weather windows, thereby providing quicker installation time.

Retrieving the line for further use on other sites can be performed as easily as laying it.

A major advantage of the flexible pipe is that it can be laid and buried in a single operation, using only one support vessel (preferably dynamically positioned). With this in mind, Coflexip has developed two types of burying machines which are operated from its own dynamically positioned vessels.

Laying vessels feature a fully redundant dynamic positioning system which permits accurate tracking of laying routes, ensures station-keeping with selection of optimum heading, enables a close approach to fixed or floating platforms, and eliminates the risk of damage by anchors in pipeline congested areas.

In addition, the vessels have the following characteristics:

- Large pipe and/or cable capacity, up to 3,600 tons
- Possibility to lay up to four line/cable bundles
- Simultaneously laying and burying when required
- Provision of deep diving facilities when required

Coflexip flowlines also can be laid with conventional, quickly mobilized supply boats equipped with a Coflexip winch.

Typical structure and characteristics of Coflexip flowlines are shown in Figures 8–5 and 8–6.

Internal Thermoplastic Sheath. Functions of the internal thermoplastic sheath are:

- To make the pipe leakproof
- To resist abrasion and corrosion

The thickness of the sheath is selected according to various factors such as internal diameter, working pressure, etc. The class of thermoplastic used depends on the type of fluid transported. The temperature limits of the end-product can be as high as 100°C (212°F) and as low as −100°C (−148°F).

Interlocked Steel Carcass. Functions of the interlocked steel carcass are:

- To resist internal pressure
- To resist external hydrostatic pressure and crushing effect
- To retain the flexibility and short bending radius of the pipe when used as an inner tube

Stainless steel strip is used for the manufacture of this carcass.

Intermediate Thermoplastic Sheath. This component is necessary whenever the pipe structure is intended for subsea service in rigorous conditions. Its purpose is to prevent the internal sheath from collapsing in the event of damage to the external sheath (by an anchor, for instance).

Double Cross-wound Tensile Armors. Functions of the double cross-wound tensile armors are:

- To resist axial load and internal pressure
- To protect the pipe from torsional strain resulting from handling and working conditions

These armors are usually made from flat steel wire. Whenever a lighter weight is required, steel can be replaced by synthetic textile or glass fiber.

The axial load results from tension and weight when the pipe is being laid or is in service. When the pipe is under pressure, the "bottom effect" also produces an axial load which depends on internal pressure and internal diameter. For example, for a 10-in. internal diameter and a 3,000 psi pressure, the bottom effect is in the order of 100 tons.

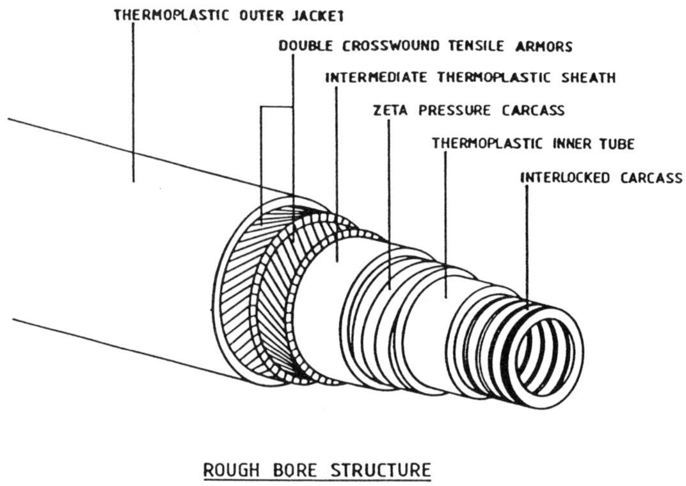

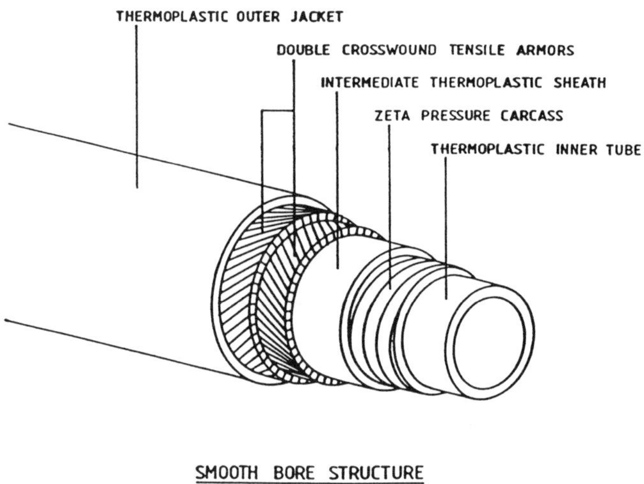

Fig. 8–5 Typical Riser Structure (Source: Coflexip)

External thermoplastic sheath. Functions of the external thermoplastic sheath are:

- To protect the metallic layers of the structure against external corrosion or abrasion
- To bind the underlying tensile armor

This sheath is made of thermoplastic as are the internal and intermediate layers. Local damage to this sheath does not endanger the pipe. Corrosion from seawater remains at a very low level, restricted to the damaged area; the lack of renewed oxygen

This list is by no means exhaustive, but indicates the characteristics of some Coflexip pipes									
I.D.		Working pressure		O.D.		Weight in air		Minimum bending radius	
in	mm	psi	bar	in	mm	lb/ft	kg/m	ft	m
1	25.4	3 000	207	2¼	57.5	4.4	6.6	1.22	0.37
1	50.8	3 000	207	3¼	81.5	7.2	10.7	1.75	0.55
2	50.8	10 000	690	4¼	108	20.7	31	2.3	0.7
2½	63.5	2 000	138	4	101.5	9.7	14.4	2.2	0.66
3	76.2	5 000	345	5	125	21	31.5	2.7	0.81
3	76.2	10 000	690	5³⁄₁₆	131	24.5	36.5	2.8	0.85
4	101.6	10 000	690	7	178	51.7	77.2	3.8	1.16
4	101.6	3 000	207	5⁹⁄₁₆	141	14.6	21.8	3	0.91
5	127	5 000	345	7¼	183.5	38.5	57.5	4	1.2
6	152.4	2 000	138	7¾	195.5	27.7	41.3	4.2	1.3
6	152.4	7 500	517	9	229	73.3	109.4	4.9	1.5
8	203.2	2 000	138	10³⁄₁₆	257	38.1	56.9	5.5	1.7
8	203.2	5 000	345	11½	292	99.5	148.5	6.3	2
10	254	1 500	103	12¼	310	55.5	82.5	6.6	2.1
10	254	4 500	310	13⅞	353	139	208	7.5	2.3
12	304.8	1 000	69	14½	368	72.1	108	7.9	2.4
12	304.8	2 500	172	15¼	388	106	158	8.3	2.55
16	406.4	750	52	20	504.5	162	242	10.8	3.3

Fig. 8–6 Characteristics of Some Coflexip Flowlines

prevents it from spreading along the pipe. Cathodic protection can be provided if required.

End-Couplings. Couplings are critical points in any piping system and need to be fully reliable. The end couplings are carefully engineered pieces of equipment, fitted onto the flexible pipe by specialized technicians and systematically tested.

They are designed to retain all the mechanical characteristics of the pipe, with a substantial safety margin. Furthermore, the couplings are internally flush and can be used with TFL (Through Flowline) tools and pigs. They can be fitted with any type of standard connection such as flanges, hubs, Unibolts, etc. The parts that are liable to corrosion from seawater undergo a special treatment, which improves their resistance to corrosion. If required, cathodic protection can be provided by means of disposable anodes tied to the couplings.

Figure 8–7 shows the basic components of a typical flowline and its end coupling

Typical 4-in. Flowline Design and Stresses (Coflexip)

The following example is based on the design of a 4-in. deep water flowline.

Water depth is an important consideration, and the most important technical characteristics to bear in mind are:

- Hydrostatic collapse at 201 bar
- A 60-ton permissible pull on a radius of 3 m (10 ft)
- A 162-ton permissible pull in straight line

The main theoretical characteristics of a 4-in. Coflexip deepwater flowline are given in Table 8–4. Stress characteristics for various diameters of flowline are given in Table 8–5.

Manufacture and Quality Control

In order to ensure constant quality and performance of the flexible pipe, a number of tests must be carried out during the manufacturing process.

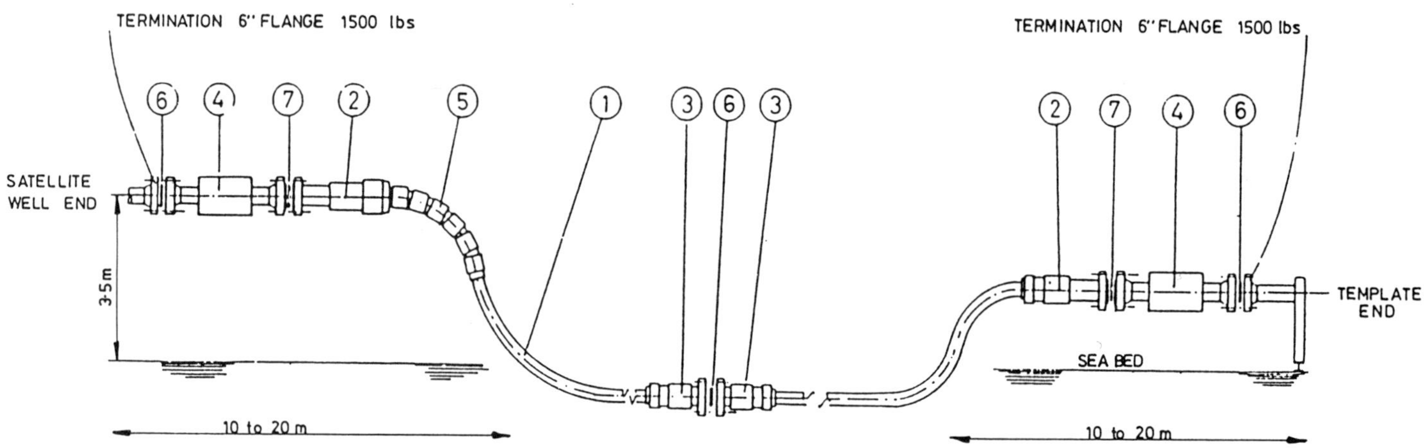

LEGEND

1 6" FLEXIBLE STRUCTURE
2 6" END FITTING
3 6" END FITTING
4 LOAD LIMITING CONNECTOR
5 ARTICULATED VERTEBRAE
6 GASKET BX 156
7 INSULATING GASKET R 46

Fig. 8–7 Components of a Typical Flowline and Its End Coupling

On arrival at the factory, raw materials and components for flexible pipe and end-fittings are subjected to a comprehensive quality control check:

- Steel components are tested for breaking strength and hardness and are checked for defects and flaws.
- Thermoplastics are tested for viscosity and moisture content.

Each step of the manufacture is accompanied by relevant controls, such as:

- Drifting of internal diameter
- Thickness of internal thermoplastic sheath
- External diameter and pitch of both layers of tensile armor
- External diameter and thickness of outer sheath

Table 8–4 Characteristics of 4-in. Coflexip Deep-Water Flowline

Characteristic	English	Metric
Inner diameter	4.0 in.	101.6 mm
Outer diameter	6.1 in.	155.0 mm
Weight in air empty	29.1 lb/ft	43.4 kg/m
Weight in air full of seawater	34.6 lb/ft	51.7 kg/m
Weight in seawater empty	16.1 lb/ft	24.1 kg/m
Weight in seawater full of seawater	21.7 lb/ft	32.4 kg/m
Design pressure	3,000 psi	206 bar
Bursting pressure	9,693 psi	668 bar
Hydrostatic collapse pressure	2,911 psi	201 bar
Damaging pull in straight line	357,265 lb	162,393 daN

Factory Acceptance Tests

The following tests are systematically performed on each individual pipe section once it has been fitted with its end-connectors:

Table 8–5 Stress Characteristics for Various Diameters of Flowline

Diameter (in.)	Bending Radius	Collapse (bar) Pressure	Pull (ton)	Breaking Pull (ton)
Tensile stress on a bending radius				
2	1.4		60	
3	1.75		50	
8	2.0		20	
Internal pressure				
1		200		
4		150		
4		250		
12		70		
Tensile strength in straight line				
1				10
2				90
3				150
4				100 (a)
12				200 (a)

(a) Test stopped before failure; pull limited by testing equipment.

- Pigging and drift test; both operations can be carried out simultaneously by using a pig and drift assembly. These are driven by fresh water which fills up the line in preparation for the pressure test.
- Pressure test.

Test pressure and test period are contractually specified. Usually, the values are as follows:

- Test pressure 1.5 × working pressure
- Test period 24 hours

All the above tests are carried out under the scrutiny of an independent certification and inspection company.

Flowline Bundles and Controlled Depth Tow Method (CDTM)

In principle, the CDTM involves the transportation of a prefabricated and fully tested flowline, or flowline bundle, suspended between a leading and a trailing tug (see Fig. 8–8). The line design requires careful consideration of strength and towing

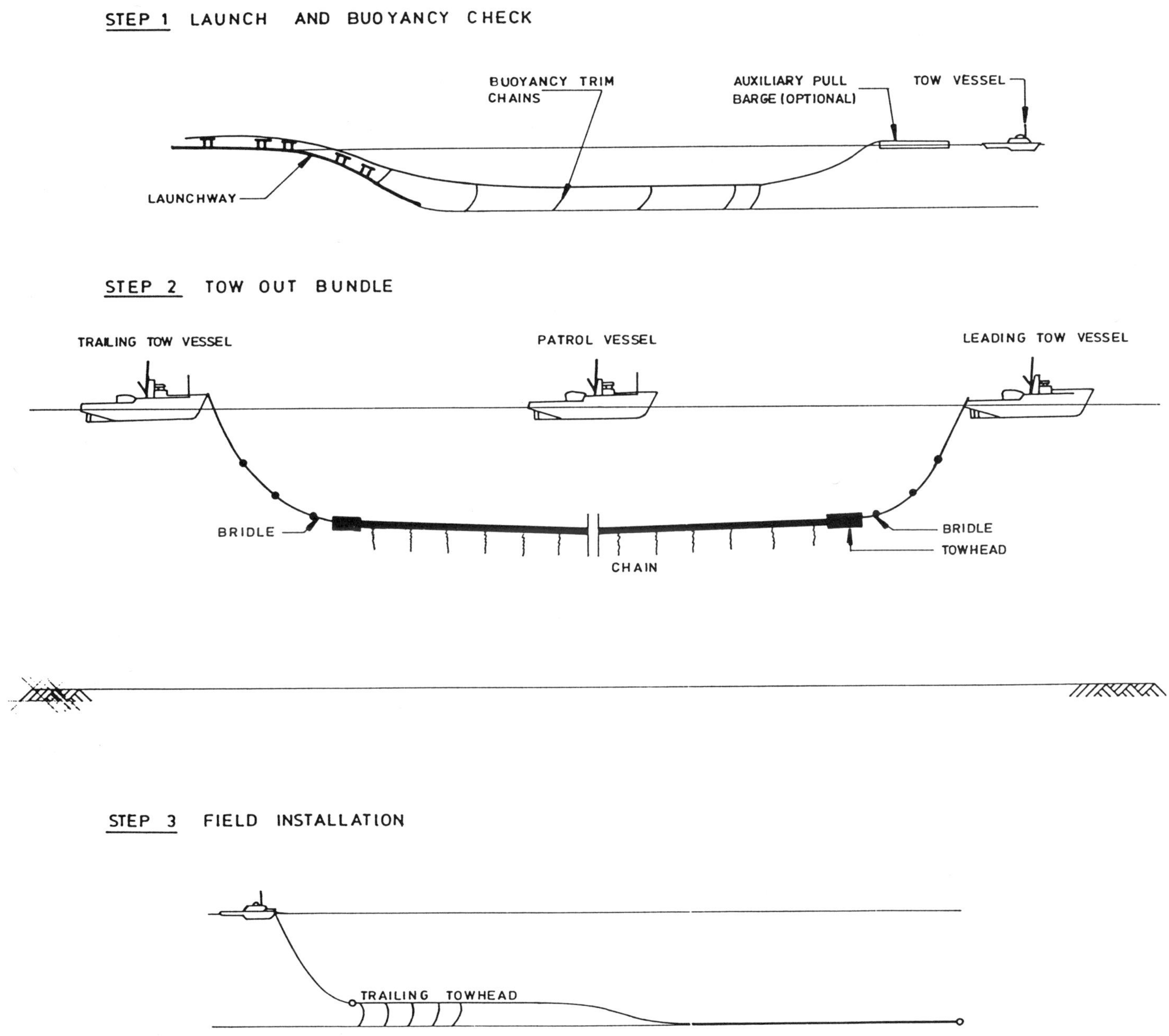

Fig. 8–8 Controlled Depth Tow Method for Flowline Bundles

characteristics, with allowance made for dimensional tolerances and the weight of its components. Another vessel accompanies the tow as a patrol vessel. Further, to maintain control during the tow, the bundle is designed and constructed within specified tolerances with respect to its submerged weight.

The bundle, excluding the chains, is designed to be buoyant. This is achieved by mounting the flowlines, control lines, etc., inside a carrier pipe. Subsequently, chains are attached so that as they touch bottom the carrier pipe floats above.

The towing speed results in a lift and straightening effect on the bundle. By controlling this speed, in combination with the tension maintained by the trailing tug (TT) at the rear, the bundle configuration and its deflections are kept under control. The towing mode of the bundle also is determined by parameters such as towing depth, deflection, stresses, tow speed, dynamic movements, etc.

Essential parameters are continuously monitored during the tow, and any necessary adjustments are made. To aid this monitoring, computer graphs, specifically prepared in advance, are available on board for each bundle. During the tow, the bundle is kept well clear of the seabed to enable a safe and unobstructed operation. Furthermore, the towheads are kept well below the water surface to minimize the effect of surface waves. Normally, this is approximately 30 m (100 ft), but this depth can be increased or reduced.

After arrival on location in the field, the bundle is gradually lowered by predetermined adjustment of the controlling parameters (tow wire lengths and tensions). The bundle settles in an off-bottom mode with part of the chains resting on the seabed. Once in this position, the bundle can easily be maneuvered to its correct position with its ends in the required target area. In the event of pull-in connections being used, the tie-in is normally performed while the pipeline bundle is still in its off-bottom position.

This tow method is especially suitable for:

- Insulated pipelines
- Lines for TFL (Through Flow Line) operations
- Deep-water operations
- Congested pipeline areas
- Bundling of flowlines, injection lines, and control lines

There is no need for anchoring the vessel in the field during the installation and to trench or bury the bundles.

In addition to this:

- The tow takes place at controlled water depth, clear from the seabed, and well below the surface.
- The complete fabrication and testing for pipe integrity is carried out onshore.
- Additional protection against damage and corrosion is achieved by the carrier pipe.

Design Philosophy

The selection of the carrier pipe is mainly determined by the following parameters:

- It should provide sufficient buoyancy to support the total system plus a specified excess which depends on the weights and weight tolerances of the bundle components. At all times, an excess positive buoyancy must be realized.
- The strength of the carrier pipe must be sufficient to withstand the combined effect of the hoop and axial stresses due to internal pressure and launch forces.
- The strength must be sufficient to withstand all hydrodynamic forces, i.e., currents and waves. Dynamic effects, especially under wave actions, must be considered including the variable forces in the towing connections resulting from vessel movements. In this context, natural frequencies of the system and its components require the necessary consideration. Computer programs have been developed to carry out these calculations.
- The effect of cross currents during tow must be considered for the carrier pipe strength. The installation forces at location will be evaluated well in advance. The bundle must withstand bending and restraining forces resulting from maximum expected cross currents on the seabed during installation.
- To counteract external water pressure on the carrier pipe, the bundle is internally pressurized with nitrogen. The internal pressure is equal to the hydrostatic pressure at the bundle's final location in the field.
- The lines are kept in their proper position by spacers at specified intervals. Thus, contact between lines and the carrier pipe is avoided during all phases of construction, tow, and installation. Type and distance of spares depends on size of flowlines and their configuration.
- At the ends of the carrier pipe, the flowlines pass through heavy end-bulkheads. As the flowlines are welded to the bulkheads (to provide air and watertight seal), the bulkheads must be designed to take all interacting forces between the flowlines and the carrier pipe, especially during pressure testing and launching.

The bundle of configuration is designed in order to prevent rotation of the bundle around its longitudinal axis during all phases.

Key Phases of a Bundle/CDTM Operation

The key phases are:

1. Operational engineering
2. Onshore construction of the bundle
3. Launch, buoyancy check, and final weight adjustment
4. Tow route survey
5. Tow initiation

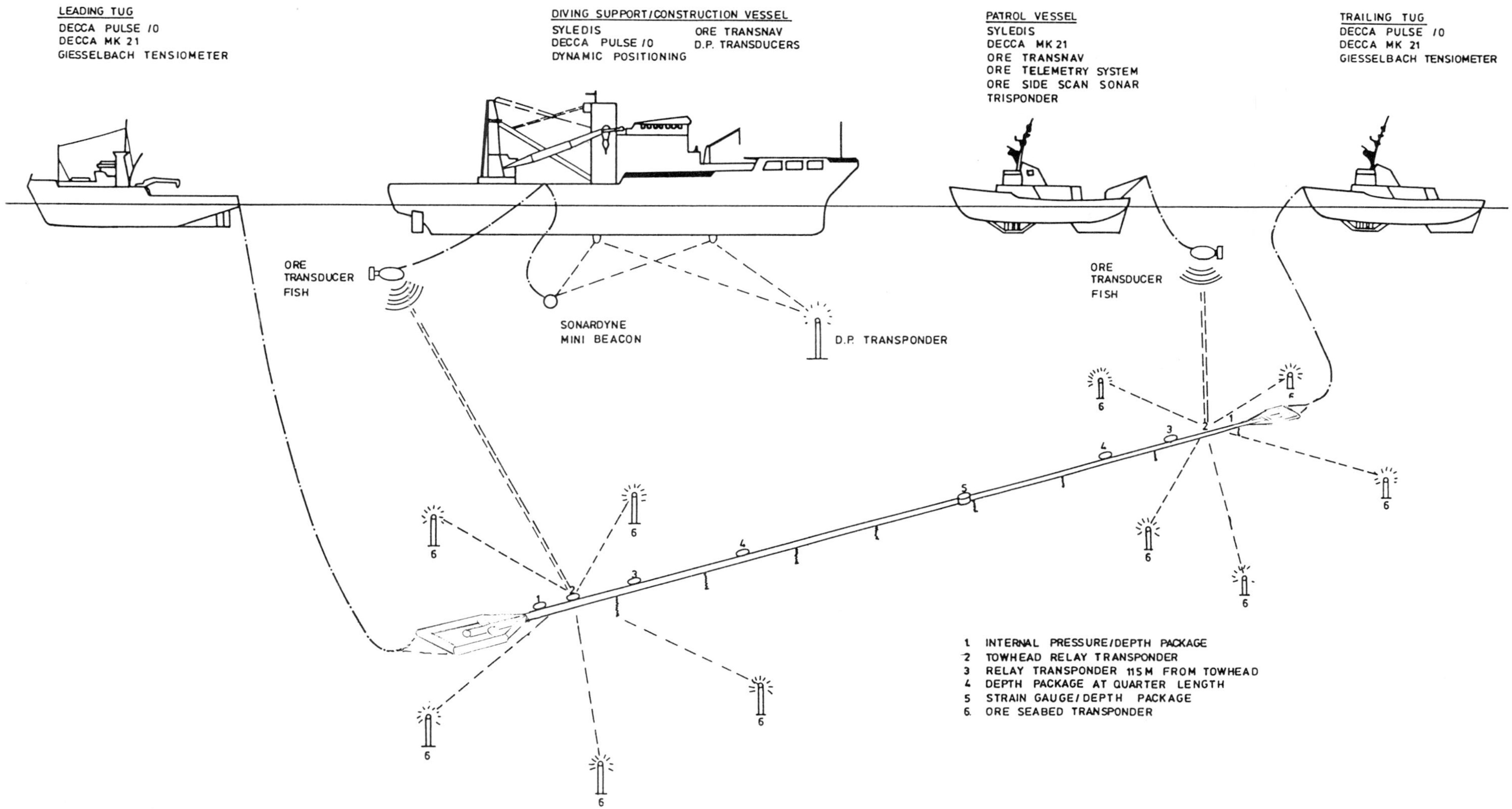

Fig. 8–9 Typical Positioning Instrumentation fcr a Flowline Bundle (or Pipeline)

6. Towing of the bundle
7. Field installation

Operational Engineering

This engineering phase provides the tools for operational control, especially during the towing phase. During the design phase, a "nominal" case is calculated which assumes that all weights and dimensions (especially outside diameter carrier) are the final values. Positive and negative tolerances then are estimated and included. Based on the results obtained, the required materials are specified. During construction, the weights of the bundle components are monitored continually. After all weight and dimension checks are completed, the final version of the operational graphs to be used during the towing operation are prepared.

By use of these graphs, parameters such as tow speed, pull-force of trailing tug, and paid out tow-wire lengths can be adjusted during the tow to ensure optimum bundle position and configuration. Continuous monitoring of tow head depths and the depths of a number of selected points along the bundle provides the necessary checks on the bundle configuration during the tow. The same approach applies to the installation of the bundle at location. All necessary graphs and calculations are prepared beforehand and, if necessary, updated on the spot.

Onshore Construction of the Bundle

The pipeline bundle is stored on bogies ready for launch with the tow heads on both ends installed. The bogies run on a narrow gauge rail track. The necessary equipment such as various instruments to supply information during tow, flowing valves, and ballast chains are mounted on the pipeline bundle. Then the carrier pipe will be pressurized.

The construction method of such a pipeline bundle varies in detail depending on the materials and contents of each bundle but follows the same pattern of events as outlined below.

- All materials are inspected on delivery to the construction site. In addition to the usual inspection, certification, and identification procedures, each pipe is measured and weighed. Any materials that eventually become part of the submerged pipe, e.g., anodes, tow heads, etc., are weighed also and the results recorded. The first construction step is to spool the pipes, i.e., to weld up three or more pipe joints into spools and complete the non-destructive testing. (This permits the use of several work areas away from the construction of the pipeline.)
- Within the fabrication shop the inner pipe spools are assembled into 450 m (1,500 ft) strings. This involves placing the required number of spools which make up the bundle section on the launch bogies. As the string-length increases, it is pulled along the track out of the fabrication shop as each spool is added.
- When each string is approximately 450 m (1,500 ft), it is pulled to its final location where a field weld is carried out between each pipe in the bundle string. Once welded and inspected, the pipes are grouped in the final bundle configuration and held in position by the spacers which are fitted along its length. The spacer also protects the inner pipelines during sleeving operations. The coating is inspected before the inner pipelines are sleeved.
- Hydraulic hoses, electric cables, or other flexible items are unreeled along the length of the pipe and fitted into the space provided in the spacer. Carrier pipe spools are made up into strings of approximately 180 m (600 ft) and threaded over the inner bundle from both ends. The carrier-pipe strings are carried by several side-boom tractors and guided over the bundle by the spacers. The carrier-pipe strings are welded together at field welding stations. Flooding valve assemblies and protection cages then are fitted to the carrier pipe along with the chain clamps and ballast chains.
- The inner lines are cleaned before pressure testing, and brush and foam pigs are air driven through the lines before gauging. Then the lines are filled against a pig with treated seawater and are pressure tested. The carrier pipe is air tested, and a leak inspection of exposed welds and fittings is carried out. The tow and installation instruments are fitted and checked, and the pipes are pressurized to complete the line ready for launch.

Launch, Buoyancy Check, and Final Weight Adjustment

After fabrication and testing, the full length of the bundle is pulled from the shore into the water. With long and heavy bundles, the required pull forces can be considerable if sliding bogies are used—typical break-out forces (to start the bundle moving) being in the order of 200–250 tons for larger diameter bundles.

The launch operation involves the presetting of pull anchors at selected locations. Then the pull barge is connected to the anchors and moved towards the shore, paying out its pull wires. When the barge is positioned close to the beach, the connection with the bundle is established. During the launch, the position of the barge is monitored and, if necessary, is corrected by a tug.

After completion of the launch, the bundle is floated above the seabed (typical 3–4 m/10–13 ft) with part of the chains resting on the seabed as shown in Figure 8–10. The height of the bundle above the seabed and the number of chain-links resting on the seabed indicates the bundle's resulting submerged weight. The height above bottom is checked with an echo-sounder survey of the bundle. A more accurate check of the number of chain-links on the seabed is performed by divers. If necessary, the submerged weight can be adjusted by adding or removing chain-links.

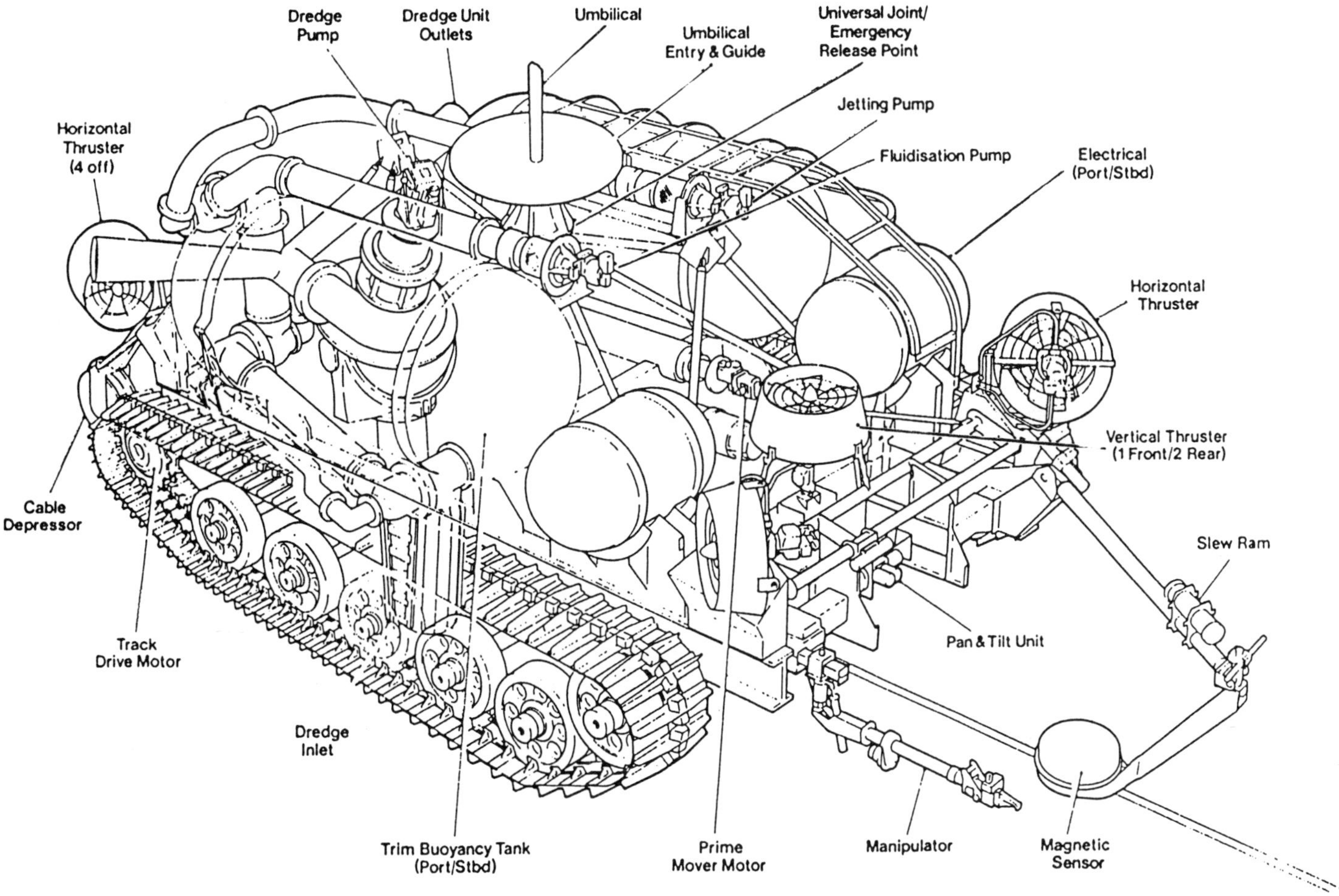

Fig. 8–10 Seadog MK1—General Arrangement

To keep the bundle in position during the buoyancy check and to avoid excessive bending under the influence of cross-currents, the bundle is kept under tension between the pull barge and a winch on the shore. During the buoyancy check, the underwater telemetry and transponders are tested to ensure proper functioning during the tow. Also, the nitrogen pressure in the carrier pipe is monitored during this period.

Tow Route Survey

Based on a study of detailed sea charts, a preliminary tow route is established, taking into account pipeline crossings, obstacles, wrecks, etc. Subsequently, the tow route is surveyed and, if necessary, changes to the route are made based on the information obtained. The survey normally covers a 600 m (1,960 ft)-wide lane surveyed by three parallel echo-sounder/ side-scan sonar runs, 150 m (490 ft) apart. It is not the intention to remove obstacles along the tow route but to adjust the tow route where necessary.

Based on the survey results, safe-landing areas can be specified to lower the bundle onto the seabed when needed, for instance if very extreme weather conditions should unexpectedly occur. Detailed surveys also are made of the launch and installation areas. Additional ROV surveys can be made of the projected location of the bundle, ensuring that no debris, boulders, etc., will interfere with the installation.

Tow Initiation

After completion of the buoyancy check and weight adjustment, the bundle is ready for the tow. The leading tug (LT) and the trailing tug (TT) are connected, and the tow can start. The tow initiation has two phases.

1. The shallow water phase: This covers the part of the tow route where water depths are insufficient for the chains to be lifted free of the seabed.
2. Lift-off phase:When sufficient water depth is available, the tow speed is increased and tensions are adjusted to "lift" the chains free from the seabed.

The moment of lift-off depends on a number of selected parameters such as tow-head depth, tensions, length of bundle, etc., but a water depth of approximately 50 m (165 ft) is sufficient in most cases. Once the bundle is completely off-bottom, adjustments to tow-lines, tensions, and tow speed will be made to establish the required bundle configuration for the voyage in deeper water. The tow heads at this stage are the required depths (typically 30 m/100 ft), and deflections are within the specified range (20–30 m/66–100 ft).

Tow of Bundle

The basic towing arrangement, shown in Step 2 of Fig. 8–8, represents the bundle suspended between the leading and trailing tugs. Its towing configuration is continuously monitored by means of an acoustic telemetry system, and all information is transmitted to the accompanying patrol vessel where the central command post is based.

The following information is generally monitored:

- Positions of the leading tug and the trailing tug
- Tow-line tensions at the leading tug and the trailing tug
- Towing speed
- Depth of bundle at five selected points (tow heads, center, quarter, and three-quarter length)
- Nitrogen pressure in the carrier pipe
- Horizontal and vertical strain.

The purpose of this control is to ensure that the following conditions are fulfilled:

- Sufficient clearance from the seabed is maintained.
- The tow heads are kept at a specified depth below waterlevel, to minimize wave effects on the bundle. This depth depends on the environmental conditions. For the North Sea, a typical depth is approximately 30 m (100 ft) below surface.

Generally, the designed tow-speed is in the range of 4–5 knots, which can be maintained during adverse weather conditions. Normally, the tow is not initiated before reasonable weather conditions are forecast.

When the tow approaches its final location, the bundle is lowered to the seabed within a preselected landing area. The tow-speed is reduced by the controlled increase of the pull force of the trailing tug and reduction of the pull force of the leading tug. The bundle ends up in an "off-bottom" mode with a part of each chain resting on the seabed. In this mode, final adjustments to the bundle's position can be made.

Field Installation

Before arrival of the bundle in the installation area, an array of transponders is installed, as shown in Figure 8–9. The patrol vessel uses this acoustic transponder array to monitor the position of the bundle.

Intermediate anchor blocks are installed to keep the bundle properly positioned in case seabed crosscurrents occur or where the final bundle route has a curved shape. The weight of these anchor blocks is designed to avoid high local stresses (typical weight is 2 tons, submerged).

After fixing the bundle, which is still in the off-bottom mode, a final check is made on its position before flooding. The anchor blocks will be removed after flooding of the bundles. As a first step, flooding the carrier involves opening flood valves installed at pre-selected locations on the bundle. Final flooding is "forced," with inhibitor added at the same time. The anchor blocks are then removed.

Tie-ins to platforms can be made by using pull-in type connectors with a receiver and pull-in sled. Either steel or flexible spool pieces can be used. This method has been successfully used on bundles installed for several oil companies, and Table 8–6 gives details of just a few of these.

EXPORT PIPELINE TIE-INS

This short section focuses attention on the importance of certain factors in the design of an export pipeline tie-in.

- Has provision already been made for tie-in on the intended trunk pipeline? It is normal practice to install tie-in points or taps during the laying of the trunk line.
- Verify whether or not provision has to be made for future tie-ins at the same tap.
- The length of time required to negotiate with the trunk pipeline operator and other parties (e.g., shore terminal operator, etc.) should not be underestimated. A period of two to three years is a realistic figure between the first discussion and the completion/commissioning of the pipeline.
- If the export pipeline requires pigging on a regular basis, launching and receiving facilities, probably subsea, must be considered at the preliminary design stage.
- In the North Sea, any tie-in valve arrangement will be of substantial dimensions. The requirement for adequate protection requires a substantial amount of offshore subsea work and between two and three weeks to install and test the whole system. Inspection and maintenance of the valves and other equipment during the years of operation adds substantially to any budget.

TRENCHING AND BURYING

Large-diameter lines and flowlines in deep waters are often left untrenched in the North Sea. However, there are many cases where trenching will still be required.

Fishing activity in the North Sea has been a significant factor in determining trenching requirements, and developments in the fishing industry over recent years have had an important

Table 8–6 Flowline Bundles Installed in the UK Sector of the North Sea (1980–82)

Oil Company and Year of Construction	Length (feet)	Meters	Bundle Configuration	Weight (ton)
Conoco North Sea Inc (1980) Murchison field	2,560	780	2 × 3.5″ × .25″ TFL line 4 × 0.875″ Hydr. lines 12.75″ × 0.25″ Carrier	102
	3,830	1,170	2 × 3.5″ × .25″ TFL line 4 × 0.875″ Hydr. lines 12.75″ × 0.25″ Carrier	149
	6,450	1,970	2 × 3.5″ × .25″ TFL line 4 × 0.875″ Hydr. line 12.75″ × 0.25″ carrier	255
Occidental (1981) Claymore field	7,000	2,130	6.625″ × 0.375″ line [(in- 5 × 0.5″ Hydr. line sulated] 14″ × 0.25″ carrier	280
	10,800	3,290	10.75″ × .438″ line [(in- 2 × 8.625″ × 0.406″ sulated] 10 × 0.5″ Hydr. lines 26.25″ × 0.344″ carrier	1,210
Shell (1982) Central Cormorant field	10,975	3,350	8.625″ × 0.6787″ line 14″ × 0.3122″ sleeve pipe 26″ × 0.406″ carrier	1,420
	10,975	3,350	8.625″ × 0.6787″ line 14″ × 0.312″ sleeve pipe 26″ × 0.406″ carrier	1,420
	10,843	3,300	2 × 4.01″ × .505″ TFL line 2 × 8.625″ × .25″ sleeve pipe 24″ × 0.375″ carrier	1,227
	10,843	3,300	2 × 4.01″ × .505″ TFL line 2 × 8.625″ × .25″ sleeve pipe 24″ × 0.375″ carrier	1,227

influence on pipeline design. Bottom trawling is the particular type of fishing which affects a pipeline, though traditional board trawls have been demonstrated to be relatively harmless to pipelines over 16-in. diameter. However, larger beam trawls are now being used in more northerly latitudes, and forces on the pipeline resulting from this activity must be considered during design. Vessels travel at up to 7 knots when trawling, and the combined weight of beams, skids, chains and nets may be up to 11 tons. Clearly, the damage potential to an exposed pipeline is considerable.

In more northerly latitudes, especially on the American continental shelf, designing for iceberg scouring must be considered also. In these areas, deep trenching may be required to give a reasonable degree of confidence that damage will not occur. In shallow waters, especially in exposed areas, pipeline stability requirements or soil movements will often dictate trenching. Various aspects of navigational activity must be a design consideration also. If ships of deep draft operate in relatively shallow water, pipelines will generally need trenching below the seabed. While there is little that can protect a line from a heavy marine anchor, any likelihood of vessels anchoring in the vicinity will generally dictate deep trenching. This is especially the case if wire-moored vessels are anticipated, as trenching will considerably reduce the risk of damage.

Spans will occur when an untrenched pipeline crosses an undulating seabed. These may be corrected with sandbags and mattresses, but there is always a danger of scour around discrete supports. Rock dumping overcomes this problem but may cause high stresses in the pipe. The most satisfactory solution will often be found in the post-trenching of high spots.

Trenching may be performed either before or after pipe-lay. Normally, pre-dredging is limited to shore approaches, but trenches have been pre-plowed in deeper water. However, the major thrust of current development work is in post-trenching techniques. These may be divided into three basic categories: jet sled, plow, and mechnical trenchers.

Jet Systems

This employs a jetting sled towed by a vessel or a seabed crawler. Jetting is now considered to be the established method of trenching submarine pipelines. However, the system is generally dependent on a large support vessel and tends to be weather-sensitive. Despite these drawbacks it is, nevertheless, the most adaptable method of trenching, suited to most types of sedimentary soil and to a wide variety of trench depths. Consequently, improvements to the sled will continue. One example is the addition of a new pipe tracking system which takes advantage of modern underwater instrumentation technology and will give considerably better steering control of the sled.

Where deep trenches are required, the jet sled's adaptability gives it a considerable advantage over other techniques, which will always require major modification in order to be effective. The jet sled's "claw" may be set to the depth required, and there is no limit to the number of passes that may be made. Additionally, if the jet sled were to be deployed from a semi-submersible support vessel, weather sensitivity would be considerably reduced. This is a natural development that must come with time. The versatility of the method will keep it alive for many years in the future.

Various contractors, based on their combined experience with flexible lines, have developed a burying plow/jet trencher for semi-soft soil conditions (sand and clay). It is equipped with an interchangeable plow-share, through which the flexible pipes or cables are guided into the trench as it is formed. In normal soil conditions, the walls of the trench collapse directly behind the machine, thus burying the lines completely. A fairly successful seabed crawler for small trenches (flow lines up to 8-in. diameter) is the BII Seadog Mk 1 shown in Figure 8–10. This new vehicle demonstrated its trenching by jet system capabilities in the Balmoral field during the summer of 1985.

Plows

A number of these have been designed to post-trench pipelines. Where these plows are used for large-diameter pipelines, a large surface support vessel is required for deployment and towing. One plow weighing over 300 tons has been successfully used for specialist applications. Current trends in plow design envisage a device which can be safely deployed and towed from a comparatively light non-specialist vessel. For smaller diameter lines, they may be deployed from a diving support vessel, and for very small lines, the plow has even been towed by a tracked underwater vehicle.

Brown & Root has developed a lightweight plow for trenching pipelines of up to 20-in. diameter. The weight has been kept to a minimum by using high-strength steels and by locating the plow beam underneath the pipeline. This reduces the bending moments in the beam by moving it closer to the line of action of force on the hares, and this in turn reduces the structural steel weight. In Figure 8–11, the UDI bottom crawler type of plow machine is shown in a typical deployment mode, combining laying and trenching operations for satellite wells control umbilicals.

Mechanical Trencher

Many underwater versions have been built over the last few years. These machines generally adapt land-based or conventional dredging technology to underwater work. The design aims to facilitate deployment from a lightweight support vessel to avoid the need for diver intervention and to operate safely on the pipeline. The various mechanical trenchers thus tend to be very different, but the principle of using cutters and eductors seems to be gaining ground.

The KBV trencher, jointly owned by Kaverner, Brown & Root, and Volker Stevin, has been under development for the past few years and has recently completed a number of successful contracts. The system is able to operate in water depth of up to 500 m (1,640 ft) and has successfully worked on span-recertification in the Norwegian trench. The trencher comprises a dredge-type basket cutter and a dredge pump soil removal system. Bogies are provided for riding the pipe and for propulsion, but it is essentially a highly maneuverable tethered ROV with positioning systems.

Future developments in trenching will be aimed at reduction in weather sensitivity and dependency on heavy support vessels, enhancement of safety, and elimination of the need for diver-assisted operations. In addition, simultaneous pipelaying and trenching and simultaneous pipe-laying and burial may become available from more contractors.

Heerema has been operating their very large mechanical trencher the Eager Beaver since 1981 with considerable success. Coflexip also has developed and now operates a successful mechanical trenching machine for hard terrain (limestone, coral, etc.). It consists of a chassis mounted on caterpillar tracks, with a vertically adjustable cutter wheel. The machine excavates trenches up to 1.5 m (5 ft) deep, into which the lines are automatically guided. It is controlled from the same vessel that lays the lines. Sensors provide continuous measurements which allow accurate positioning of the machine and recording of the "as laid" track. Subsea television cameras ensure complete video-monitoring of the operations.

Coflexip is proposing a new machine combining the jet and plow techniques. The design is based on the well-proven previous machines which have been extensively used in the past for burying flexible lines. More than 250 km (155 miles) of flexible lines have been buried to date at a trench depth varying from 1 to 2.5 m (3–8 ft).

The machine is a seabed crawler that uses high-pressure jets and a suction dredge to dig the trench. This solution is the more cost effective for medium to very soft soil conditions. The machine can be safely used anywhere in the field, even in the large pockmarks, close to templates, wellheads, or other pipelines.

It is fitted with a burying shoe which forces the pipeline to the bottom of the trench. By taking advantage of the flexibility of the Coflexip pipe (bending radius less than 2 m), the length

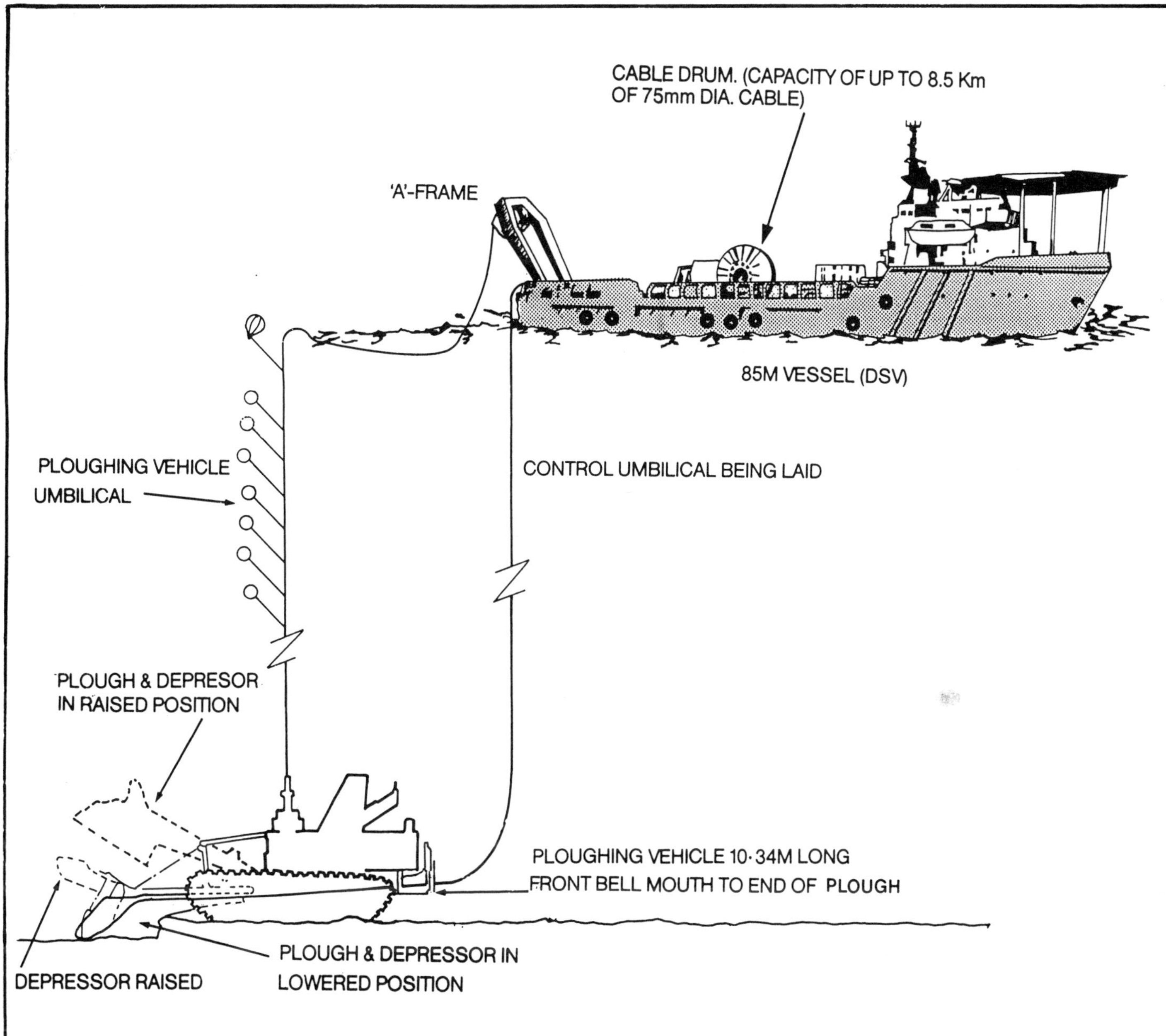

Fig. 8–11 Typical Deployment of a Bottom Crawler Plow Machine

of the shoe is minimum, therefore reducing the size of the machine as well as the friction of the soil along the shoe and, consequently, the power of the machine.

Owing to the small length of the burying shoe, the pipeline is in the bottom of the trench before trench collapse occurs. This allows a trench to be dug with vertical walls, therefore reducing the quantity of material removed and increasing the burying speed. This method ensures a complete back-filling and hence the required protection and thermal insulation of the flowlines.

FLOWLINE INSULATION

The reason for applying insulation materials to subsea flowlines is to ensure that the temperature of the fluid product does not fall below an acceptable level at the subsea-manifold, at the riser-level end of the first stage separator.

Because of the relatively low ambient temperatures of seawater at depth, fluids in submarine pipelines tend to lose heat progressively as they flow down the line. This is mainly due to heat loss through the pipe wall. The rate of loss depends on a number of factors including inlet temperature, environment temperature, and the length of line, but the single most critical factor is the heat conductivity of the pipe components.

Heat loss in pipeline fluids can create serious problems such as unacceptable levels of emulsions, hydrates, and wax deposits. In the production of waxy oils with a high pour point, the need for effective insulation against heat loss becomes even more critical.

In interrupted flow conditions involving the shut in of export or loading systems, it is particularly important to ensure that in-line fluids remain at a sufficiently high temperature to allow flow to resume. This problem may be overcome by use of fluid-return lines, heat-tracing methods, or thermal insulation. Frequently, thermal insulation provides the best long-term solution, but it has to be remembered that static fluids will ultimately reach the environmental temperature.

Thermal insulation of a flowline is achieved by use of a material with low heat conductivity which can enclose the external pipe for its full length, including field joints. Thus, selection of the most appropriate insulation material is crucially important for any thermally efficient system.

Three proven insulation techniques are available:

1. Hutchline system
2. Webco system
3. Flexible line insulations

Each is reviewed below.

Hutchline System (Refer to Fig. 8–12(1))

The basic components are:

- Rubber—A polychloroprene which has been tested and already used in marine environments for many years
- Insulation material—Typically a P.V.C. foam (brand name Klegecell), a closed-cell waterproof foam with high compressive strength (Other types of insulating material may be used, depending on the application.)

The system's key characteristics are as follows:

Adaptability

The design of the system allows the adjustment of the insulation level to meet client requirements.

This adaptability is obtained by:

- Using different numbers of layers of insulating materials
- The possibility of utilizing layers of insulating material of densities decreasing with distance from the steel pipe (Since the compressive strength of the insulating material improves at lower temperatures.) Refer to Figure 8–12 (2).

Excellent Thermal Efficiency

The system is designed to present neither radial nor longitudinal joints, which are always sources of significant thermal loss.

Working Depth

Using what is currently the highest available insulating material density, the system can be used at depths of up to 200 m (650 ft), with a very limited long-term creep.

Corrosion and Strength

No water ingress is possible between the different layers, owing to their very strong adherence after vulcanization. Thus the system is fully protected against corrosion. After vulcanization, the strong bonding between the different layers gives the system a very high shear resistance.

Webco Flowline Insulation System

The link between Shell's Underwater Manifold Center and the Cormorant A platform is the first application for a new elastomer pipe-coating system called Linecote IPC that combines corrosion protection with a high degree of thermal insulation. Produced by Webco, the system was devised after extensive investigation of the properties of various insulation materials, particularly their ability to withstand hydrostatic pressure. Refer to Figure 8–12 (3).

Information on the thermal and mechanical properties of the wide variety of insulation materials available leads to the conclusion that relatively low thermal conductivity values (referred to as K values) are attained with low-density polyurethane.

But, while polyurethane foam presents the best low-conductivity characteristics, it too has disadvantages. The fact that the material is not of the 100% closed-cell-type means that water ingress to all of the foam components may occur if even one cell section is damaged. Another major disadvantage is that compressive strength limitations of the polyurethane foam restricts its application to water depths of less than 30 m (100 ft) unless the material is itself enclosed by a high-pressure-resistant jacket such as steel. After an in-depth investigation of the properties of various insulation materials, Webco concluded that PVC foam offered the best combination of characteristics.

Webco Linecote IPC system

PVC foam for the Linecote IPC line insulation system has been chosen mainly because of its 100% closed-cell construction.

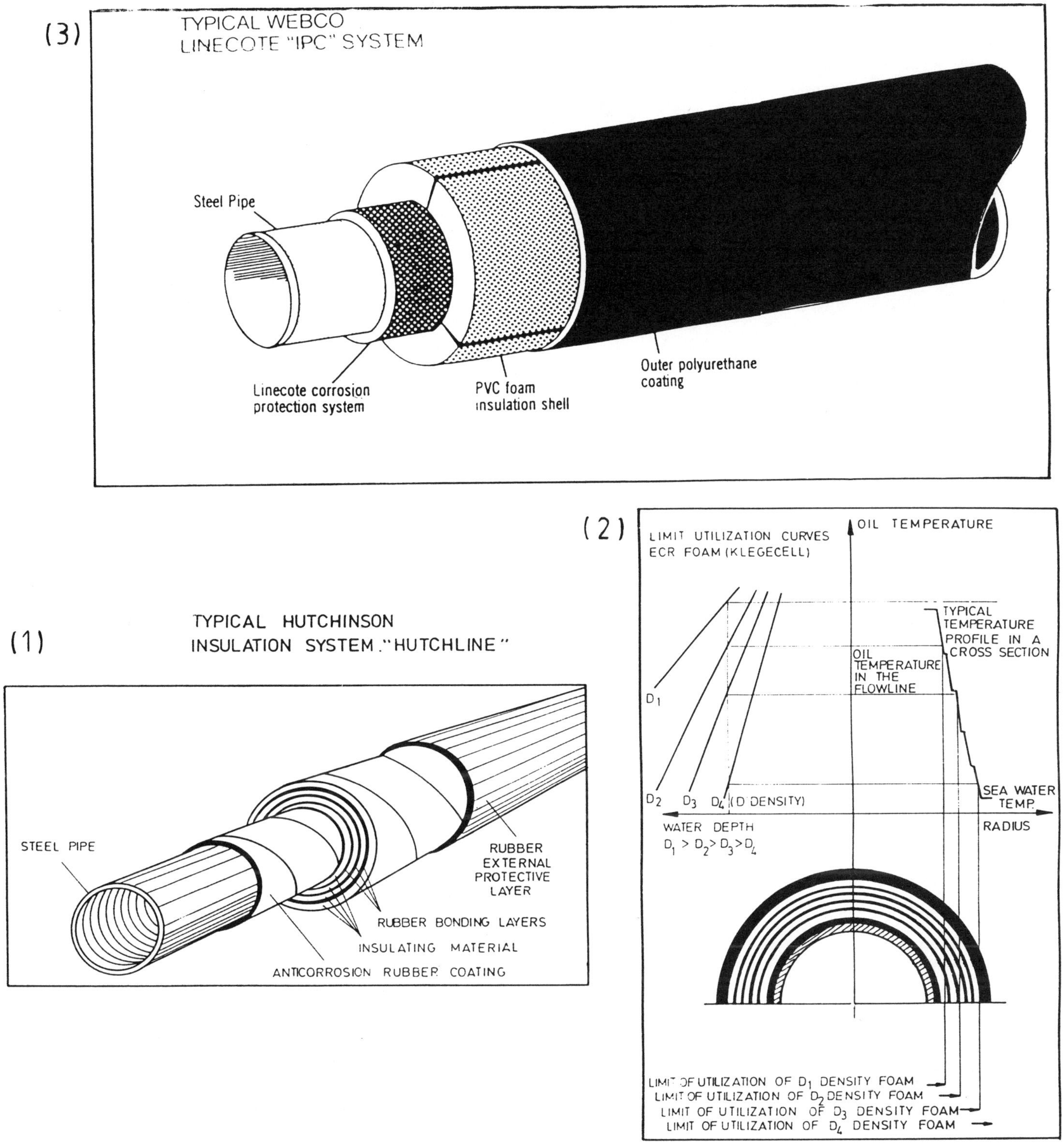

Fig. 8–12 Flowline Insulation Techniques

This makes the insulation coating impermeable to seawater. The ability of the foam to withstand hydrostatic pressure is another of its major advantages.

From the results, even a relatively low-density PVC foam can provide a hydrostatic pressure safety margin of twice the seawater pressure at a depth of 200 m (650 ft). This safety margin increases progressively as insulation densities are increased (see Fig. 8–13, Tables 1 and 2).

Webco has devised the Insu-Calc system, a computer program to determine the correct density of insulation required for any subsea pipeline or flowline. Essentially, it calculates the heat loss through the pipe wall and hence the temperature drop along a length of line of given dimensions. Factors used to provide base data for each computer run include wellhead input temperature, acceptable oil-output temperature, length of line, volume of flow, water depth, and seawater temperature.

The insulation thickness is designed for various pipe diameters needed to achieve a minimum platform temperature of 26°C (79°F), starting with wellhead temperature of 48.8°C (120°F) along a 3,650 m (12,000 ft) flowline with a flow rate of 6,000 B/D.

By combining the insulation properties of 100% closed-cell PVC foam with the Linecote corrosion coating, Webco has produced a system which combines a high degree of corrosion resistance with high thermal insulation. The composite consists of:

- Linecote corrosion coating
- 100% closed-cell PVC foam insulation
- An outer coating of polyurethane abrasion-resistant material

Linecote itself is an established flexible elastomer coating providing excellent physical properties and a chemical bond which is itself greater than the tensile strength of rubber (see Fig. 8–13, Table 3).

Field Joints and Laying Limitations. An effective technique has been developed to coat field joints with the same three-stage system, ensuring uniform thermal insulation properties over the jointed area. This in turn ensures a uniform temperature profile along the whole length of the coated line. One most important characteristic of the system is the ability of the various components to absorb laying stresses. This allows pipes coated with Linecote IPC to be laid by any method, including reel ship.

(1)

TYPICAL PHYSICAL PROPERTIES OF PVC FOAM INSULATION

	TYPE 1 Density 130kg/m^3	TYPE 2 Density 200kg/m^3	TYPE 3 Density 360kg/m^3
Compressive strength at 20°C (MPa)	2.5	4.5	> 9.0
at 70°C	1.4	3.0	> 6.0
Compressive modulus at 20°C (MPa)	89	153	650
Tensile strength at 20°C (MPa)	3	5	> 9.5
Flexural strength at 20°C (MPa)	3.5	4.9	12
Flexural modulus at 20°C (MPa)	80	120	200
Thermal conductivity (watts/mk)	0.035	0.04	0.06

Higher and lower density materials are available to allow for differing water depths, and the physical properties of the foam vary accordingly.

(2)

HYDROSTATIC TESTING OF PVC FOAM

	TYPE 1 Density 130kg/m	TYPE 2 Density 200kg/m	TYPE 3 Density 360kg/m
Failure pressure (bar)	45 - 50	60 - 70	> 75

Water Depth of 200 metres = 20 bar

TABLE 3: INSULATION THICKNESS FOR MINIMUM PLATFORM TEMPERATURE OF 26°C

PIPE DIAMETER	6 INCH	8 INCH	14 INCH
Flow rate (b/d)	6,000	6,000	6,000
Oil input temperature °C	48.8	48.8	48.8
Oil output temperature °C	27.0	28.6	25.9
Required insulation (mm)	20	30	40

(3)

WEBCO LINECOTE CR60/4A-8 - PHYSICAL PROPERTIES

Hardness	65
Specific gravity	1.36
Min tensile strength (psi)	2,500
Min elongation at break (%)	400
Min tear strength (lb/inch) (Crescent)	250
Max compression set (%)	15
Max Akron abrasion resistance (m/s/1,000 revs)	0.3
Resistance to cut growth	Excellent

Linecote offers excellent chemical properties and resistance to the degradive effects of seawater, ozone, oils, waxes, gases, as well as most aliphatic hydrocarbons.

Fig. 8–13 Webco Linecote IPC System—Tables

Coflexip Flowline Insulation

Two systems are possible:

1. Standard Coflexip pipe with double Rilsan outer sheath and a thermal conductivity of 4.1 Kcal/hour/meter/°C.
2. Flexible Coflexip pipe with an additional foam insulation between layers; the thermal conductivity of this pipe is mainly dependent on the thickness of foam which is incorporated.

The thickness of foam can be adapted to meet the required insulation level. For an equivalent pipe, the thermal conductivity can be as low as 1.1.

However, it is important to note that, while insulation of flexible flowlines can be obtained by deep burial (5–6 ft) the insulation obtained for a trenched pipe is dependent on the various parameters as indicated below:

Standard Coflexip pipe with double Rilsan layer.

Soil thermal conductivity (Kcal/hm°C)	Trench depth (meter)	Overall insulation coefficient (Kcal/hm°C)
0.9	2.5	1.1
0.9	2.0	1.15
0.9	1.0	1.4
1.3	2.5	1.4
1.3	2.0	1.5
1.3	1.0	1.75

Standard Coflexip pipe with additional foam insulation

Soil thermal conductivity (Kcal/hm°C)	Trench depth (meter)	Overall insulation coefficient (Kcal/hm°C)
0.9	2.5	0.63
0.9	2.0	0.65
0.9	1.0	0.72
1.3	2.5	0.72
1.3	2.0	0.73
1.3	1.0	0.8

Chapter 9

Operations and Maintenance

This chapter covers the overall concept of producing crude oil and gas from a reservoir. It includes:

- Start-up
- Normal operation (production, workover, and drilling)
- Turndown
- Routine shutdown
- Weather shutdown
- Emergency shutdown

Concepts and philosophies are outlined rather than operational details. The aim is to ensure that the various systems proposed are compatible with each other and the existing systems with which they will interface.

The following applies to a floating production vessel.

RESERVOIR FLOW RATES

Single and multi-well developments are considered below.

Single Well

A single well development (a satellite well) has three main advantages, namely:

1. It allows field development without extensive use of directional drilling methods.
2. It facilitates early production.
3. It allows the satellite wells to be connected to the template by flowlines, where the control module and chokes for these wells are situated on the template for ease of access.

Multi-Well Template

The flow rate from each well is controlled by a subsea choke and directed to one or two production headers. Normally, wells are open to the two production headers, which operate in parallel and feed the two production risers. These risers connect the subsea headers to the two-stage separation facilities.

For the start-up of a well, it is possible to isolate the headers and operate one at high pressure and one at a lower pressure. In this mode, high pressure fluids are directed to the first-stage separator and then on to the second-stage separator, where they commingle with the low pressure reservoir fluid. The latter enters directly into the second-stage separator.

RESERVOIR PRESSURE MAINTENANCE

Reservoir pressure is maintained by water injection. The total flow of water is measured by a flow meter on the discharge side of the water injection pump.

Flow distribution between individual wells is controlled by the adjustable subsea chokes, provided for each of the injection wells. The subsea chokes can be adjusted by divers, ideally by an ROV or remotely from the surface.

These chokes can be coarsely calibrated for variations in choke opening and choke pressure differential by occasional checks when all the flow is directed individually to each of the wells. The pumps are sized for the maximum need of the reservoir.

Having set the subsea chokes to give the required flow of fluid from each well, the operation of the first-stage separator is fully automatic and should require no operator intervention. The pressure is controlled at 250 psi by a back pressure controller releasing all gas to flare—surplus to fuel gas and gas lift requirements. Crude oil level is controlled conventionally by means of an offtake control valve.

Control of the second-stage oil and gas separation is similar to that of the first-stage. The normal pressure of 159 psi is controlled by a back pressure controller releasing associated gas to flare. The required pressure in the second-stage separation is dependent on the separator temperature and will require occasional adjustment as changes in well flow rates, seawater temperature, etc., result in separation temperature changes.

Depending on the nature of the reservoir fluids, a means of

heating or cooling crude oil can be provided in the oil/gas separation system.

In normal operation, the crude oil export flow is controlled by the oil level in the second-stage separator. The level controller resets the setpoint of a flow controller on the discharge side of the crude oil pipeline pumps.

Pumps are often provided in pairs, sized for the maximum production rate. The crude oil is metered for fiscal purposes by a meter located downstream of the booster pumps. To avoid meter errors, any spillback from the pipeline pumps is cooled and recycled downstream of the meters.

Fuel gas is taken from the first-stage separator and its pressure controlled by the separator pressure controller. It should normally operate at 250 psi, with a lower limit of 200 psi. Below about 200 psi, it is necessary to switch to diesel oil firing for the gas turbines, which takes place automatically.

The system provides for all the normal duties (cooling, water supply to desalination, washdown, pressuring of firemain, and sanitary water) anticipated in the transport mode, plus any process plant cooling duties and the supply of injection water.

The controlling duty is water injection and for this reason the cooling duties are arranged in series with the injection water supply, so that the latter receives warmed water. The two systems are regulated by having a small pressure-controlled spillback to the sea upstream of the water injection system. For stability, there must be a small spillback at all times, with the pumps automatically started or shut down as dictated by system demand to maintain this.

The system is regulated by a sequence timer, water levels, and flow rates.

A microprocessor-based energy management system can be used to control the generation and use of all onboard power supplies, to ensure that the generation system operates within its design capacity and an optimum use of machinery. It also would control functions such as the restart of motors and automatic load sharing.

START-UP

This brief description outlines the sequence of operations required to start up the production vessel from an on-station, blacked out condition, with the rise and manifold onboard.

Any other starting condition (e.g., normal start-up from a temporary shutdown) will only require some of the operations in this schedule. The sequence of operations is as follows:

1. Purge emergency diesel room with battery-driven fan.
2. Start diesel-driven air compressor.
3. Start emergency diesel generator to provide minimum of services for safety. These services include:
   - Emergency and navigation lighting
   - Radio and telecommunications
   - Purge fans for main generators and switchrooms
   - Battery charges for continuous power supply
4. Start main diesel generator to provide power for essential and other utility services, to include:
   - Seawater for cooling
   - Cooling water, high temperature, and low temperature systems
   - Potable water
   - Hot water
   - Sewage
   - Lighting
   - Cooking/laundry
   - Instrument and service air
   - Bilge/ballast pumps
5. Shut down emergency diesel generator as soon as the emergency switch board is supplied by a main diesel generator.
6. Start other main diesel generators as required to meet demand for:
   - Anchor winches
   - Thrusters
   - Draw works
   - Rotary table
   - Slush pumps, etc.
7. Ballast the semi-submersible down to operating draft—if a floating production platform is used as the main platform.
8. Use divers (or other relevant means) to connect guidelines to template guide posts.
9. Lower manifold to template, and use divers to make all necessary subsea connections. Lower control modules to template, and use divers to make all necessary connections. (This activity will continue for several weeks.)
10. Connect riser to template.
11. Adjust diesel generator power output to meet normal demand.
12. Start inert gas system and purge the entire process system. Start continuous purge to flare header.
13. Carry out system check on all subsea functions.
14. Carry out system check on all process controls.
15. Use propane bottles to ignite flare pilots.
16. Line up selected well to test separator.
17. Open well choke. Kick-off selected well using start-up/booster compressor and gas lift compressor with compressed inert gas via the second-stage separator. (This will take approximately 30 hours or longer.)
18. As crude enters the test separator, set the pressure controller to the separator pressure (typical valve, 250 psi).
19. Start one gas turbine generator on diesel. Shut down diesel generator.
20. Increase flow of well fluid and commission level controller. Set back-pressure controller on second-stage separator.
21. When an adequate level is obtained in the second-stage

separator, start booster pump on recycle and completely fill pump/metering system.

22. Start pipeline pump on recycle. Increase well flow and when normal level is achieved in second-stage separator, commission flow to export line.
23. With first well at normal flow, gradually divert flow to first-stage separator and adjust pressures to normal operation conditions.
24. Commission fuel gas system. Transfer flare pilots to fuel gas. Switch gas turbine generator to gas firing.
25. Commission additional wells as required using associated gas for kick-off. Bring on additional turbine generator as required, and shut down diesel generators when the system is stable. Bring on additional cooling water pumps as required.

Start-up of Water Injection

The start-up of the water injection system is set out below:

1. Gradually open valves to admit water to the water injection system through to the de-aerator. Maintain a small flow through the system to drain, and check out operation of filters.
2. Open valves around booster and injection pumps and fill pump system with water. Close discharge valves.
3. Start vacuum pumps, and then start booster and injection pumps at minimum flow discharging overboard.
4. Continue until water quality meets required specification.
5. Open water injection well, and slowly open discharge valve to divert flow to well. Carry out calibration check.
6. Open other injection wells as required, starting additional cooling water pumps and the second injection pumps as demand increases. Check calibration of each well in turn.

TURNDOWN

Either as a preliminary operation before shutdown or as a means of reducing production for whatever reason, it is necessary to be able to turndown the plant. The extent to which this is carried out is dependent on the nature of the problem, but the following is a typical procedure:

1. Close in injection wells one at a time until flow rate is reduced sufficiently to allow one injection pump to be shut down. Shut down any surplus cooling water pumps and trim cooling water system. (This is a major power-shedding operation and might be the total extent of turndown required.) If necessary, close in remaining injection wells, shut down second injection pump, and trim cooling water system.
2. Close in producing wells one at a time until flow rate is reduced sufficiently to allow one pipeline pump and its booster to be shut down.
3. Check and trim all ancillary systems.

This is a typical minimum stable turndown condition.

NORMAL SHUTDOWN

From the turndown condition, a normal shutdown would proceed with the major activities as follows:

1. Switch one gas turbine generator to diesel oil firing.
2. Switch flare pilots to bottled propane.
3. Gradually shut in all except one producing well, checking that pipeline pump minimum flow recycle device is operating correctly.
4. Decrease level in first-stage separator to minimum.
5. Decrease level in second-stage separator to minimum.
6. Start a diesel generator.
7. Shut down gas-fired turbine.
8. Close in the last well. Close all subsurface safety valves. Shut down oil export pump and shut all riser isolation valves.
9. Trim power consumption.
10. Start a second diesel generator, and then shut down the final gas turbine.
11. Extinguish flare pilots.

The system is now in a stable shutdown state.

Weather Shutdown

In advance of a weather shutdown, the production will be turned down and shut down as already described. Given time, there will be a "hold" at each major step in the program in turn permitting a "go" or "no go" decision.

As the weather deteriorates, preparations will be made to disconnect the riser and, if necessary, to stack it. A likely procedure would be:

1. Use kill pump to purge the subsea manifold and all production risers with seawater via the kick-off riser back to the test separator.
2. Slacken the anchor chains.
3. Commence deballasting of semi-submersible to increase air gap.
4. Operate thrusters to minimize loads on anchor chains.
5. Release risers from production vessel.
6. Meanwhile, if weather continues to deteriorate, all dry and wet muds are dumped to reduce weight. The contents of the first-stage separator, second-stage separator, and test separator can be routed into the slop oil tank or alternatively can be burned via the oil burner.

This completes the weather shutdown procedure.

ON-BOARD INCIDENT SHUTDOWNS

All critical items are protected by safety devices which in the event of an on-board incident shut down equipment to avoid injury, damage, or spillage.

Fire

The main processing area is divided into two distinct zones with regard to fire. The first, where structural damage could occur in the presence of a fire, will be protected by an automatically initiated deluge system supplemented by manually operated monitors, foam generators, and hoses.

The second area will be protected by manually operated monitors and supplemented by foam generators and hoses.

In the less hazardous indoor areas, protection will be by an automatic sprinkler system, supplemented by hoses and hand-held extinguishers.

Each gas turbine enclosure will be protected by an automatic Halon system. Each enclosure will be independent of the others so that a fire alarm in one will not shut down all the turbines.

The accommodation area will be protected by an automatic fresh water sprinkler system. When the supply of pressurized fresh water is nearly exhausted, the sprays will automatically receive salt water. This system will be supplemented by hoses and hand-held extinguishers.

The helideck will be protected by manually operated foam monitors.

The fire main will be pressurized by a jockey pump. If the pressure in the main falls, a fire pump will be automatically started. If the pressure continues to fall, a second pump will start. These two pumps are powered through the main switchboard if a main generator is working.

If, for whatever reason, the power through the emergency switchboard to the fire pumps fails, then the emergency generator will start automatically if no main generator is working.

Catastrophic Incident

In the event of a catastrophic incident, a manually operated button will start the following operations:

1. Close all downhole safety valves
2. Trip out water injection and oil pipeline pumps
3. Cut off gas to all gas turbines, boilers, and heaters
4. Close all riser isolation valves
5. Close valves between on-board riser manifold and first-stage, second-stage, and test separators
6. Open depressing valves to flare
7. Shut down all non-essential process equipment
8. Purge flare system with inert gas

The emergency diesel generator will be started manually, as will one diesel generator. The gas turbine generators will shut down. All non-essential power users will be shut down.

At this stage, fire fighting and evacuation can take place under the safest conditions for the maximum length of time.

Given time, the well kill unit will be used to purge out the risers with seawater so that the riser can be released without causing pollution.

Also given time, all unnecessary solids and non-hazardous fluids will be dumped.

INTRODUCTION TO DOWNHOLE OPERATIONS

As long as step-out satellite wells are used as part of subsea completion systems, there will be a need to service them.

Of course it will always be possible to relocate over a satellite with the same type of rig as was originally used to drill and complete the well, but this is not as economic as it used to be. Increasing operating costs and inflation have made it difficult to justify routine re-entries for maintenance or logging.

Certain operations (TFL) exist and may eventually become routine, but this technology has not yet claimed that TFL tools will ever be able to duplicate the abilities of rotating pipe in drilling operations, pulling the tubing string, or any of the spot services provided by coiled tubing. The TFL technique at present also suffers from a degree of experimental uncertainty and has been known to lead to downhole fishing operations. The difficulties of positive location and retrieval, and the usually justified caution of the offshore oil industry, will probably inhibit full acceptance of TFL. For the purpose of this section, only direct re-entry/workover operations are considered, on the basis that they are simply more economical and lead to a suitable design for small field developments.

In a review of economic considerations over downhole operations, there is a crossover point at which non-essential re-entry using a drilling rig becomes economically unacceptable. The benefit of production logging, desanding, acidizing, chemical and nitrogen injection, and zone testing, etc., can be outweighed by the commercial disadvantages of hiring a full-sized drilling rig to provide the necessary support facilities.

To illustrate the point using a particular well, assumptions can be made that 30 days of downhole work per year could improve production by as much as 6 %. The true cost of operating a semi-submersible exploration drilling rig is, of course, a variable factor, as is the selling price of oil.

However, without planned downhole maintenance, the longer a well produces the sooner it will need unplanned workover, and, of course, inflation will ensure that this is more expensive.

Initially, drilling and completion followed by hookup will ensure a period of productivity, thereafter which is associated with minimal maintenance costs. Finally, a period will follow when depletion and workover costs will eventually exceed the value of oil produced. Productivity vs. maintenance cost is explained by the bathtub curve in Figure 9–1. It shows that to increase the net benefit, point x, the crossover point, must either be moved upwards or to the right.

Production is basically a function of a specific well and so are costs, but the workover has positive influence on both.

Downhole operations can be classified in three main categories, namely:

1. Major workover operations
2. Wireline workover operations
3. Well testing, circulation, and stimulation

Major workover operations normally cover the following tasks.

- Christmas tree removal/change out
- Tubing pulling/change out
- Casing downhole interventions
- DHSV change out (depending on completion design)
- Gas lift mandrel change out
- Downhole pressure/template sensors change out

Workover operations using an integral workover riser for major workover must be run through a marine riser, BOP stack configuration. This arrangement is illustrated in Figure 9–2(1) and is common to both template and satellite wells.

The use of an integral workover riser is not mandatory. It is common practice by many operators to use a multi-purpose tool run-on tubing. In all cases, a terminal head is required to provide a secure hydrocarbon path when circulating.

Light workover is achieved by entering the tree via the tree mandrel or re-entry hub. Again, the equipment used is common for both satellite and template trees. A typical arrangement is show in Figure 9–2(2) whre the importance of the tree running tool is shown in achieving the connection to the production/annulus bores and the tree control functions.

The light workover operations are numerous and include:

- Setting and retrieving plugs
- Opening/closing sliding sleeves
- Performing bottom hole pressure/temperature surveys
- Sampling
- Running and retrieving wireline retrievable subsurface safety valves

In addition to the above completion activities, other basic well maintenance and workover activities are:

- Replacement of malfunctioned SCSSV (Surface Controlled Subsea Valve)
- Locking into tubing retrievable SSSVs a valve to extend the life of a well and to avoid a major workover
- Setting and pulling plugs to secure a well
- Reperforating production tubing

As drilling operations have moved into deeper waters, wireline operations from a floating rig have become common practice.

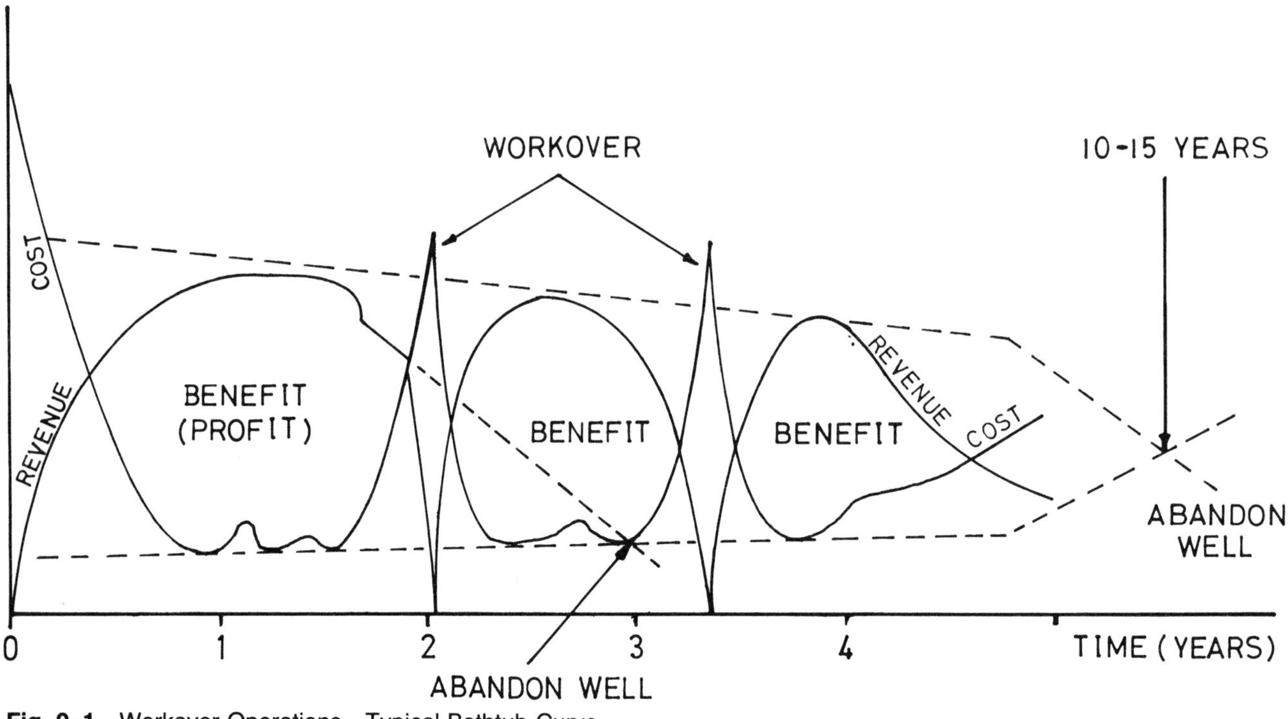

Fig. 9–1 Workover Operations—Typical Bathtub Curve

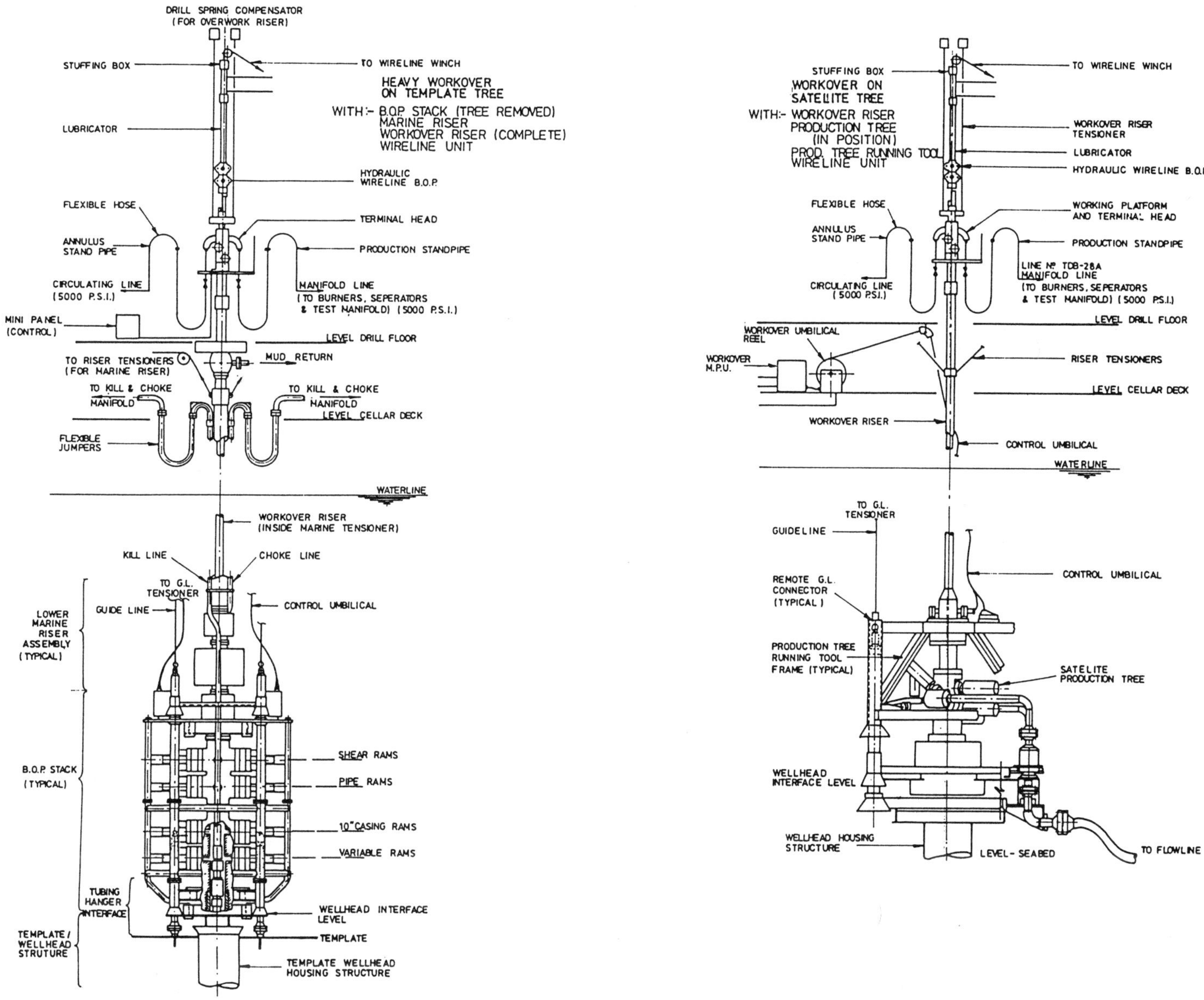

Fig. 9–2 Workover Interfaces and Equipment 9–2 (1) and (2)

Over the last decade, work from rigs has escalated in the North Sea with many companies conducting wireline work on exploration wells, during well testing operations, and on completion and workover of subsea wells.

Equipment utilized on the surface, as well as downhole, is exactly the same as that used on a fixed platform. The remotely operated hydraulic stuffing box and wireline blowout preventors is equipment sometimes not used on a platform but considered essential on a floater. These two assemblies, when controlled remotely, help minimize the number of times the wireline personnel need to climb the lubricator stack-up. With very few exceptions, all operations carried out on a platform can be accomplished on a floating rig.

Wireline units utilized on platform and floating rigs are generally skid-mounted and are either one or two piece units (see Fig. 9–3). The one-piece unit has the power pack mounted beside the reel unit. The two-piece unit consists of a separate reel unit and power pack, the two units being connected by hydraulic hoses or electric cable depending on whether the power pack has a diesel engine or an electric motor. Units can be either single or double drum, with a double drum having the following normal capacity:

- 7,620 m (25,000 ft) of 0.108-in. wire
- 6,096 m (20,000 ft) of 0.187-in. braided wire or single conductor line

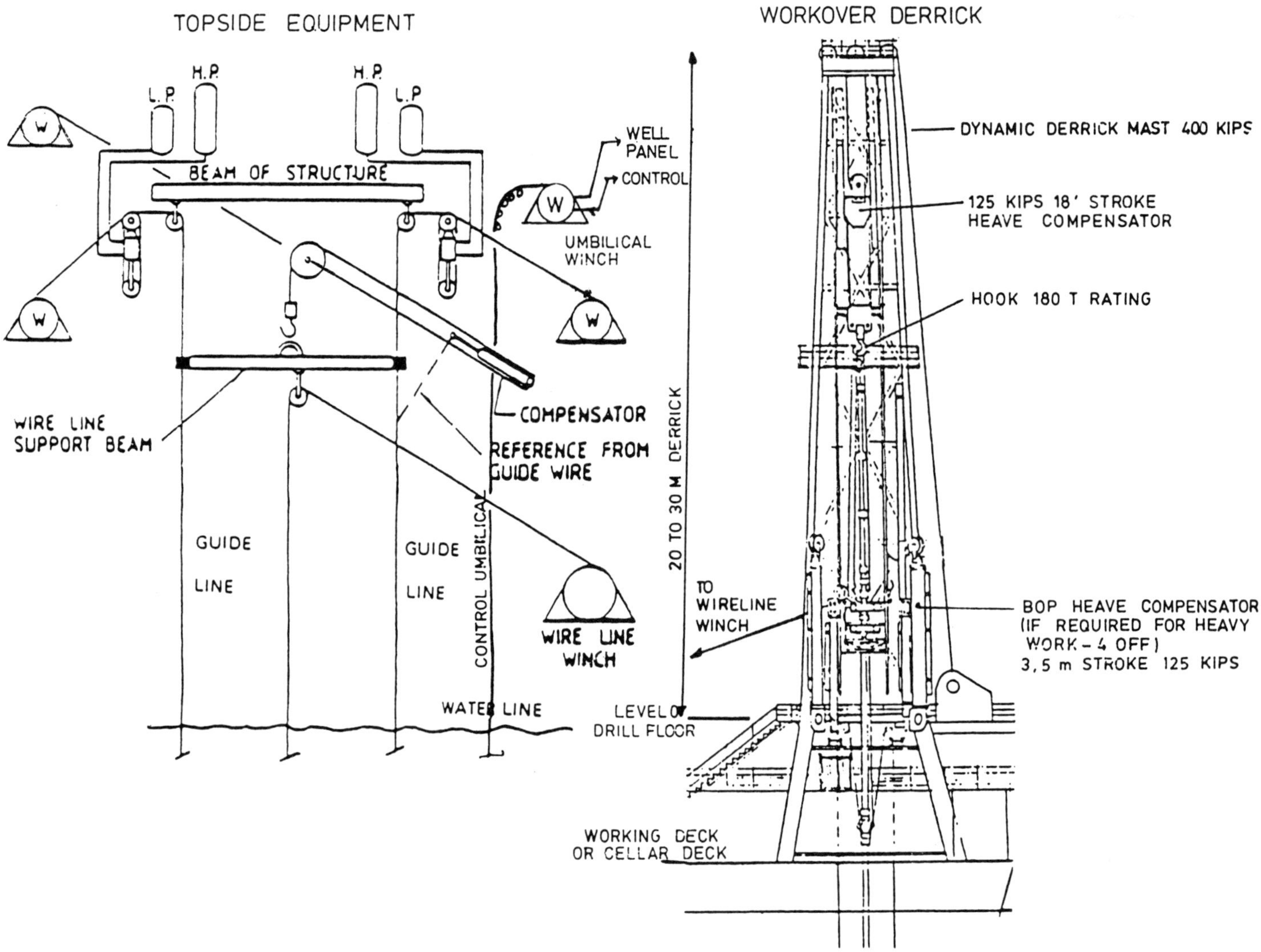

Fig. 9–3 Wireline Units 9–3 (1) and (2)

The lubricator is derived from a method of applying grease to pump rods, etc., while these items are operating under pressure. The wireline lubricator is used as a method of introducing the necessary subsurface wireline tools into the conduits to perform an operation under pressure. With the tools installed, the wellhead valve then is opened to allow entry into the wellbore. Lubricator assemblies are available in pressure ranges from 2,000 psi to 22,500 psi working pressures. They conform to specifications set by API 6A. A typical layout is shown in Figure 9–3(2).

Running and setting tools are used to place flow control devices into the tubing string. Refer to Figure 9–4(1). These devices are landed into nipples and placed within the string at predetermined depths during the running of the tubing. Numerous designs of landing nipples exist, which means that the flow control device's profile must match the nipple profiles. The tool illustrated in Figure 9–4(2) has nipple selectivity designed into the tool and can be landed and locked into any of several landing nipples placed at various depths. A third type of tool may be used to lock the device at any depth using slip type collets. Refer to Figure 9–4(3).

To pull the flow control devices, a second series of pulling tools is utilized. The three main types are shown in Figures 9–4(4) and 9–4(5).

The service tools required to activate these devices are referred to as positioning or shifting tools and will vary in design to conform to the design of the sliding side-door. Three main types are illustrated in Figure 9–5.

The gas lift mandrel is a typical flow control device, which generally incorporates a valve in a side pocket. The kick-over (gas lift valve) tool is used to install or retrieve flow control devices from side pocket mandrels. Being set in the offset portion

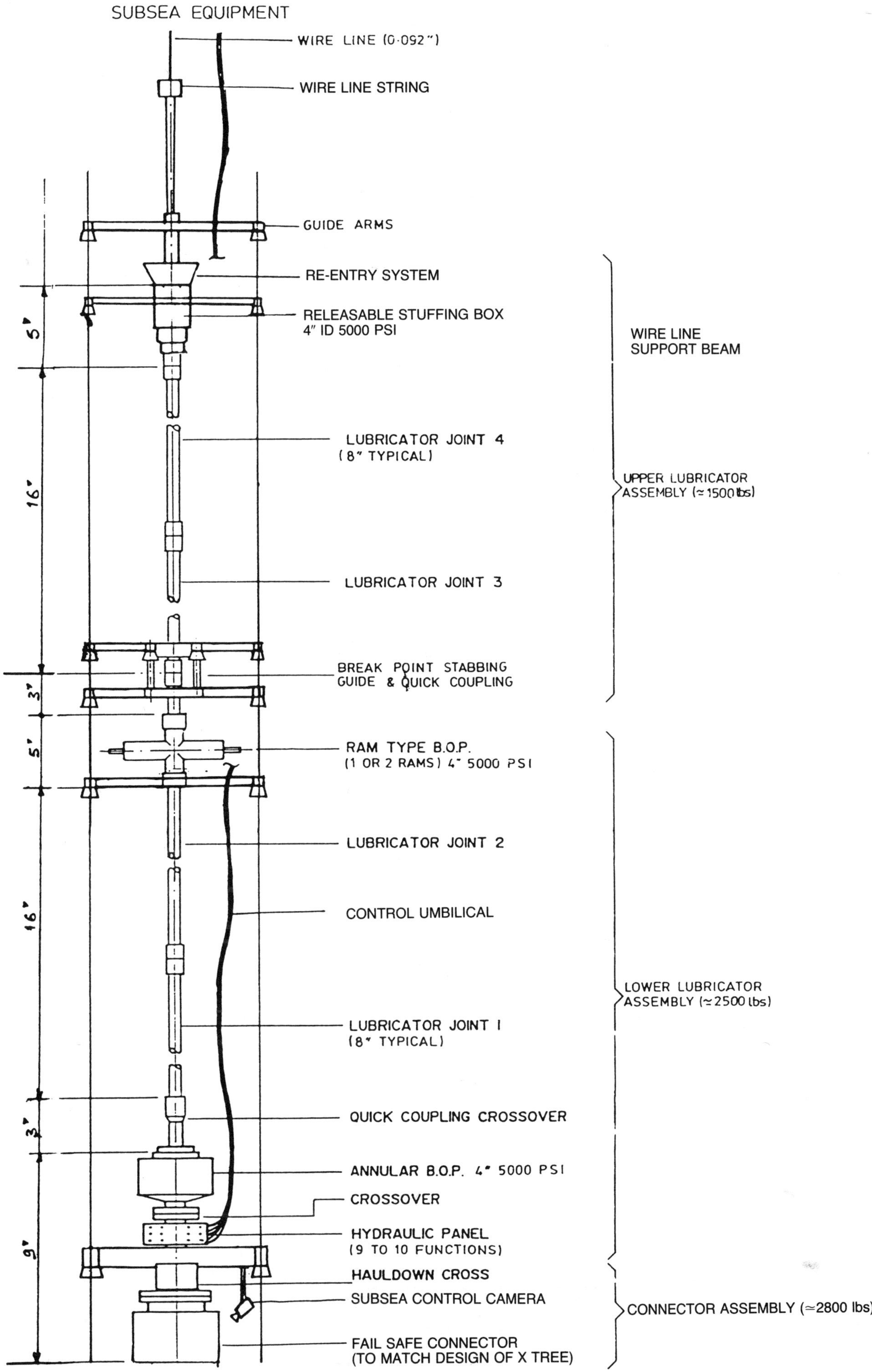

Fig. 9–3 Continued

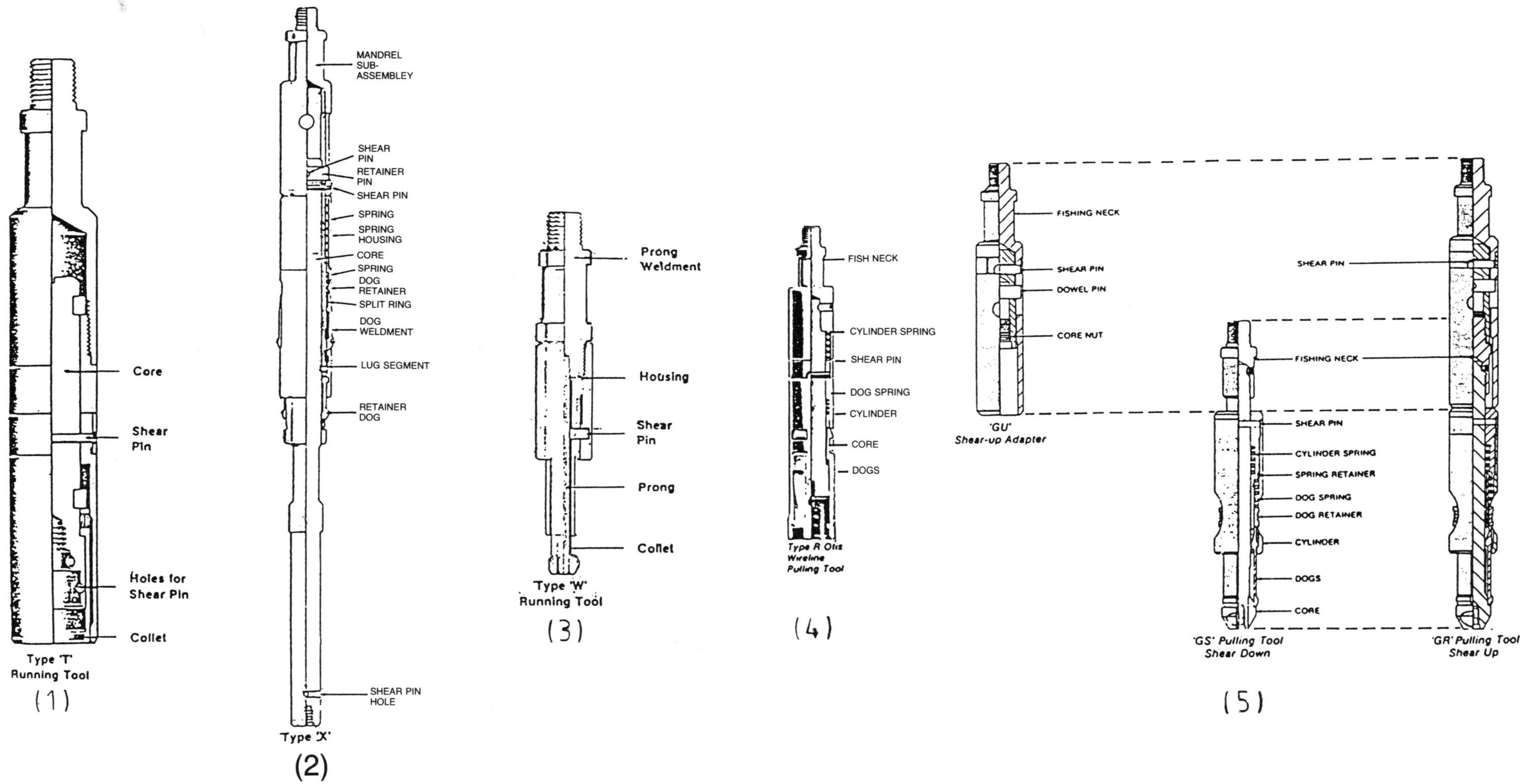

Fig. 9–4 Wireline Running and Setting Tools 9–4 (1), (2), (3), (4), and (5)

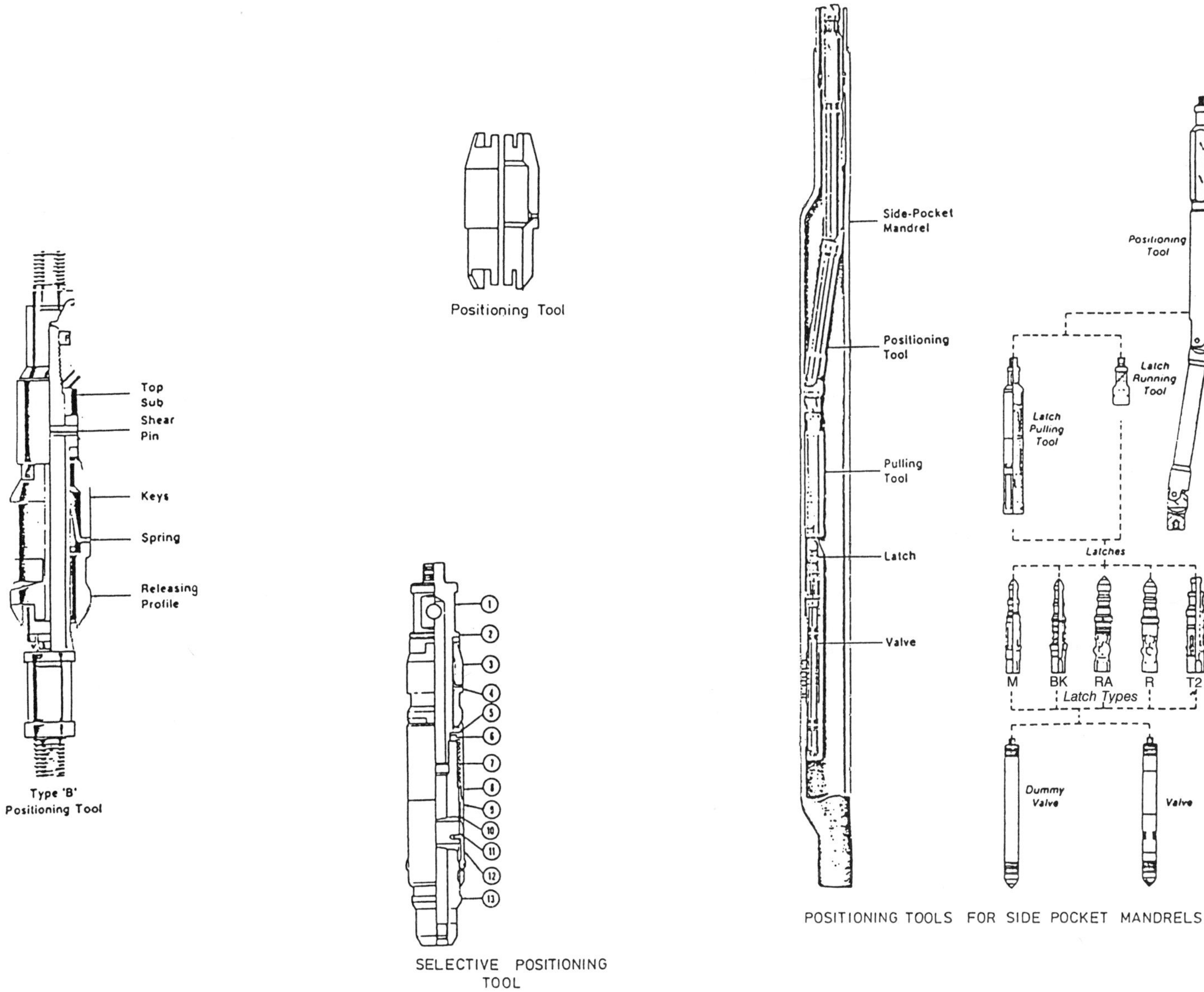

Fig. 9–5 Sliding and Positioning Tools

of the mandrel means that these devices can be pulled and reset without distributing any other valve in the tubing. Refer to Figure 9–5(4).

The tool is lowered just below the mandrel and is pulled back through. When the kick-off tool contacts a shoulder in the upper end of the mandrel, it is placed in a position that forces the pulling tool to a slight angle to the vertical axis. Then, as it is lowered again, it is directed towards the top of the valve in the side-pocket, where it engages the pulling neck of the valve. An upward jar action shears a pin holding the latch of the valve and frees it to be pulled.

Running and setting the valve is accomplished in much the same manner as the pulling procedure.

In some instances, all of the valves in the tubing string will be pulled and replaced with dummy valves. In deep wells, there may be 10 or 12 valves in the tubing.

SUBSEA WIRELINING

Wireline operations are normally carried out from a surface vessel, drilling rig, or floating production vessel, through the workover/completion riser. Refer to Figure 9–3(2) for more details. The wireline unit is mounted on the surface vessel, inside the derrick. Wireline operations were carried out by Flopetrol-Schlumberger first in Zakum field, and then in various oil fields around the world. These used a subsea wireline lubricator and associated equipment, keeping the wireline winch, hydraulic power unit, and remote control panel on the support vessel. The operations have been very successful.

Improved technology permitted another company, Otis, in conjunction with Comex Houlder, to carry out a successful test for Hamilton Brothers in 1984. This was in the North Sea. The objectives of the test on a real wellhead were:

- Position keeping ability of a medium size DSV type vessel
- Lubricator make-up, rigging, and handling methods on the deck without derrick but using only a heave compensated crane
- Subsea handling and installation with ROV and/or divers
- Wireline system release and retrieval
- Ultimately develop a method of performing wireline work from a Dynamically Positioned Diving Support Vessel (DSV with DP)

The work was carried out from Seabex One, a medium size, modern DSV with 6,000 ton displacement. The method used avoided the use of guide wires, but this is not recommended. The test was successful though the need for a handling derrick was strongly demonstrated.

A typical deployment operation from a DSV is shown through five key figures on Figure 9–6, and typical equipment required on deck and subsea are presented in Figures 9–3(1) and 9–3(2).

Operations carried out by subsea wireline can be:

- Running and pulling the wireline retrievable DHSV (depending on completion design)
- Bottom hole pressure survey
- Bottom hole temperature survey
- Production logging
- Tubing services, workover only
- Replacement/serving of gas lift valves

The only sequences of operation requiring ROV/diver assistance for convential trees are:

- Setting of guide lines
- Removing Christmas tree cap protector
- Installation of special flange for hydraulic connection

The special releasable stuffing box allows various sequences to be performed without having to pull the complete lubricator and BOP assembly.

The subsea lubricator can operate either with slick line for wireline operation or with a conductor cable for logging of pressure read-out. Typical dimensions of the equipment are indicated on Figure 9–3.

Service Tools

There are a variety of service tools for purposes other than running or retrieving downhole flow controls. All are run on a basic tool string even though they serve a special function. They fall into three main groups:

1. Fishing tools
2. Miscellaneous service tools
3. Service tools for recommended work

Conventional overshot or *internal-neck dog type pulling tools* may be used to retrieve lost tool string assemblies, since most items have a fishing neck of some sort on their upper end, which may have a shear-up or shear-down safety release.

Wireline grab tools are used to extract broken wirelines or cable from tubing or casing. These are run down into a lost wire, are jarred down, and then pulled up to a trap-loop of wire on the tools' barbs.

Overshot tools are used to retrieve items without fishing necks and generally utilize a collet mechanism with serrated inner surfaces. The collet may be sleeve-supported for retrieval of heavy items.

Spears are used to retrieve items without fishing necks such as large internal diameter stubs, mandrels, springs, etc. Generally, they have a collet mechanism with a serrated outer surface. The collet is usually supported by an inner core for lifting heavy items.

Magnetic fishing tools are used to remove small particles of ferrous material from the top of tools, such as shear pins or broken keys. These include a skirt to prevent "fish" from being dragged off while being pulled from the well. Refer to Figure 9–7(1).

Alligator grab, used to retrieve small items too irregular in shape for either overshoot or spear type tools, such as pins, dogs, broken springs, etc. It consists of two or more "jaw" segments, run in an open position, which closes when the obstruction (fish) is encountered.

Blind box, used for heavy downward jarring, such as driving a "fish" downhole. It is basically a solid metal tool with flat bottom, hardened to prevent damage. Refer to Figure 9–7(2).

Gauge cutter, used to verify that the tubing I.D. is sufficient to run the desired flow control. It has a sharp lower end to permit the cutting of paraffin, scale, or other deposits from the tubing wall. The large I.D. and fluted areas permit fluid or cuttings to bypass the tool. Refer to Figure 9–7(3).

Tubing swage, used to remove mashed spots and large obstructions in the tubing. It is basically a wedge, jarred down through any tight place to force the tubing out far enough for flow control items to pass. Refer to Figure 9–7(4).

Tubing broach, used to cut burrs in tubing I.D. caused by perforating, rust, bent tubing, etc. It usually consists of a series of spools with increasing O.D. so that only a small portion of the obstruction is removed each run. This tool is used cautiously to prevent damage to the tubing I.D.

Impression tool, used to take an impression of hard or metallic objects so their shapes or relative positions may be determined.

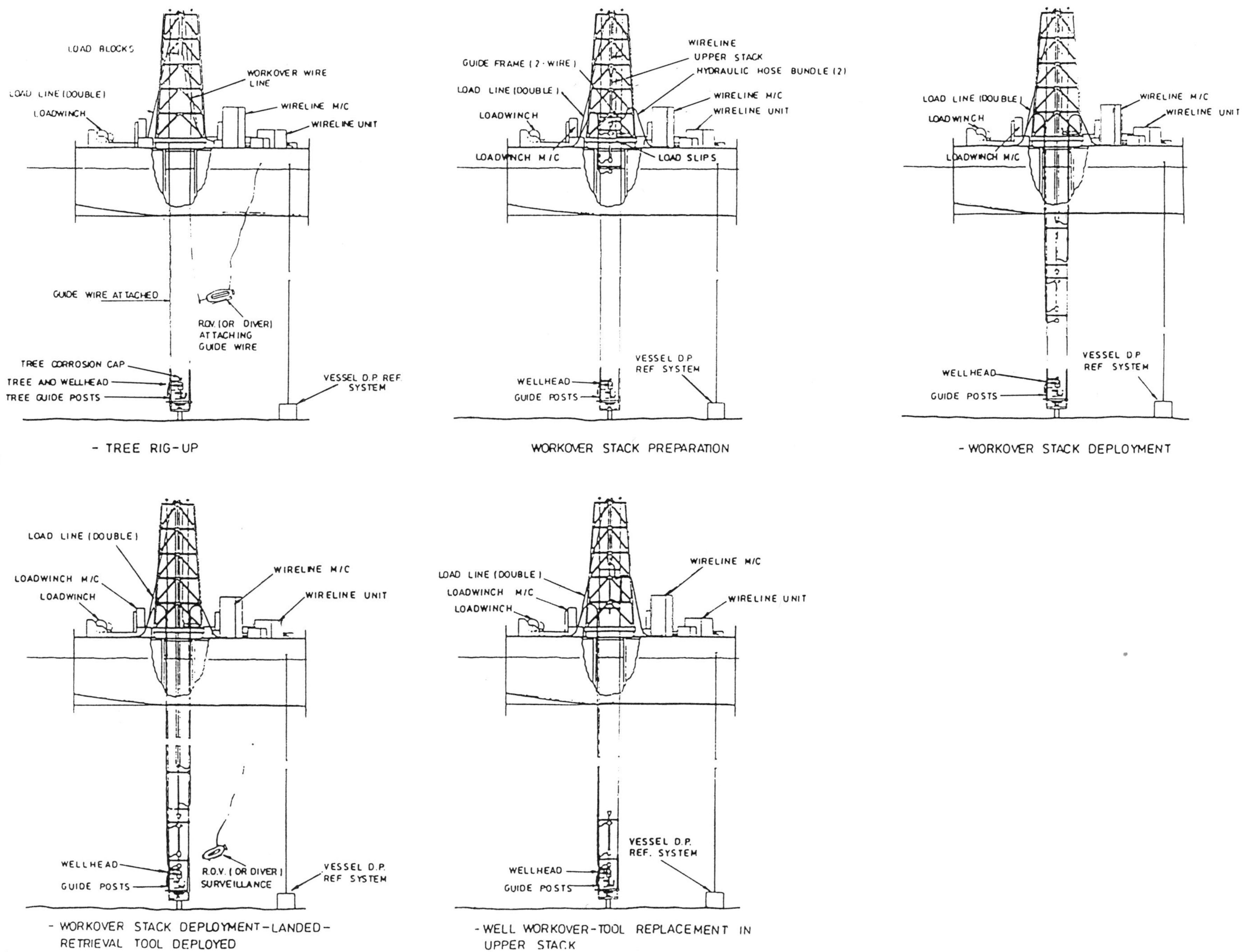

Fig. 9–6 Typical Deployment From a Support Vessel (DSV)

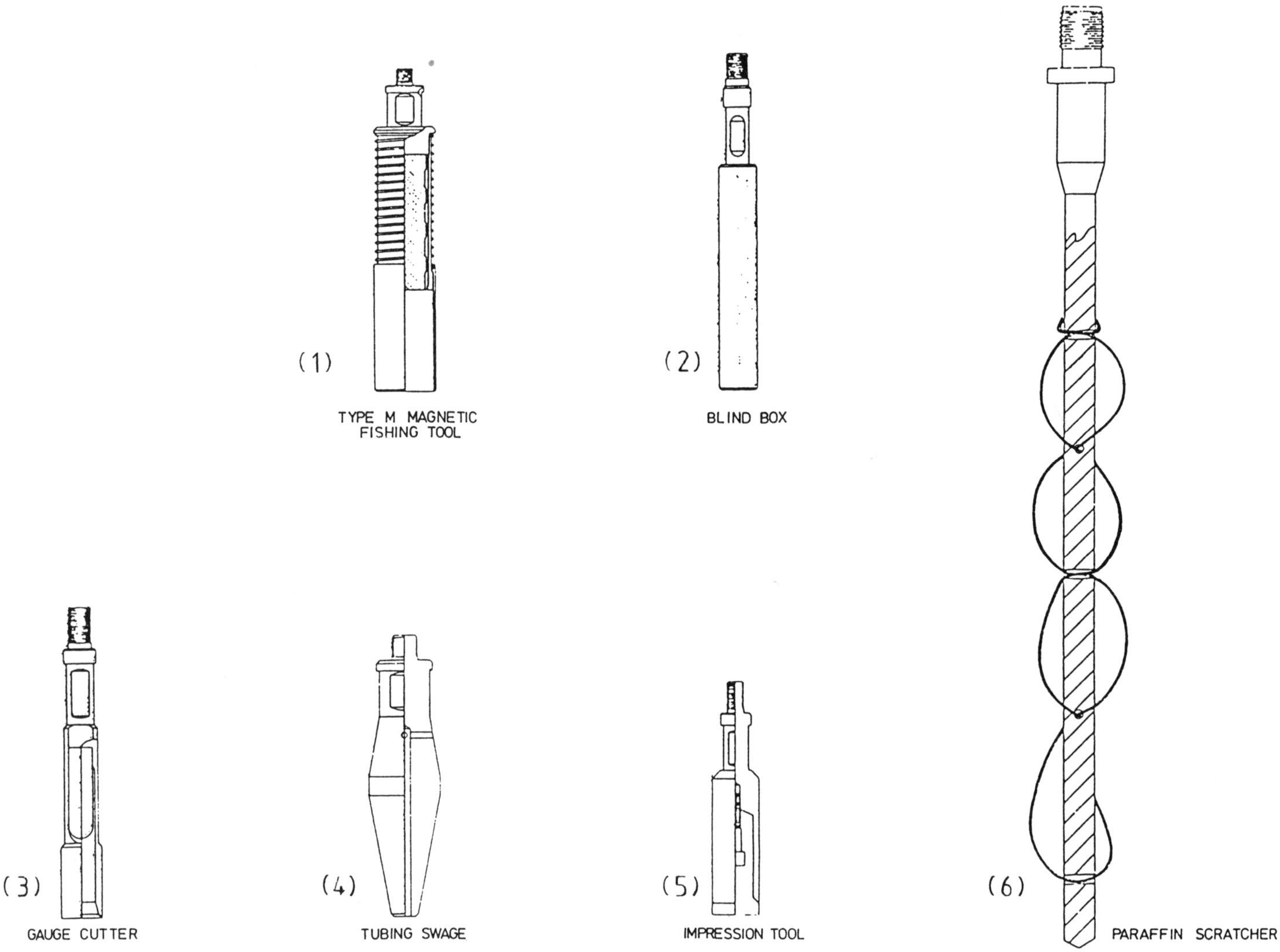

Fig. 9–7 Remedial Work Service Tools

It is a lead-filled cylinder with the lower end depressed prior to each run to effect a usable impression. Refer to Figure 9–7(5).

Paraffin scratchers are used to remove light deposits of paraffin downhole. The design is similar to a round wire brush. Sometimes it may be found necessary to flow the well at a reduced rate to prevent the tool from clogging. Refer to Figure 9–7(6).

Paraffin cutters are used to remove paraffin or wax deposits downhole. They have a series of spring-loaded knives of O.D. comparable to tubing I.D.

Pump type bailers are used to retrieve sand or other trash and carry it back to the surface. The tool consists of an outer cylinder to carry sand after it has been retrieved. A piston is used to suck sand into the clyinders. Check valves retain the sand in the cylinder after it has been sucked in. Refer to Figure 9–8(1).

Hydrostatic bailer, sealed at the surface and run into the well with the internal chamber of the bailer at atmospheric pressure. When the bailer reaches the object to be bailed, downward jarring shears a small sealing disc to admit fluid and trash into the chamber. A check valve retains the trash in the cylinder. A release valve is opened at the surface to bleed off the well pressure in the chamber prior to draining the bailer. Refer to Figure 9–8(2).

A series of pressure surveys are used to monitor pressure decline and secure accurate flowing and static bottom hole information for more effective gas lift calculations. They also may be used to determine a point of production inflow (open perforation).

Perforators provide access to casing annulus to circulate or kill a well. In addition, they can be used to produce additional zones and to install pack-off type gas lift or chemical-injection

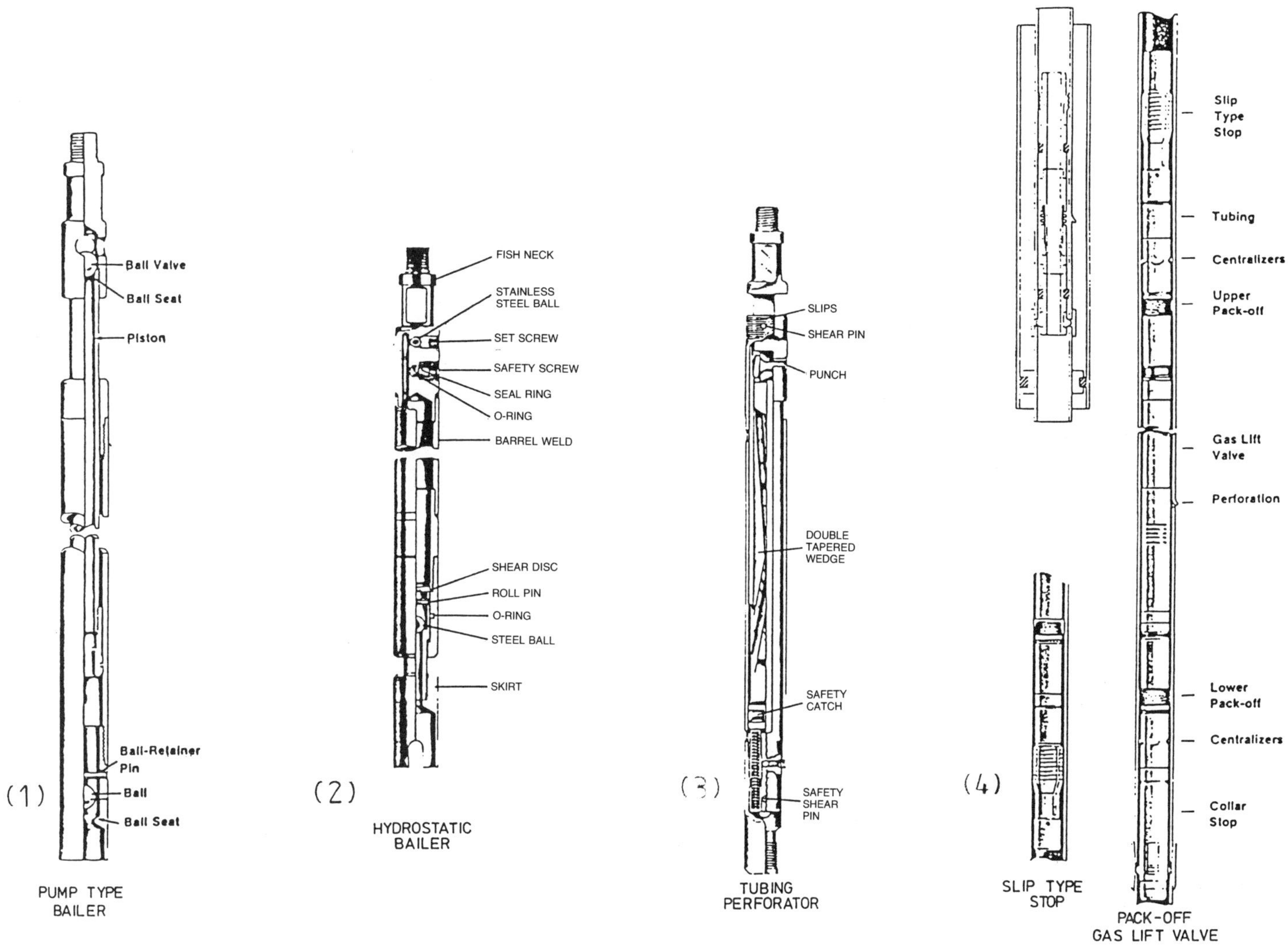

Fig. 9–8 Other Service Tools

valves. They permit production through a tail pipe that has been plugged and cannot be re-opened.

Two types of perforators are available—mechanical which are jarred down to force tapered wedges through the tubing wall or explosive which uses a charge to perforate the tubing. Refer to Figure 9–8(3).

Pack-offs are used to isolate holes in the tubing or leading connections and have large I.D./O.D. ratio with compression-set seal elements in tubing string. The lower pack-off may use slip-type stops or collar stops. The upper pack-off uses slip-type stops. Refer to Figure 9–8(4).

Coiled Tubing Methods

It is becoming increasingly common for North Sea operators to turn to the use of coiled tubing for workover operations to solve many common well problems. Many of these problems may be solved using alternative methods; however, compromises usually have to be made to achieve satisfactory results.

Most operators, when running coiled tubing from a floater, have a subsea coiled tubing/wireline BOP in addition to a surface BOP. This is considered a necessary precaution in the event of an emergency occurring with the coiled tubing in the hole. The subsea tree valves could not be closed to secure the well with the coiled tubing across them. With a subsea BOP, the shear rams can cut the coiled tubing allowing it to fall clear of the tree valves which permits the upper section of coiled tubing to be pulled clear.

Some of the problems that may be tackled using coiled tubing methods are:

- Kicking-off difficult wells.
- Backflushing water injectors to clean them up.
- Spotting acid or scale treating chemicals across a specific zone.

- Circulating scale/fill from top of a stuck wireline plug.
- Circulating at tubing retrievable sub-surface safety valve should valve fail. This may dislodge the debris causing the valve to fail.
- Spotting cement or water shut-off chemicals should one particular section of a production zone be producing all the water.
- Sand clean-outs, should any of the gravel packs fail.
- Other treatments or stimulations that may be required, should satellite wells be brought on stream from different reservoirs.

Unfortunately, previous use of floaters for this purpose has been somewhat restricted due to the lack of an efficient heave compensation. Recent experiences with the system shown in Figure 9–9 have shown that this problem can be overcome. In addition, further work is being carried out using integral risers fitted with a coiled tubing injection head. In this instance, the heave compensation requirements are catered for by the workover riser arrangements through tensioners and the heave compensated derrick to which the handling/terminal head is attached. The tubing head replaces the conventional wireline BOP and lubricator head.

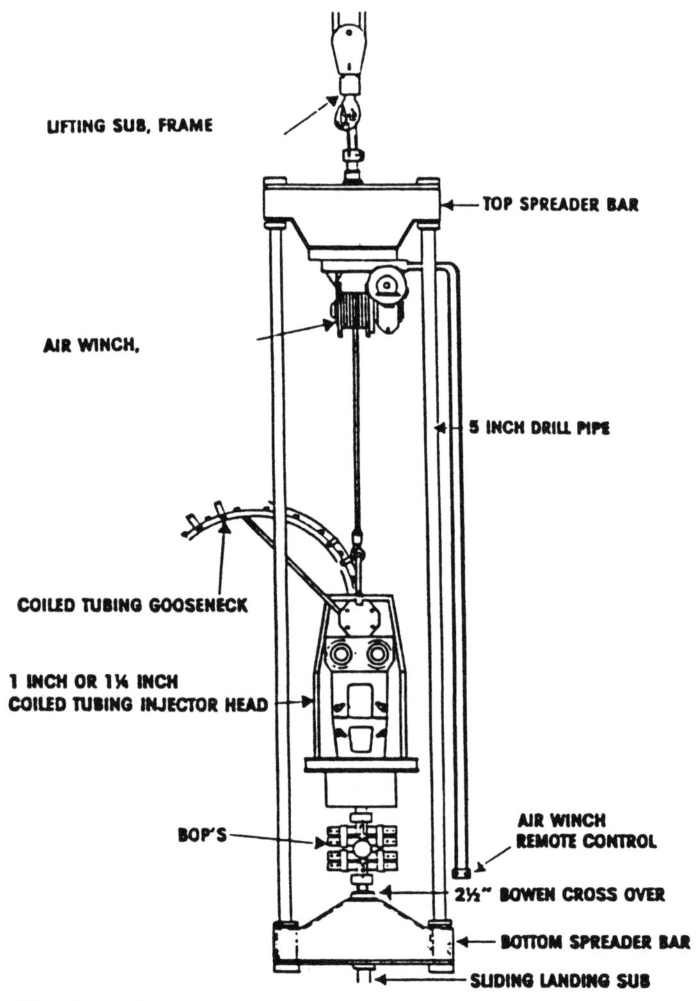

Fig. 9–9 Coiled Tubing Head and Frame

Areas for Improvement

Wireline operations, using the conventional lubricator mounted directly on the wellhead, plus a swell-compensating device, have proved feasible and perfectly practical in limited water depth. In deeper water, it will be necessary to develop alternative means of carrying-out wireline work. The simplest and most adaptable method, evaluated in the Zakum scheme in the 1970s, appears to be the remotely controlled submersible wireline winch. More development work would be required to bring this to a reliable and useful state.

For use in deeper water (where divers or guidelines are not available), it would be necessary to operate the winch with a search propulsion system in order to locate and move to the wellhead. A remotely operated connector for latching onto the wellhead can be made available, and adequate internal power and electronic capability would accommodate the propulsion and search systems. Two ideas for the development of subsea wireline apparatus are shown in Figure 9–10, one of which envisaged diver intervention.

There is also room for new support vessels such as small semi-submersibles, or large DSV monohulls, purpose-designed for defined workover duties. Projects are currently being studied by a number of operators and contractors, with various proposals under consideration. The great advantages of such vessels are that they have all the benefits of their bigger brothers, the exploration rigs, in providing large deck areas and low motion characteristics compared with equivalent monohulls. Additionally, they can be tailor-built for their intended purpose without expensive over-capacity.

A small semi-submersible, the 75 m (246 ft) long and 50 m (164 ft) wide Amethyst, with a movable deckload of 1,600 tons, is planned to be operational shortly. Most interesting is its design philosophy:

- Full drilling capacity but only in 200 m (650 ft) of water
- Full well completion capability
- Full workover capability
- Full wireline capability
- Full well stimulation and other injections
- Full well testing (including long-term)
- Full saturation diving for subsea construction

The concept provides a truly multi-functional support vessel to a producing field at an attractive cost. This project will be watched by the industry with great interest. Refer to Figure 9–11 for more details on this vessel.

Other vessels can be improved or developed. Of all the current possibilities for semi-submersible hull configurations, the streamlined column Swath or semi-submerged ship is probably

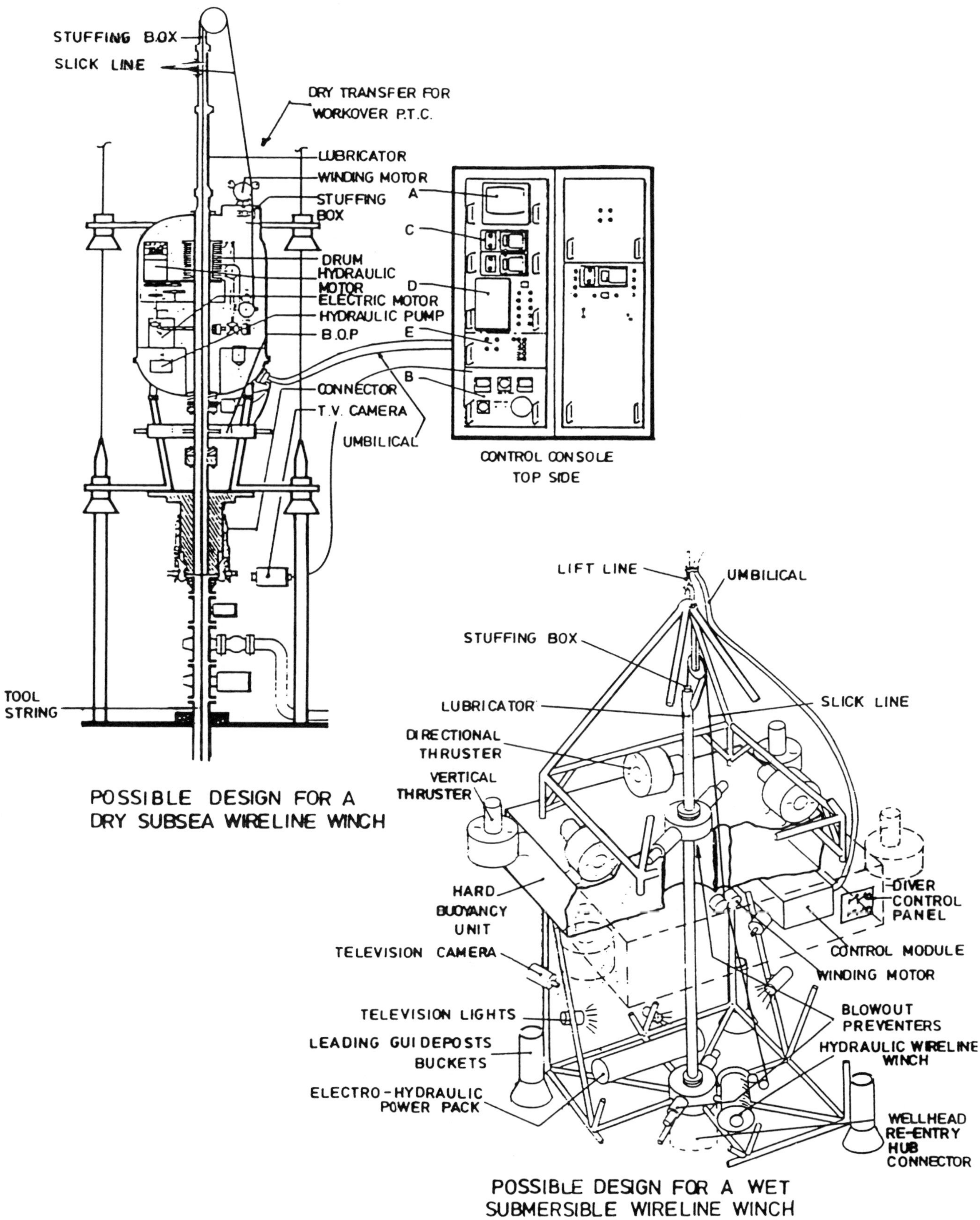

Fig. 9–10 Subsea Wireline

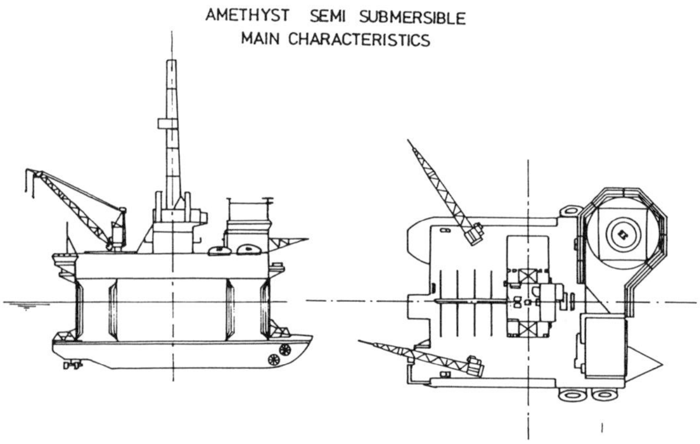

BASIC DATA		
LENGTH OF PONTOONS	75,0 m	246 ft
BREADTH OVERALL OF PONTOONS	50,4 m	165 ft
DEPTH OF PONTOONS	7,95 m	26 ft
LENGTH OF DECK (APPROX.)	73,3 m	240 ft
BREADTH OF DECK	44,4 m	146 ft
DEPTH OF DECK	7,95 m	26 ft
MOONPOOL	8×12 m	26×40 ft
PAYLOAD	2600 t	5732 kips
DECKLOAD - MOVABLE	1600 t	3527 kips
4 AZIMUTH THRUSTERS AFT	4 × 1400 kW	4 × 1904 hp
4 TUNNEL THRUSTERS FORWARD	4 × 750 kW	4 × 1020 hp
SPEED IN TRANSIT	10 knots	
ACCOMMODATION FOR 124 PERSONS		
4 POINT MOORING WITH :- 2 × 1.100 m OF WIRE FORWARD		
2 × 2.200m OF WIRE AFT		
(CONVERSION TO A 8 POINT MOORING SYSTEM IS POSSIBLE)		
CAPACITIES		
FUEL OIL	1430 m^3	8995 bbl
FRESH WATER	270 m^3	1698 bbl
DRILL WATER	450 m^3	2830 bbl
BULK MATERIAL	264 t	528 kips
MUD	223 m^3	1403 bbl
SACK MATERIALS	150 t	330 kips
OPERATIONAL CHARACTERISTICS		
DRAFT, TRANSIT	7,75 m	25 ft
DRAFT, OPERATIONAL I	18,95 m	62 ft
DRAFT, OPERATIONAL II	20,95 m	69 ft
DRAFT, SURVIVAL	13,15 m	43 ft
AIRGAP AT OPERATIONAL DRAFT I	7,0 m	23 ft
AIRGAP AT OPERATIONAL DRAFT II	5,0 m	16 ft
DYNAMICALLY POSITIONED:		
WATER DEPTH	200 m	656 ft
MAXIMUM DRILLING DEPTH		
WORKOVER	4600 m	~15000 ft
DRILLING	3000 m	~10000 ft

Fig. 9–11 Workover—New Semi-Submersible

the most favorable for workover. It has the ability to transit without deballasting, and this leads to time saving. It is comparable to a monohull vessel of the same approximate displacement in fuel efficiency in calm water. However, it has a considerable edge in rough seas.

Its construction is simple in that it has no tubulars or cylindrical braces and nodes, and it can be built using conventional flat plate construction with stressed skin techniques achieving considerable savings in structural weight and cost. The technology is proven, such vessels have been built, and classification/certification is not a problem.

The typical equipment installed for workover missions would include:

- A small drilling derrick complete with motion compensated block, draw works, rotary table, guide wires, cellar deck, and control and tensioning equipment, etc.
- A mud circulating system and storage.
- Drilling fluids storage.
- Pipe skid and lay down area.
- Dynamic positioning and integrated anchoring system.
- Saturation diving system.

Such a design has been developed by SSCO, Semi-Submerged Ship Corporation of California, and Worley Engineering and is available to shipyards and operators.

Stena U.K. Ltd, have two "Stena Workwell" class under construction for well workover operations and also well stimulation. The vessels have more than 12,000 ton displacement.

A typical equipment configuration and flow diagram for well testing is presented in Figure 9–12.

UNDERWATER SERVICES AND SUBSEA INSPECTION/MAINTENANCE AND REPAIR

Inspection requirements are specified in order to ensure safe and continuous operations of the offshore installations above standard certification requirements. Such inspection also serves the purpose of assessing and minimizing maintenance and repair requirements.

Structural inspection programs generally include:

- Metal thickness measurements
- Marine growth measurements
- Corrosion potential measurements
- Detailed inspection of caissons
- Detailed inspection of boat bumpers

Operational requirements also embrace subsea production systems, risers, flowlines, and pipelines.

The inspection of an offshore structure is a demanding task in terms of time and cost. Baseline surveys of a structure take place during construction and installation and then continue during its operational life.

In the North Sea, companies agree on the importance of planning and scheduling the inspection and maintenance programs. These provide a constant assessment of the integrity of the structure, as well as reducing the need for a major survey to be carried out in order to qualify for recertification. The procedure adopted for an underwater structural inspection begins with a full visual survey, followed by a detailed visual inspection and cleaning, and non-destructive testing of selected critical areas. The survey is planned on an annual basis and is conducted under the surveillance of the appropriate certifying authority.

When inspecting steel platforms, the program includes:

- A detailed visual inspection of selected nodes
- Non-destructive testing of selected nodes and members
- Metal thickness measurements
- A survey of the corrosion protection system

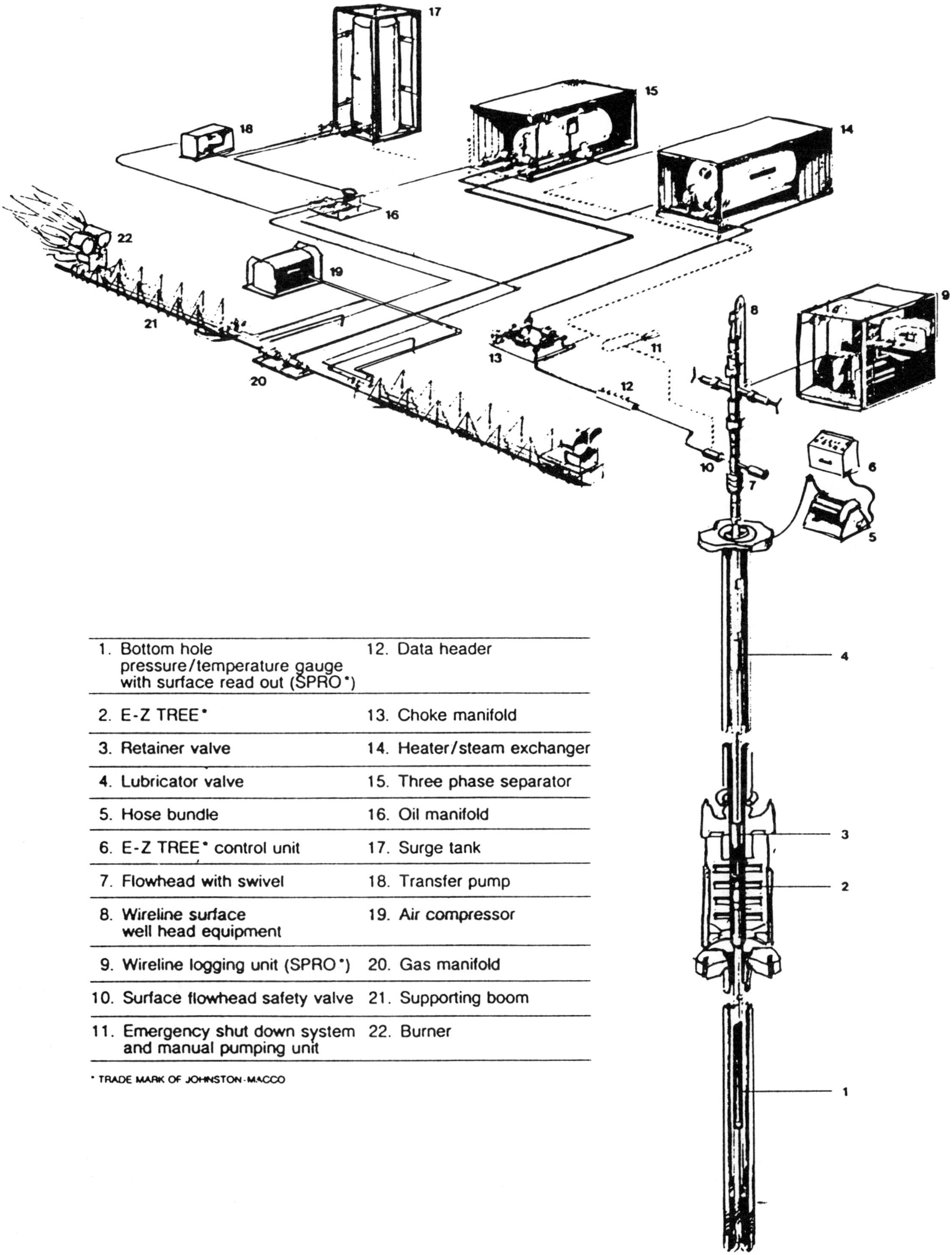

1. Bottom hole pressure/temperature gauge with surface read out (SPRO*)	12. Data header
2. E-Z TREE*	13. Choke manifold
3. Retainer valve	14. Heater/steam exchanger
4. Lubricator valve	15. Three phase separator
5. Hose bundle	16. Oil manifold
6. E-Z TREE* control unit	17. Surge tank
7. Flowhead with swivel	18. Transfer pump
8. Wireline surface well head equipment	19. Air compressor
9. Wireline logging unit (SPRO*)	20. Gas manifold
10. Surface flowhead safety valve	21. Supporting boom
11. Emergency shut down system and manual pumping unit	22. Burner

* TRADE MARK OF JOHNSTON-MACCO

Fig. 9–12 Workover—Well Servicing Operation

- A survey to determine the effects of sea bottom scouring
- A detailed inspection of all risers
- Marine growth measurements

The inspection of both steel and concrete structures is a complicated problem, as they are often situated in hostile environments which produce a variable loading factor. In addition, non-destructive testing of the entire reinforced and pre-stressed concrete platform is unsatisfactory, as surface areas to be inspected are vast (often over 130,000 m^2/1,400,000 ft^2). In the light of these problems, a suggested inspection program for concrete structures would include:

- A detailed inspection of the splash zone
- The installation of structural monitoring equipment to give continuous stress analysis
- An initial inspection of the whole surface area, reducing after the first two years to a program which will concentrate only on selected and critical regions
- A scour survey
- A survey of the structure's base
- A detailed inspection of all risers

Other seabed installations, for example subsea completions and spar buoys, usually require only visual inspection for leaks, damage, and marine corrosion. Inspection programs for these secondary structures are generally uncomplicated and relatively short in duration.

Subsea Production Systems Inspection

Subsea systems for marginal fields tend to incorporate high reliability components which require little or no scheduled maintenance during the life of the field. Exceptional maintenance due to accidents, environmental damage, or defective components should be accommodated by the design. Where possible, individual wells or components can be isolated to permit individual well shut in and repair without necessitating a complete shut down of the system.

The following general procedures are taken into account for operation planning:

1. All components are to be functionally tested at time of installation.
2. Periodic pressure and/or functional testing will be required. In general, periods between tests should not be more than one year.
3. Visual inspection should be carried out on a frequency to be defined. In general, one annual visual inspection of all components should be required. More frequent inspection can be performed at the operator's discretion.
4. Special inspection or non-destructive examination requirements should be defined in the detailed design of the system.

Inspection and maintenance guidelines are set out below.

Risers

1. Visual inspection by ROV. Risers will be routinely monitored for gas permeation and venting.
2. Inspection on daily basis to detect lean and general conditions. Detailed annual inspection in accordance with warranty.
3. Periodic functional testing to verify operation; ability of release system during period of shutdown.
4. Annual refurbishment of connectors and control will be performed by supplier's personnel.

Template

A detailed inspection and N.D.T. examination of template piping and structural components should be conducted on a planned basis.

This should include:

1. General visual inspection of piping, connectors, and structure
2. Ultrasonic inspection of selected components
3. Check of the cathodic protection system and anode inspection

All template piping should be designed for access by either divers, ROV, or both.

When designed for diver access, piping connectors are usually flanged to permit diver make-up using torque inverted bolt tensioning devices.

The in-situ repair of valves is not recommended. Occasional seabed injection may be required prior to removal of piping spools and valves. The affected lines are flushed with seawater and depressurized prior to diver operations. Divers will periodically blast-clean the template area to remove marine growth and sediments.

Manifolds should be maintained in-situ wherever possible. Valves will be replaced where found defective. The manifold could be recovered for major overhaul or inspection.

Chokes in service should be subject to visual inspection only. Chokes should be oily body pressure tested. Complete shut off of chokes should not be tested. Function tests should be performed periodically.

Chokes should be of suitable trim and material as to minimize replacement for normal wear. Repair of chokes and hydraulic actuators should be done on surface. The choke and actuator should be a unitized assembly to minimize removal and re-installation time.

Trees

Routine visual inspection should be by divers or ROV only. Periodic functional and pressure testing of trees is recommended to be conducted during workover periods.

When required, the tree should be retrieved to the surface for repairs or refurbishment. The in-situ repair of components other than controls plumbing is not recommended.

Subsea Controls

Visual diver or ROV inspection of subsea control modules and lines is to be conducted. Other inspection and testing should be scheduled.

It is imperative that the subsea control level of cleanliness is maintained. This must be ensured through regular maintenance of filters and fluid mixing system and by continual inspection/testing of fluid and water quality. Flexible hoses should be periodically replaced.

Maintenance of the electronic equipment will be scheduled and conducted with back-up system in use, to minimize shut-in periods.

The maintenance of the subsea components, namely the modules, junction boxes, and interconnections, should be conducted in the supplier's shore-based support facility, under clean room conditions.

The surface to subsea control umbilical may have to be replaced within the life of the field. This operation will be planned to coincide with other shutdowns, such as export pipeline maintenance.

Periodic surveys should be conducted on an annual basis as dictated by statutory requirements.

No general maintenance of satellite wells and flowlines will be required.

Pipeline Inspection

In the North Sea, the majority of pipelines are buried, and this has resulted in inspection work being extremely specialized.

Major surveys include:

- How to assess the pipe metal condition when it has a concrete and bitumen coating
- How to assess the state of cathodic protection system on buried pipelines

Current density measurements give a curve showing current densities along the pipeline whether it is buried or not.

Cathodic protection readings give measurements in contact with the base metal areas using separate reference cells.

Visual, video, and sonar measurements are used in pipeline tracking to locate buried lines. Seabed profiling by echo sounders and pressure sensors and trench profiling techniques are used also.

A survey of pipeline routes is used in the assessment of spanning (caused by erosion of soil underneath the pipe). Damage to exposed pipeline lengths from trawls and anchors is obtained from surveys also.

Pipeline movements on/off the seabed are obtained from detailed surveying.

Using flowmeters within a leak detection system allows continuous measurements of the hydro-carbon content of the seawater.

This system can be combined with an acoustic hydrophone to monitor the noise level. Together, they allow very small leaks to be accurately pin-pointed.

NDT TECHNIQUES

The role of NDT techniques include:

- Integrity monitoring
- Periodic inspection of critical areas
- Measurement and assessment of known defects

The results from an NDT inspection can be used to illustrate the presence or absence of significant defects in a given component. It also will permit proper assessment of any defect to be visually recorded during the assembly phase. Good reliability will be achieved when each factor in the complete chain of activities within the inspection system is correctly designed and controlled.

Various techniques have been developed for specific application.

Magnetic Particle Inspection (MPI) is highly sensitive to surface breaking defects of a linear nature in ferromagnetic steels. Limiting dimensions for detection where the surface condition is good are of the order of 2 to 3 mm (depth and length). The technique is primarily sensitive to defects which lie perpendicular to the direction of the magnetic field. Generally, two perpendicular directions of magnetization are considered adequate to find all orientations of flaw. However, on complex geometries, the directions of magnetization must be considered further. MPI is suitable for underwater inspection.

Dye Penetrant Inspection (DPI) can be applied to both ferromagnetic and non-ferromagnetic surfaces. it is sensitive to very small defects of both plane and curved types. The method can be applied on clean, relatively smooth surfaces and is more time consuming than MPI. DPI is suitable for dry inspection only.

Eddy current techniques for surface inspection use small hand-held probes and various types of instruments which give either a meter read-out or a screen display. Although relatively simple in concept, it is very sensitive on good surfaces where sub-millimeter defects are found quite reliably. Eddy current testing is much less easy to apply on welded structures. This arises because the technique is very sensitive to the geometrical and material property changes which occur at welds, namely weld cap roughness, hardness changes, and material composition changes. It is sensitive to probe-to-workpiece spacing (lift-off) also.

AC potential drop instruments have been developed for the

detection and measurement of surface breaking cracks. They have been applied recently to measurement of fatigue cracks in platform structures. The instruments use a high frequency, 5kHz, so that the impressed current is carried in a thin layer at the surface. The presence of a crack leads to an increase in the potential drop between a pair of contacts on the surface compared with that between a similar pair of contacts remote from the crack.

Radiography is capable of detecting both surface breaking and buried cracks. The method is widely applied to examination of welds and castings. It is best suited to detection of flaws such as gas pores and slag particles. Also, it is able to detect cracks, provided that the crack is not tightly closed and that the radiation beam can be directed approximately parallel to the plane of the crack. An obvious disadvantage of radiography is that non-essential personnel have to be excluded from a substantial area around the test zone.

In principle, *ultrasonics* have the capability of detecting and sizing cracks at or near the surface and within the wall thickness of a component. However, these capabilities are dependent on the procedures employed, on the equipment used, on the nature of the cracks themselves, and on the skill of the operator.

Acoustic Emission (AE) has an apparent suitability for continuous monitoring of a structure for defect growth but is less likely to be useful for periodic inspection (except in conjunction with hydro-testing).

The capabilities of AE to detect crack growth are dependent on the material micro-structure, the cracking mechanism, and the details of the AE technique, namely transducer type and location, electronic processing, thresholds, etc. AE techniques should be demonstrated prior to application by means of control procedure tests on suitable specimens to establish its effectiveness and limitations.

Manned Intervention

A number of techniques may be used for manned intervention, classed under two general headings:

1. *Man in the water*—working under the pressure at the depth of the water surrounding him.
2. *One atmospheric diving*—with the man inside a pressure proof suit or submersible.

Man in the Water

Diving is the traditional way of working underwater. Nowadays it offers a number of well-researched techniques which can give a capability down to depths in the order of 450 m (about 1,500 ft).

The diver is a maneuverable tool to the offshore oil industry and his "hands-on" ability and dexterity cannot be equalled by any other manned intervention method. However, the diver does have certain drawbacks which will be reviewed together with other diving techniques that may be used.

The Decompression Requirement. Decompression, as a method of avoiding *decompression sickness* (the bends), is a way of life to divers. On a project level however, it can be a time consuming and costly business. Furthermore, diving at greater depths offers less useful bottom time and requires longer decompression, which can make the planning of continuous operations quite difficult.

Surface Orientated vs. Saturation Diving. In surface orientated operations, the diver leaves the surface, completes his work, and then fulfills his decompression commitment in the water before surfacing. In a variation, called *surface decompression* (of North American origin), he surfaces directly on reaching 10 m (33 ft) his decompression stop and is compressed again in a deck chamber to complete his decompression period. This technique probably carries a higher than normal risk of the bends.

The method usually put forward by contractors involves the diver descending with an umbilical. This typically consists of a breathing gas hose, hot water hose (for the hot water suit), communications, and a safety line. On his back he wears an emergency bale-out cylinder. In the United States, he would probably make a through water descent or use a stage, having a statutory working maximum of 68 m (220 ft). In the North Sea, a wet bell would probably be used, an inverted weighted cupola filled with breathing gas, offering a safe haven, emergency gas, and good communications. Here the maximum working depth allowed is 50 m (163 ft).

For greater working depths, a saturation diving technique is used. This involves the diver remaining under pressure at a given depth long enough for all body tissues to become fully saturated with the inert gas in the breathing mixture. No matter how long he now stays under pressure at that depth, his decompression requirement will not alter.

Saturation periods of 4–5 weeks are not uncommon, with divers living in surface chambers and descending to work up to 10-hour shifts on the work-site by way of a diving bell. Two divers then lock-out of this bell in turn for a recommended five-hour maximum in-water period.

Whereas air can be used in the shallows, its nitrogen content becomes increasing narcotic with depth. Mixtures of oxygen and helium, a far less narcotic gas, are used in deeper diving. However, contractors may well nominate to use this *heliox* mix from as shallow as 30 m (100 ft).

Helium, being one-seventh the density of air, eases respiratory effort at depth, but it has five times air's thermal conductivity. Also, it produces an upward frequency shift in the voice, the "Donald Duck" effect on communications.

In very deep diving, say over 300 m (100 ft), mixtures of

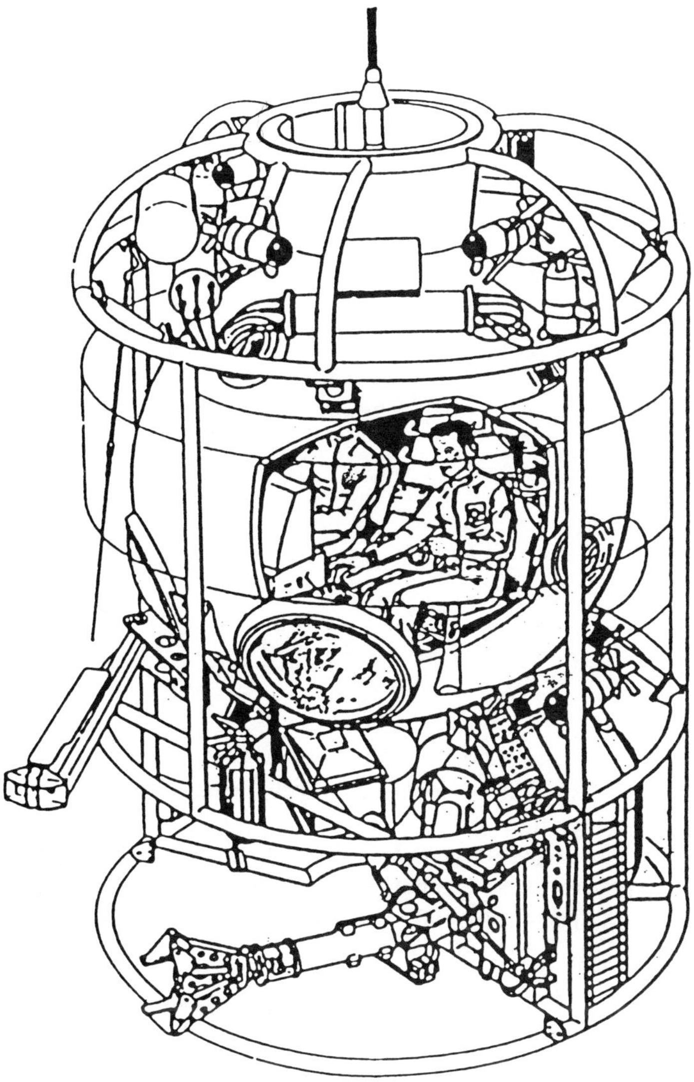

Fig. 9–13 Mobile Observation Bell (Source: Comex)

oxygen, helium, and hydrogen are coming into use—hydrogen being the least dense and narcotic gas available.

The modern diving support vessel (DSV), with its onboard saturation system and deck cranes, can prove a cost-effective unit within a well managed program. As fields move into deep water however, different manned intervention methods must be considered.

One-Atmosphere Diving

Where an operator can descend to a work site within a pressure proof enclosure at one atmosphere pressure, then decompression becomes unnecessary. The unit can be used as and when necessary, but it does have drawbacks. Of primary importance is recovery in event of any accident and the limitations with regard to working within structures. Finally is the fact that all work has to be conducted remotely via manipulators. In some ways, the modern purpose-adapted remote controlled vehicle can offer almost as much as one-atmosphere diving in certain situations.

Typically, all one-atmosphere systems have a depth potential of 300 m (1,000 ft) or more.

Articulated Diving Suits

Jim, the modern version of a 1930s suit, actually walks across a solid bottom while WASP, OMAS, and SPIDER use thrusters. All are lowered by cable, are self-contained, and have proved successful on various operations.

One-Atmosphere Bells

An adaption of the standard diving bell, these are lowered down guide wires trailing a service umbilical and have powerful enough thrusters to swim horizontally for considerable distances (see Fig. 9–13).

Manned Submersibles

Free ranging 2–3 man miniature submarines (some with diver lock-out capability) have, in the past, proved very successful in the survey and observational mode on pipelines. Nowadays, the number of contractors who offer them has decreased (see Fig. 9–14).

Tethered Submersibles

A miniature one man submersible with thrusters, which differs from the articulate suits in that it is designed to work horizontally.

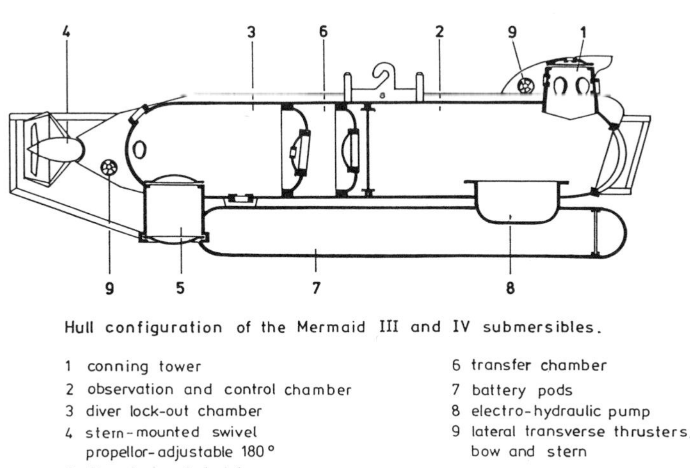

Fig. 9–14 Manned Submersible

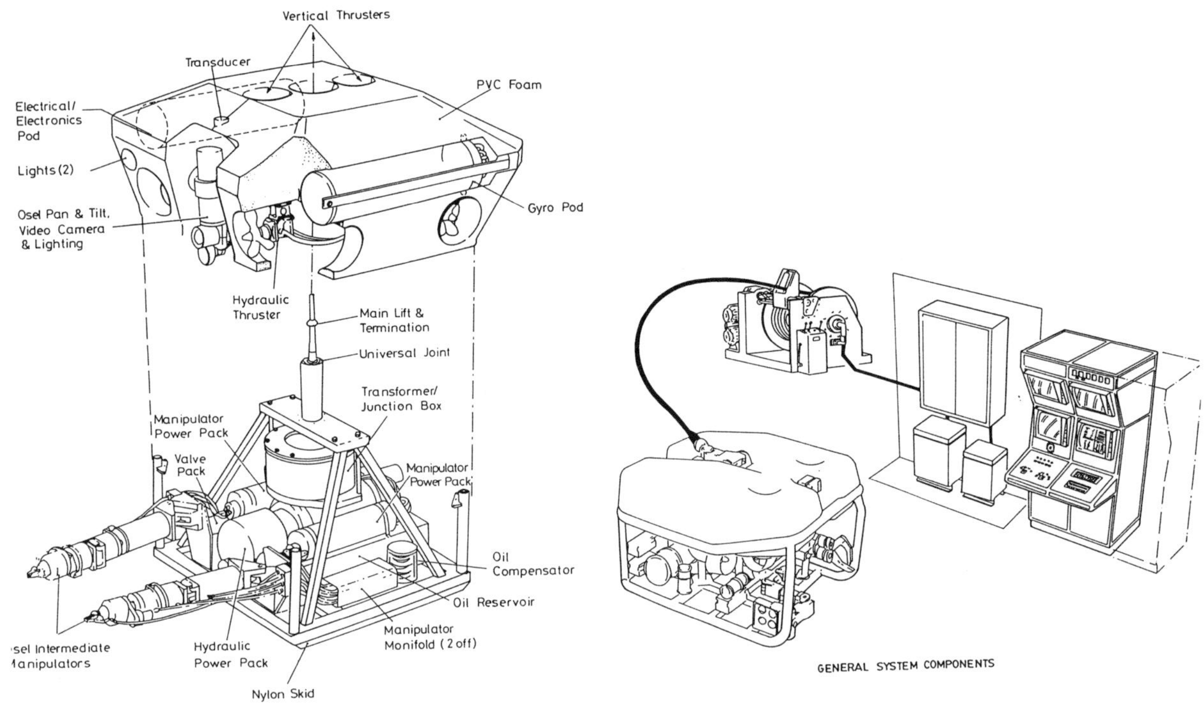

Fig. 9–15 Remote Operated Vehicles (Source: OSEL)

Remote Operated Vehicles (ROV)

The industry has seen the evolution of three main types:

1. *Observation vehicles*—Carrying a TV camera, these have identified a slot in inspections or working in conjunction with divers. They are typified by being lightweight and having relatively weak thruster power. That is, they have difficulty maintaining station in any real continuous current. Even so, they are useful tools.
2. *Specialized ROV*—Larger more powerful vehicles whose main role is in pipeline survey operations. A range of additional equipment including sub-bottom profilers, side scan sonar, etc. is offered by contractors to back up to the two or more TV cameras the vehicles carry. Once deployed, it is not uncommon for them to stay submerged for a week or more on the job.
3. *Modular Multi-Task ROV*—Whereas the previous two ROV types are standard models, these are usually purpose-designed for a specific job and named accordingly (see Fig. 9–15).

Although free swimming, they are intended to carry out their work load while attached or seated in a set position on a structure. They are used to deploy dedicated modular work packages which carry out specific tasks such as maintenance. In one application, North Cormorant UMC, the ROV moves on rails.

In at least one case, the Chevron's Montanaza D2 well in 742 m (2,434 ft) of water off Spain's Mediterranean Coast, a specified ROV was planned to be adapted to service the wellhead using modular style packages.

Index

D

E

F

G

H

I

J

K

L

M

P

Q

R

T

U

X

Y

Z